W0230185

CANNABIS

CANNABIS
The Genus *Cannabis*

Edited by

David T. Brown
School of Pharmacy, University of Portsmouth, UK

CRC PRESS

Boca Raton   London   New York   Washington, D.C.

FIRST INDIAN REPRINT, 2012

This book contains information obtained from authentic and highly regarded sources. Reprinted material is quoted with permission, and sources are indicated. A wide variety of references are listed. Reasonable efforts have been made to publish reliable data and information, but the author and the publisher cannot assume responsibility for the validity of all materials or for the consequences of their use.

Neither this book nor any part may be reproduced or transmitted in any form or by any means, electronic or mechanical, including photocopying, microfilming, and recording, or by any information storage or retrieval system, without prior permission in writing from the publisher.

Direct all inquiries to CRC Press LLC, 2000 N.W. Corporate Blvd., Boca Raton, Florida 33431.

© 1998 CRC Press, LLC

Trademark Notice: Product or corporate names may be trademarks or registered trademarks, and are used only for identification and explanation, without intent to infringe.

Visit the CRC Press Web site at www.crcpress.com

Printed and bound in India by
Replika Press Pvt. Ltd.

ISBN 10 : 90-5702-291-5
ISBN 13 : 978-90-5702-291-3

FOR SALE IN SOUTH ASIA ONLY.

CONTENTS

PREFACE TO THE SERIES

There is increasing interest in industry, academia and the health sciences in medicinal and aromatic plants. In passing from plant production to the eventual product used by the public, many sciences are involved. This series brings together information which is currently scattered through an ever increasing number of journals. Each volume gives an in-depth look at one plant genus, about which an area specialist has assembled information ranging from the production of the plant to market trends and quality control.

Many industries are involved such as forestry, agriculture, chemical, food, flavour, beverage, pharmaceutical, cosmetic and fragrance. The plant **raw** materials are roots, rhizomes, bulbs, leaves, stems, barks, wood, flowers, fruits and seeds. These yield gums, resins, essential (volatile) oils, fixed oils, waxes, juices, extracts and spices for medicinal and aromatic purposes. All these commodities are traded world-wide. A dealer's market report for an item may say "Drought in the country of origin has forced up prices".

Natural products do not mean safe products and account of this has to be taken by the above industries, which are subject to regulation. For example, a number of plants which are approved for use in medicine must not be used in cosmetic products.

The assessment of safe to use starts with the harvested plant material which has to comply with an official monograph. This may require absence of, or prescribed limits of, radioactive material, heavy metals, aflatoxin, pesticide residue, as well as the required level of active principle. This analytical control is costly and tends to exclude small batches of plant material. Large scale contracted mechanised cultivation with designated seed or plantlets is now preferable.

Today, plant selection is not only for the yield of active principle, but for the plant's ability to overcome disease, climatic stress and the hazards caused by mankind. Such methods as *in vitro* fertilisation, meristem cultures and somatic embryogenesis are used. The transfer of sections of DNA is giving rise to controversy in the case of some end-uses of the plant material.

Some suppliers of plant raw material are now able to certify that they are supplying organically-farmed medicinal plants, herbs and spices. The Economic Union directive (CVO/EU No 2092/91) details the specifications for the **obligatory** quality controls to be carried out at all stages of production and processing of organic products.

Fascinating plant folklore and ethnopharmacology leads to medicinal potential. Examples are the muscle relaxants based on the arrow poison, curare, from species of *Chondrodendron*, and the antimalarials derived from species of *Cinchona* and *Artemisia*. The methods of detection of pharmacological activity have become increasingly reliable and specific, frequently involving enzymes in bioassays and avoiding the use of laboratory animals. By using bioassay linked fractionation of crude plant juices or extracts, compounds can be specifically targeted which, for example, inhibit blood platelet aggregation, or have antitumour, or antiviral, or any other required activity. With the assistance of robotic devices, all the members of a genus may be readily screened. However, the plant material must be **fully** authenticated by a specialist.

The medicinal traditions of ancient civilisations such as those of China and India have a large armamentaria of plants in their pharmacopoeias which are used throughout South East Asia. A similar situation exists in Africa and South America. Thus, a very high percentage of the World's population relies on medicinal and aromatic plants for their medicine. Western medicine is also responding. Already in Germany all medical practitioners have to pass an examination in phytotherapy before being allowed to practise. It is noticeable that throughout Europe and the USA, medical, pharmacy and health related schools are increasingly offering training in phytotherapy.

Multinational pharmaceutical companies have become less enamoured of the single compound magic bullet cure. The high costs of such ventures and the endless competition from me too compounds from rival companies often discourage the attempt. Independent phytomedicine companies have been very strong in Germany. However, by the end of 1995, eleven (almost all) had been acquired by the multi-national pharmaceutical firms, acknowledging the lay public's growing demand for phytomedicines in the Western World.

The business of dietary supplements in the Western World has expanded from the Health Store to the pharmacy. Alternative medicine includes plant based products. Appropriate measures to ensure the quality, safety and efficacy of these either already exist or are being answered by greater legislative control by such bodies as the Food and Drug Administration of the USA and the recently created European Agency for the Evaluation of Medicinal Products, based in London.

In the USA, the Dietary Supplement and Health Education Act of 1994 recognised the class of phytotherapeutic agents derived from medicinal and aromatic plants. Furthermore, under public pressure, the US Congress set up an Office of Alternative Medicine and this office in 1994 assisted the filing of several Investigational New Drug (IND) applications, required for clinical trials of some Chinese herbal preparations. The significance of these applications was that each Chinese preparation involved several plants and yet was handled as a **single** IND. A demonstration of the contribution to efficacy, of **each** ingredient of **each** plant, was not required. This was a major step forward towards more sensible regulations in regard to phytomedicines.

My thanks are due to the staff of Harwood Academic Publishers who have made this series possible and especially to the volume editors and their chapter contributors for the authoritative information.

Roland Hardman

PREFACE

Cannabis sativa is a dioecious, bushy plant, probably originating from central Asia, but now considerably more widely disseminated and enjoying a truly international reputation. Records indicate that cannabis was used by man some 2–3,000 years before Christ. Then, as now, it provided a source of fuel, textiles, paper, rope, medicines and intoxication.

The plant exudes a resin containing psychoactive compounds called cannabinoids. Trichomes which secrete the resin are most abundant in the flowering heads and surrounding leaves. There are over 60 cannabinoids, the most familiar of these being delta-9-tetrahydrocannabinol (THC). The amount of resin produced, and its cannabinoid content are strongly influenced by plant gender, cultivation conditions and time of harvest. In addition to naturally occurring cannabinoids, the search for new 'phytopharmaceuticals' has led to the development of synthetic or semisynthetic derivatives with enhanced medicinal properties and reduced side effects. This in turn has led to a greater understanding of cannabinoid pharmacology and pharmacokinetics in addition to providing several promising lead medicinal compounds.

Cannabis and its derivatives are used medicinally in a range of disorders; although often illicitly so. Traditional uses such as the relief of pain, have been extended to include the reduction of intra-occular pressure in glaucoma, relief of spasticity in multiple sclerosis, treatment of chemotherapy-induced nausea and vomiting, and stimulation of appetite in AIDS patients. This book reviews evidence for the justification of these uses. Cannabis also has its darker side; it is the most commonly abused illicit substance on the planet. A body of evidence suggests that cannabis can cause both physical and psychological harm, although the extent of this is the topic of hot debate, extended in this volume.

There have been at least 23 international symposia on cannabis which are referenced, together with over 700 other citations, in this book. In addition, a search on the World Wide Web reveals a burgeoning number of correspondents wishing to air their views on all aspects of cannabis use. This serves to illustrate the interest which cannabis still generates.

Cannabis has had a long and chequered history spanning some 5,000 years, as the source of substances of abuse, of medicines, and products useful in manufacturing industry. Study of derivatives from the cannabis plant are still providing us with surprises and possibilities which sustain our fascination with the plant. This book provides a detailed review of the use and abuse of cannabis and the national and international problems which surround them. It provides a body of evidence, in one volume, from which the reader can obtain a clear view of where society stands in its relationship with cannabis and the likely paths which that relationship may take in the future.

David T. Brown

CONTRIBUTORS

David T. Brown
Senior Lecturer in Clinical Pharmacy
School of Pharmacy
University of Portsmouth
St Michael's Building
White Swan Road
Portsmouth PO1 2DT
UK

Alpana Joshi
Research Associate
National Center for the Development
  of Natural Products
University of Mississippi
MS 38677
USA

William G. Notcutt
Consultant Anaesthetist
The James Paget Hospital NHS Trust
Lowestoft Road
Gorleston, Great Yarmouth
Norfolk NR31 6LA
UK

Roger G. Pertwee
Department of Biomedical Sciences
Institute of Medical Sciences
University of Aberdeen
Foresterhill
Aberdeen AB25 2ZD
UK

Geoffrey F. Phillips
62 Parkhill Road
Bexley
Kent DA5 1HY
UK

Mario A.P. Price
Senior Pharmacist
The James Paget Hospital NHS Trust
Lowestoft Road
Gorleston, Great Yarmouth
Norfolk NR31 6LA
UK

Amala Raman
Lecturer in Pharmacognosy
Department of Pharmacy
King's College London
Manresa Road
London SW3 6LX
UK

Richard B. Seymour
Information and Education Director
Office of the President
Haight Ashbury Free Clinics Inc.
409 Clayton Street
San Francisco CA 94117
USA

David E. Smith
President and Medical Director
Haight Ashbury Free Clinics Inc.
409 Clayton Street
San Francisco CA 94117
USA

Simon Wills
Head of Drug Information Service
Pharmacy Department
St Mary's Hospital
Portsmouth PO3 6AD
UK

1. CANNABIS USE AND ABUSE BY MAN: AN HISTORICAL PERSPECTIVE

SIMON WILLS

Head of the Drug Information Service, St Mary's Hospital, Portsmouth, UK

ORIGINS

The hemp plant, *Cannabis sativa*, is native to central Asia north of the Himalayas. It was initially confined to an area stretching from Turkestan in the west, to Pakistan in the east. Southern China probably marked the northernmost boundary of this original domain. Hemp has subsequently become much more widespread, largely due to the intervention of man. Cannabis, a dioecious species, is a member of the Cannabidaceae family, which contains only one other genus – *Humulus*. The hop plant, *Humulus lupulus*, is used to preserve and flavour beer.

Throughout history, the hemp plant has been widely used: the seeds can be eaten and also produce oil for lamps or cooking; the stems produce fibres for textiles or rope; the flowering heads and leaves produce resin used as medicine or for intoxication.

ANCIENT CHINA

The ancient Chinese, and others inhabiting the plant's native region of central Asia, must have discovered the properties of cannabis centuries before it came to the attention of other more distant civilisations. In the right conditions the plant grows quickly to maturity, and the multiplicity of potential uses made it too valuable to be ignored. Copies of a Chinese herbal, thought to have been originally written in the 3rd millennium BC by the emperor Shen Nung, show that cannabis was used medicinally (Mechoulam, 1986). This is the most archaic written record of the uses of cannabis. The original does not survive, but later copies reveal that the conditions treated included rheumatism, gynaecological disorders, absentmindedness and malaria. In this herbal, and in others written much later, excessive use is described as causing symptoms akin to intoxication, usually described as the 'appearance of spirits'. Hua Tu (115–205 AD) was a renowned surgeon in ancient China. He is believed to have used cannabis as a form of anaesthetic (Guthrie, 1946). Following administration of the drug to patients, he performed a variety of operations including laparotomies and splenectomies.

A biography of the Chinese physician Hoa-tho, who practised around 220 AD, reveals a knowledge of the anaesthetic and analgesic effects of cannabis which was generally administered in a drink of wine:

...he administered a preparation of hemp (Ma-Yo) and, in the course of several minutes, an insensibility developed as if the patient had been plunged into drunkenness or deprived of life.

Then, according to the case, he performed the opening, the incision or the amputation and relieved the cause of the malady; then he apposed the tissues by sutures and applied liniments. After a certain number of days or the end of the month the patient finds he has recovered without having experienced the slightest pain during the operation.

(Walton, 1938a)

The oldest preserved specimens of hemp are portions of cloth from a Chinese burial site dated to around 1200 BC (Richardson, 1988). However, clay pots at earlier sites sometimes show markings which may be impressions of the woven or twisted fibres. Hemp was widely used as a basic clothing material by the majority of Chinese society – who could not afford silk – and this persisted until cotton was introduced in the 10th or 11th century. Hemp was also used to make paper. During the period 33–7 BC, Fan Sheng-chih, a consultant to the Emperor, wrote a manual on farming techniques which included a detailed discussion of the method for raising cannabis and other plants. So important was the crop that when the population lost confidence in the state coinage during the reign of Emperor Wang Mang (9–23 AD), it became one of the basic commodities which was used as currency in its place (Twitchett and Loewe, 1986).

At the turn of the century the British botanist, Ernest Wilson, visited China and reported on the method of preparation of cannabis for fibre production (Wilson, 1913). He described a process which had not changed for thousands of years:

Several plants yielding fibres valued for textile and cordage purposes are grown in China. In Szechuan the most important of these is the true Hemp (Cannabis sativa), colloquially known as "Hou-ma". This crop is abundantly cultivated around Wenchang Hsien and P'i Hsien. It is a spring crop, the seeds being sown in February and the plants harvested the end of May and beginning of June, just as they commence to flower. The stems are allowed to grow thickly together and reach 8 feet in height. The culms are reaped, stripped of their leaves, and often the fibre is removed there and then. More commonly, however, the stems are placed in pits filled with water and allowed to ret for a few days; they are then removed, sun-dried, stacked in hollow cones, surrounded by mats, and bleached by burning sulphur beneath the heaps. After these processes the fibrous bark is stripped off by hand. The woody stems that remain after the bark has been removed are burned, and the ashes resulting, mixed with gunpowder, enter into the manufacture of fire-crackers. Hemp, or "Hou-ma," is the best of fibres produced in Western China for rope-making and cordage purposes generally. It is also used locally for making grain-sacks and coarse wearing apparel for the poorer classes. Quantities are used in the city of Paoning Fu for these latter purposes. It is in great demand on native river-craft and is largely exported down river to other parts of China. It is this hemp that is principally exported from Szechuan. True Hemp (Cannabis) is an annual and is grown as a summer crop in the mountains for the sake of its oil-containing seeds. Hemp oil is expressed and used as an illuminant and is said not to congeal in the coldest weather.

There appears to be a dearth of information on the abuse of cannabis in China. It is unlikely that deliberate intoxication was not described through ignorance of this effect. The tradition of Chinese medicine has been to investigate and classify the properties of an enormous range of natural products. Some have speculated that the effects of intoxication with cannabis might be incompatible with the Chinese temperament, sense of public dignity and attitude to life (Walton, 1938; Mechoulam, 1986).

ANCIENT EGYPT

The oldest surviving original document which mentions cannabis is the Ebers papyrus of Egypt, which dates from the 16th century BC. A hieroglyphic symbol (pronounced "shemshemet") has been assumed to represent cannabis because the plant referred to in the text is cited as a source of both fibre and medicine (Mechoulam, 1986; Nunn, 1996). It was administered orally to treat "mothers and children" for an unstated purpose, and was also employed as an enema, eye preparation and medicated bandage. Cannabis is mentioned briefly in a number of later medical papyri in which additional methods of administration are described including vaginal application and use as a fumigant (Nunn, 1996). However, the infrequency of reference and lack of therapeutic detail suggest that cannabis was not commonly used medicinally. There is also no explicit description of the intoxicating effects of cannabis. Its use in incense may reveal an association with religious ritual – perhaps to produce hallucinations of a quasi-religious nature – but this must remain speculative. Analysis of hair from Egyptian mummies dating back to 1070 BC has revealed surprisingly high levels of cannabis: 800–4100 ng/g compared to 2–1000 ng/g for present day German drug addicts (Parsche et al., 1993). Only the higher echelons of Egyptian society were mummified and these sections of society often had important religious functions. Cannabis was not native to the area, and so it might have been available in limited supply such that it was used mainly by the more prosperous elite.

THE BIBLE AND JUDEA

There are no obvious references to cannabis in the Bible. However, the neighbouring Assyrians are known to have used cannabis widely for a variety of medicinal purposes. Cannabis was administered orally for the treatment of impotence and depression, topically for bruises and by inhalation for a disease assumed to be arthritis. The drug was also used in various forms to ward off evil (Mechoulam, 1986). Until the late seventh century before Christ, the Jewish and Assyrian peoples were in close contact (Mechoulam et al., 1991) and because of their geographical proximity it is most unlikely that the Jews were unaware of the existence of cannabis. Mechoulam has speculated that following the decline of the Assyrian civilisation, Jewish kings such as Josiah may have sought to purge their culture of Assyrian influences which may have been seen as pagan or immoral by the orthodox. Any existing references to cannabis in the Old Testament could have been removed at this time. Despite this, Mechoulam cites one instance where the ancient word for cannabis might have been preserved in the Bible. In the Old Testament, the prophet Ezekiel mentions trade in a product called *pannag*.

Judah and the cities in what was once the Kingdom of Israel sent merchants with wheat, minnith and pannag, and with honey, oil and balm.

(Ezekiel, Ch. 27 v. 17)

The letters 'p' and 'b' are often interchangeable in Hebrew, and *pannag* or *bannag* is thus very similar to the Sanskrit word for cannabis *bhanga*, the Hindu *bhang* and the Persian *bang*. It has even been speculated that some of the more vivid personal experiences, or revelations, recounted in the Bible may have been descriptions of cannabis intoxication. Examples include the madness of King Saul, and the almost psychedelic visions of the prophet Ezekiel (Creighton, 1903).

Cannabis was certainly in use in Jerusalem during the later stages of Roman occupation. The remains of a fourteen year old girl and a full-term baby were found near Jerusalem in 1993 (Zlas *et al.*, 1993). These were dated to the 4th century AD by coins buried with the bodies. The remnants of burned cannabis were also found, leading to speculation that drug fumes may have been inhaled as an aid to childbirth – either as an analgesic or to aid uterine contractions. But this interpretation has been questioned (Prioreschi and Babin, 1993). The burning of cannabis might equally have represented part of the burial ritual, or simply a popular habit amongst the indigenous people.

PREHISTORIC EUROPE AND SCYTHIA

Hemp seeds are the most durable part of the plant, and so are most likely to be preserved at archaeological sites. Seeds have been found associated with Neolithic habitations in Germany, Switzerland, Austria and Rumania (Rudgley, 1993). Rudgley speculates that hemp may have initially grown as a weed around prehistoric settlements, particularly since the rubbish piles which accompany many dwellings would typically have been rich in nitrogen, and *Cannabis sativa* is a nitrophilic species.

Objects known as 'polypod' bowls have been found in eastern Europe from the early third millennium onwards (Rudgley, 1993). These may have been braziers used to burn cannabis for the purpose of intoxication. Some of these bowls are decorated with the impression of coiled rope. This is most likely to have been hemp since it was the most commonly used material for the preparation of fibre. The earliest examples of polypods are found in the east, suggesting an east-west migration of this culture. This conforms to the known direction of the spread of cannabis use. A grave in present day Rumania, dated with certainty to the third millennium BC, was found to contain a cup of charred hemp seeds.

In the mid 5th century BC, the Greek historian Herodotus described a technique for burning cannabis when he travelled through the Black Sea area of the region known as Scythia (Herodotus, tr. Rawlinson, 1949). Scythia covered a large area stretching from the Ukraine to the borders of present-day India. Its peoples were largely nomadic. Whilst there, Herodotus described the funerary customs of the Scythians:

After the burial, those engaged in it have to purify themselves, which they do in the following way. First they well soap and wash their heads; then, in order to cleanse their bodies, they act as follows: they make a booth by fixing in the ground three sticks inclined towards one another, and stretching around them woollen felts, which they arrange so as to fit as close as possible: inside the booth a dish is placed upon the ground, into which they put a number of red-hot stones, and then some hemp-seed.

Hemp grows in Scythia: it is very like flax; only that it is a much coarser and taller plant: some grows wild about the country, some is produced by cultivation: the Thracians make garments which closely resemble linen; so much so, indeed, that if a person has never seen hemp he is sure to think they are linen, and if he has, unless he is very experienced in such matters, he will not know of which material they are.

The Scythians, as I said, take some of this hemp-seed, and, creeping under the felt coverings, throw it upon the red-hot stones; immediately it smokes, and gives out such a vapour as no Grecian vapour-bath can exceed; the Scyths, delighted, shout for joy, and this vapour serves instead of a water-bath; for they never by any chance wash their bodies with water.

In the late 1940s, archaeological excavations in Pazyryk in Siberia vindicated Herodotus' observations of Scythian customs. Two copper vessels were unearthed containing the remains of burned cannabis, together with stones used to heat them, and a tent-frame. The practice was clearly widespread because Siberia, although still part of Scythia, is a considerable distance from the Black Sea. Herodotus seemed to interpret the exposure to burning cannabis as a form of cleansing, but the Scythians' shouting for joy suggests that intoxication occurred during the funeral rite that he witnessed. Perhaps, cannabis caused them to see 'spirits', as certain of the ancient Chinese herbals have recorded. Although Herodotus states that cannabis seeds were used, this part of the plant actually contains very little psychoactive component and so cannot have been used to produce the effects described. It is most likely that either lumps of resin were burned and he assumed that these were the seeds of a plant, or that the whole plant was burned and only the seeds survived intact for later inspection.

A section of Dacian society was known as the *kapnobatai*, or "smoke walkers", which may indicate use of cannabis intoxication. Dacia covered part of present day Transylvania and eastern Hungary, and cannabis is known to have been cultivated in neighbouring Thrace. Cunliffe has suggested that these elite may have been priests (Cunliffe, 1994). However, Dacia was annexed by the Roman empire in the first century AD and these religious traditions were rapidly suppressed.

ANCIENT GREECE AND ROME

The use of cannabis in ancient Greece and Rome has been reviewed in detail by Brunner (1973). 'Cannabis' is the Latin word for the hemp plant; in ancient Greek it was written $\kappa \acute{\alpha} \nu \nu \alpha \beta \iota \zeta$. None of the ancient Greek or Roman writers have described the intoxicating effects of cannabis upon their own citizens. This leads to the conclusion that the populace was either unaware of the intoxicating effects or chose to abstain from them for some reason. It seems unlikely that either civilisation could have been completely ignorant of this property of cannabis, since the plant was widely used at the height of each civilisation for the production of rope and coarse fabric. Perhaps, like the Chinese, cannabis did not suit the Greek or Roman temperament. Both peoples consumed large amounts of wine, and perhaps cannabis was viewed as a less desirable substitute or was taken only occasionally and in private by a select few. If cannabis was used widely it is not credible that it should go unmentioned by all classical writers. Even the more salacious or meticulous classical authors, who do not hesitate to report the depths of debauchery, do not describe it. By contrast, intoxication with alcohol is described by many.

In modern times, the most popular method of taking cannabis is via smoking, usually mixed with tobacco. The practice of smoking was completely alien to all ancient Europeans, and tobacco, of course, only reached Europe in the sixteenth century. The Scythian manner of burning cannabis in an open brazier under a tent was clearly not very sociable or elegant. It may have been viewed as a primitive custom by the ancient Greeks and Romans who regarded the Scythians as barbarians. For whatever reason, it seems unlikely that this custom was ever widely recognised or copied. The drug could have been taken orally and there seems to be a long tradition of this practice in Persia, Arabia and Egypt, but it is unpleasant to take in this way, unless the taste is heavily disguised with other substances. Furthermore it can take a considerable time for the intoxication to begin after taking cannabis orally; alcohol on the other hand is rapidly absorbed from the stomach. In addition, it is pertinent to recall that cannabis does not provide such a reliable method of intoxication as alcohol, which was very freely available in ancient Rome and Greece and very cheap. The eastern reaches of the Roman Empire included cultures where cannabis was used widely (e.g. Scythia and Arabia). However, there is evidence that even the Scythians eventually found alcohol more to their liking once it was introduced to them (Rudgley, 1993).

The only definite description of the taking of cannabis for pleasure or intoxication in the classical literature is given by the Greek historian Herodotus during his travels through Scythia (see above). However, in the Odyssey, as related by Homer, Helen adds *nepenthe* to the wine of her guests after the siege of Troy:

But Jove-born Helen otherwise, meantime,
Employ'd, into the wine of which they drank
A drug infused, antidote to the pains
Of grief and anger, a most potent charm
For ills of ev'ry name. Whoe'er his wine
So medicated drinks, he shall not pour
All day the tears down his wan cheek, although
His father and his mother both were dead,
Nor even though his brother or his son
Had fall'n in battle, and before his eyes.

                                                              (Homer, 1992 edn)

The narrative continues by explaining that the drug had been given to Helen by an Egyptian, "*For Egypt teems with drugs*". There has been much speculation as to the identity of this substance; some have suggested that it may have been cannabis (Singer and Underwood, 1962; Burton, 1894 edn [a]; Walton, 1938). As the story continues the guests do not become sedated or start to hallucinate, and so presumably the *nepenthe* was given to promote relaxation and discourse rather than heavy intoxication, sedation or psychotomimetic effects. It has also been suggested that Greek warriors may have taken nepenthe as a courage-boosting intoxicant before charging into the violence of combat (Cooper, 1995). Others have speculated that the 'wine of the condemned' cited by the Greek writer Amos in about 700 BC as a method of reducing the pain of a slow death was also cannabis (Walker, 1954). In reality it is impossible to determine the identity of substances such as these from the vague descriptions given. Frequently the events described in the Odyssey, for example, are clearly completely fabricated, and consequently the drugs depicted may also not have been based upon actual substances. The Greeks were aware of a number of psychoactive and sedating preparations apart from alcohol including opium, henbane and mandragora. Any one of these could have been the unknown drugs described by Homer or Amos.

Most other references to cannabis in the Greek and Roman literature describe the medicinal value of cannabis or its use for producing rope or material. Galen and Ephippus describe how the seeds may be cooked and eaten as a delicacy. Galen possibly hints at the intoxicating powers of cannabis in his description of those who enjoy eating the seeds:

There are some who fry and consume [the seed] together with other desserts. I call "desserts" those foods which are consumed after dinner in order to stimulate an appetite for drinking. The seed creates a feeling of warmth, and – if consumed in large amounts – affects the head by sending to it a warm and toxic vapour.

(Brunner, 1973)

It is possible that Galen misinterpreted what he saw or had described to him (it does not seem to be a first hand account). Some of his contemporaries may have burned cannabis like the ancient Scythians to produce the heady vapour that he mentions. Like Herodotus, he also may have described the parts of the plants used as "seeds" through ignorance, when resinous material or the whole plant may actually have been used.

Pliny the Elder (Gaius Plinius Secundus, c. 23–79 AD) mentions cannabis in his masterwork *Naturalis Historia* of 77 AD (Pliny, 1950 edn [a]). He classifies hemp as belonging to the "fennel class" since like fennel, dill and mallow, the hemp plant is a tall, upright, rapidly-growing shrub. Lucius Junius Moderatus Columella, also writing in the first century AD, classifies hemp differently. In *De Re Rustica* he describes it as a "pulse or legume", together with plants such as the bean, millet, flax and barley (Columella, 1960 edn [a]). Columella carefully describes the method for cultivating hemp:

Hemp demands a rich, manured, well-watered soil, or one that is level, moist, and deeply worked. Six grains of this seed to the square foot are planted at the rising of Arcturus, which means toward the end of February, about the sixth or fifth day before the Calends of March [ie February 24th or 25th]; and yet no harm will be done in planting it up to the spring equinox if the weather is rainy.

(Columella, 1960 edn [b])

He later explains that although it is possible to estimate the time and manpower required to plant, tend and harvest many related plants, for hemp "the amount of expense and attention required is not fixed". Presumably this was because the rate of growth and maturity of the plant is greatly affected by climate which varied considerably across the Roman Empire. Pliny also explains briefly how the plant is cultivated, before describing the harvest and preparation of the plant:

Hemp is sown when the spring west wind sets in; the closer it grows the thinner its stalks are. Its seed when ripe is stripped off after the autumn equinox and dried in the sun or wind or by the smoke of a fire. The hemp plant itself is plucked after the vintage, and peeling and cleaning it is a task done by candle light. The best is that of Arab-Hissar, which is specially used for making hunting-nets. Three classes of hemp are produced at that place: that nearest to the bark or the pith is considered of inferior value, while that from the middle, the Greek name for which is "middles", is most highly esteemed. The second best hemp comes from Mylasa. As regards height, the hemp of Rosea in the Sabine territory grows as tall as a fruit-tree.

(Pliny, 1950 edn [b])

Pliny explains that esparto, made from a species of coarse grass, was used as a basic source of fibre at sea *"on dry land they prefer ropes made of hemp"*, and later he mentions hemp again, reiterating that it is *"exceedingly useful for ropes"*. Several other Roman and Greek authors mention hemp and its usefulness in the production of rope, including Athenaeus, Apsyrtus, Lucilius and Varro (Brunner, 1973).

In volume twenty of the *Historia*, Pliny describes the medicinal uses of cannabis:

Hemp at first grew in woods, with a darker and rougher leaf. Its seed is said to make the genitals impotent. The juice from it drives out of the ears the worms and any other creature that has entered them, but at the cost of a headache; so potent is its nature that when poured into water it is said to make it coagulate. And so, drunk in their water, it regulates the bowels of beasts of burden. The root boiled in water eases cramped joints, gout too and similar violent pains. It is applied raw to burns, but is often changed before it gets dry.

(Pliny, 1950 edn [c])

Dioscorides and Galen both mention the ability of cannabis seed to reduce sexual potency and to treat earache. Galen also comments on its value in reducing flatulence. Pseudo-Apuleius advocates use of the herb mixed with grease to treat swelling of the chest, and cannabis mixed with nettle seeds and vinegar for cold sores (Brunner, 1973).

PERSIA AND ARABIA

Cannabis has had a long association with Persia and Arabia. Indeed the term "hashish" is Arabian and is taken from the phrase *hashish al kief* ("dried herb of pleasure"). Several early manuscripts describe the popular use of cannabis for intoxication or medicinal purposes. In the *Makhsanul aldawaiya*, an ancient Arabic drug formulary, cannabis is described as "a cordial, a bile absorber, and an appetizer, and its moderate use prolongs life. It quickens the fancy, deepens thought and sharpens judgement" (Chopra and Chopra, 1957). In the *Herbarium amboinence* written in 1095 AD, Rumphius reported that the followers of Mohammed used cannabis to treat gonorrhoea and asthma. Cannabis was also claimed to reduce bile secretion and diarrhoea, and to alleviate the distress of a strangulated hernia (Chopra and Chopra, 1957).

Sylvester de Sacy collected a series of early medieval Arabian manuscripts which describe the use of hashish. In these the Garden of Cafour, near Cairo, is cited as an infamous location for hashish smoking by fakirs, who wrote poetry to praise the intoxicating properties of the plant. An example of such poetry is given below:

The green plant which grows in the Garden of Cafour, replaces in our hearts, the effects of a wine old and generous,
When we inhale a single breath of its odour, it insinuates itself in each of our members and penetrates through the body,
Give us this verdant plant from the Garden of Cafour, which supersedes the most delicate wine,
The poor when they have taken only the weight of one drachm, have a head superb above the Emirs.

(Walton, 1938a)

In 1251 the garden was destroyed and the destruction was viewed by many as the work of God, since the taking of cannabis was widely viewed as a form of debauchery.

The writer Ebn-Beitar wrote about cannabis in the early thirteenth century. He described the low esteem with which cannabis users were viewed:

People who use it habitually have proved its pernicious effect, it enfeebles their minds by carrying them to manic affections, sometimes it even causes death... I recall having seen a time when men of the vilest class alone dared to eat it, still they did not like the name of 'takers of hashish' applied to them.

(Walton, 1938a)

Ebn-Beitar describes cannabis as *"a revolting excrement"*, and in order to illustrate the repugnance with which it was viewed by the ruling classes, he recounts the story of one local leader who attempted to rid his people of the practice of taking hashish:

L'Emir Soudon Scheikhouni to whom it pleases God to be merciful, wishing to destroy this abuse made investigations in a place named Djoneina...he had dug up all that he found of this abominable plant in these places and arrested the dissolute people who ate this drug; he ordered that the teeth of those who had eaten it be pulled and many were subjected to this ordeal.

(Walton, 1938a)

However, the historian explained that since this event the taking of cannabis became so common, that it was impossible to control either its public usage or the antisocial behaviour that he believed it engendered.

Hasan about 1260 described how Haider, leader of a holy order of fakirs, happened to eat a sample of the hemp plant whilst out walking because he was hungry. He returned to his brethren with *"an air of joy and gayety quite contrary to what we were accustomed to see"* and he subsequently encouraged all of his followers to *"take little nourishment but chiefly to eat of this herb."* (Walton, 1938a). Another Arabic writer in 1394 described the widespread use of cannabis in the Timbaliere region:

The use of this cursed plant has become today very common; libertines and feeble-minded people occupy themselves with it to excess and strive to exceed each other in its immoderate usage...without any shame.

(Walton, 1938a)

In traditional Mohammedan medicine, or *Tibbi*, the properties of cannabis have been described as: promoting insanity, causing unconsciousness, weakening the heart, annulling pain, inhibiting secretion of semen and enabling the individual to gain control over ejaculation (Chopra and Chopra, 1957).

Cannabis is mentioned in Sir Richard Burton's 1885 translation of the *1001 Tales of the Arabian Nights*. This series of mythological tales dates back to at least the 10th century, and is centred on Persia, Arabia and China. In one story, King Omar sedates Princess Abrizah, in order to seduce her, with *"a piece of concentrated bhang, if an elephant smelt it he would sleep from year to year"* (Burton, 1894 edn [b]). In another tale, a thief named Ahmad Kamakim drugs the eunuchs guarding the Caliph's valuables with hemp fumes (*Ibid.* [e]). The intoxicating and sedative effects of cannabis are described in several of the other tales (*Ibid.* [a], [d]). For example in the "Tale of the Kazi and the Bhang-eater", the Sultan and his vizier observe the Kazi and the Bhang-eater becoming intoxicated with cannabis (*Ibid.* [a]). The vizier explains to the Sultan:

"I conceive that the twain are eaters of Hashish, which drug when swallowed by man garreth him prattle of whatso he pleaseth and chooseth, making him now a Sultan, then a Wasir and then a merchant, the while it seemeth to him that the world is in the hollow of his hand... but whoso eateth it (especially an he eat more than enough) talketh of matters which reason may on no wise represent" Now when they had taken an overdose, they got into a hurly-burly of words and fell to saying things which can neither be intended nor indited.

In a footnote to the "Tale of King Omar and His Sons", Burton reveals that cannabis has been used in Arab medicine as an anaesthetic "for centuries before ether and chloroform became the fashion in the civilised West" (*Ibid.* [e]). He provides further information:

...[An] anaesthetic administered before an operation, a deadener of pain, like myrrh and a number of other drugs. For this purpose hemp is always used... and various preparations are sold at an especial bazar in Cairo.

(*Ibid.* [a])

Burton also provides details of the colloquial names for cannabis preparations and fascinating insight into the range of preparations of cannabis used in Arab countries:

The Arab "Banj" and Hindu "Bhang" (which I use as most familiar) both derive from the old coptic "Nibanj", meaning a preparation of hemp.

(*Ibid.* [a])

The Arab "Barsh" or Bars [is] the commonest kind [of hashish]. In India it is called Ma'jun (=electuary, generally); it is made of Ganja or young leaves, buds, capsules and florets of hemp (C. sativa), poppy-seed and flowers of the thorn-apple (Datura) with milk and sugar-candy, nutmegs, cloves, mace and saffron, all boiled to the consistency of treacle, which hardens when cold... These electuaries are usually prepared with "Charas," or gum of hemp, collected by hand or by passing a blanket over the plant in early morning; it is highly intoxicating. Another aphrodisiac is "Sabzi," dried hemp-leaves, poppy-seed, cucumber-seed, black pepper and carda-moms rubbed down in a mortar with a wooden pestle, and made drinkable by adding milk, ice cream etc. The Hashish of Arabia is the Hindustani Bhang, usually drunk and made as follows. Take of hemp-leaves, well washed, 3 drams; black pepper 45 grains; and of cloves, nutmeg, and mace (which add to the intoxication) each 12 grains. Triturate in 8 ounces of water, or the juice of water melon or cucumber, strain, and drink. The Egyptian Zabibah is a preparation of hemp-florets, opium, and honey, much affected by the lower orders, whence the proverb: "Temper thy sorrow with Zabibah." In Al-Hijaz it is mixed with raisins (Zabib) and smoked in the water-pipe.

(*Ibid.* [d])

In 1298, Marco Polo prepared the story of his 26 years of travelling through the Orient. In it he recounts the middle eastern legend of the "Old Man of the Mountain" who lived in Mulehet in Persia (Bellonci, 1984 edn). He probably passed through this area in 1271–2 near the fortress of Alamūt where the Old Man is reputed to have lived. The Old Man was called Alaodin. He created a magnificent, secret garden in an inaccessible valley between two mountains, the entrance to which was guarded by an impregnable fortress. He based his garden on the prophet Mohammed's vision of Paradise – it was filled with luxurious plants and fruit, golden palaces, splendid artwork, beautiful women and streams of wine, milk and honey. Alaodin selected young men with proven fighting ability and drugged them. They were then carried into the garden and, upon awakening, each believed he was in Paradise. However, a second dose of the drug was used to sedate them and return them to the fortress. Believing him to be a great prophet, the young men begged to be taken back to Paradise and, knowing that they would do anything to return, Alaodin would order them out of the fortress to murder one of his opponents before allowing them back. Any that died in the attempt knew that, after death, they would return to Paradise anyway, and all were happy to do his bidding.

Many believed that Alaodin's drug was cannabis. The Alaodin referred to by Marco Polo was one of a number of men who bore the 'Old Man' title. All of them were leaders of a sect known as the Neo-Ismailites which was founded in the eleventh century by a Persian, al-Hasan ibn-al-Sabah (d.1124). He became the first to bear the title "Old Man of the Mountain" when his sect captured the mountain fortress of Alamūt in 1090. The Neo-Ismailites were not popular with neighbouring cultures, especially devout Muslims, and were given a variety of derogatory names including *Hashshāshūns* (literally "those addicted to hashish" in Arabic) or *Hashishīyya* ("smokers of hashish" in Syriac). These may have reflected the social practices of the sect's members, or may have been simply insults (Boye, 1968). Interestingly, in Burton's translation of the *Arabian Nights*, cannabis users are always portrayed in a derogatory light. In 1809, Silvestre de Sacy concluded that the garden paradise did not really exist and that it was simply an illusion created by consuming hashish. He further advanced a theory that since the drug used by the Old Man was likely to have been hashish, his group of murderers was known as hashishins. A corruption of this word, de Sacy claimed, gave rise to the western word "assassins" (Rudgley, 1993; Boye, 1968), but this has since been contested. Nonetheless this story was later to have some impact amongst the artistic elite in nineteenth century France.

INDIA

The earliest record of cannabis in Indian literature is found in the *Atharva Veda* which may have been written as early as 2000 BC. 'Bhang' is referred to briefly. However, it is unclear whether this is a direct reference to cannabis or another sacred plant (Chopra and Chopra, 1957). Cannabis was known in India at least as early as 1000 BC because "bhanga" is mentioned in the Susruta, which dates to this period. It is advocated for the treatment of catarrh accompanied by diarrhoea, excess production of phlegm and biliary fever (O'Shaughnessy, 1839; Walton, 1938; Chopra and Chopra, 1957). The medicinal qualities of cannabis are described in more detail in the *Rajanirghanta* edited by Narahari Pandita in 300 AD. The drug was recommended as a soothing, astringent preparation which could reduce the production of phlegm, stimulate the appetite, boost the memory and alleviate flatulence (Chopra and Chopra, 1957). Indian surgeons may have used cannabis as an anaesthetic, as did those in ancient China. Another very early use for cannabis was to promote valour and allay fear in warriors about to do battle.

The Tajni Guntu, an early materia medica has been ascribed to the fourteenth century AD (O'Shaughnessy, 1839). Cannabis is described in this source as a promoter of success, and a giver of strength. Other effects described include a reeling gait, laughter and excitement of sexual desire. Garcia da Orta, a Hindu physician, gave an account of the social use of cannabis in 1563. He described the use of the plant to produce intoxication, cause hallucinations, increase appetite, allay anxiety, promote merriment and induce sleep. Other works of the sixteenth and seventeenth centuries, such as the *Dhurtasamagama* and *Bhavaprakash* mention similar indications. Cannabis formed an important part of the herbal armamentarium of traditional Hindu medicine, known as Ayuverdic. It has been used for treating the conditions already alluded too, namely: bowel disorders, reduced appetite, insomnia, reticence of speech and sadness (Chopra and Chopra, 1957).

However, cannabis has also been taken to induce intoxication for many centuries. In 1659 Aurangzeb, Emperor of India, attempted to deal decisively with the abuse of cannabis which he regarded as a vice. He regarded himself as a champion of pure Islam and a censor of public morals, and shortly after his coronation he forbade the cultivation of "bhang" throughout his realm (Dodwell, 1974). However, this was a short-lived prohibition which was virtually impossible to enforce.

In 1839, Dr W.B. O'Shaughnessy published a monograph *"On the Preparations of the Indian Hemp, or Gunjah"*. O'Shaughnessy was employed at the Medical College of Calcutta by the British East India Company. His paper was a summary of all the data that he could gather about the plant, and much of this information concerned usage in India. The various types of cannabis used in India are described below:

Sidhee, subjee, and bang (synonymous) are used with water as a drink... 540 troy grains are well washed with cold water, then rubbed to powder, mixed with black pepper, cucumber and melon seeds, sugar, half a pint of milk, and an equal quantity of water... *Gunjah* is used for smoking alone – one rupee weight, 180 grains, and a little dried tobacco are rubbed together with a few drops of water... The *Majoon*, or Hemp confection, is a compound of sugar, butter, flour, milk and sidhee or bang.

In 1893–4, the British Government, conscious of the powerful influence of cannabis in Indian Society, commissioned the seven volume Indian Hemp Drug Commission Report. This described in detail the history of cannabis use in India and the utilisation of the drug by natives at the time of publication. The report concluded that:

The evidence shows the moderate use of ganja or chara not to be appreciably harmful, while in the case of bhang drinking, the evidence shows the habit to be quite harmless. The excessive use does cause injury... [it] tends to weaken the constitution and to render the consumer more susceptible to disease... Moderate use of hemp drugs produces no injurious effects on the mind... excessive use indicates and intensifies mental instability.

(Walton, 1938b)

The government of India took the view that cannabis was so much a part of Indian society that it would be impossible and unreasonable to ban it.

In a very detailed account of the uses of cannabis in India in 1957, Chopra and Chopra reviewed the medicinal applications of the drug. Traditional indications for the use of cannabis by the various indigenous populations of India included: digestive disorders, diarrhoea, cholera and dysentry, colic, malarial fever, nervous diseases, insomnia, mania, epilepsy, hysteria, gonorrhoea, urethritis, reduced appetite, gout, rheumatic diseases, migraine, dysmenorrhoea, cough, asthma, bronchitis, oliguria, dysuria, pain and heat stroke. In addition, cannabis was applied externally to treat open wounds, haemorrhoids, conjunctivitis, orchitis, erysipelas and toothache. Cannabis was also used as a prophylactic against malaria and cholera. The authors described the method for smoking cannabis:

The equipment for smoking differs in various parts of the country, most widely used being a simple earthenware chillum similar to that used by the poorer classes for smoking tobacco, resembling a funnel with a wide base and a long neck. In addition to this, the smoker must have a brazier, a pair of tongs and a piece of cloth to be wrapped round the neck of the funnel. The method is simple. The ganja is first moistened with a little water to soften it and is then placed in the palm of the left hand and kneaded with the thumb and forefinger of the right hand to a pulpy mass. An amount of tobacco, a little less than the ganja, is then placed inside the chillum, the prepared ganja being placed on top of it. The usual practice is to put the kneaded ganja (or charas) between two thin pieces of broken earthenware, thus preventing the rapid combustion of the drug by the ignited charcoal and helping to reduce the temperature of the smoke, which might otherwise be too hot. A piece of glowing charcoal or smouldering cow-dung cake is placed with a pair of tongs on the chillum thus prepared. A piece of moistened cloth is then wrapped round the neck of the chillum, which is held between the palms of the hands. The mouth is applied to the opening formed between the thumb and forefinger of the right hand and the smoke is inhaled deeply into the lungs. The smoke is retained in the lungs for as long as possible and is then allowed to escape slowly through the nostrils, the mouth being kept shut. The longer the smoke is retained, the more potent are the effects obtained. Experienced smokers are able to retain the smoke for quite a long time.

Apart from the chillum method of smoking described above, ganja and charas are also smoked in the ordinary hookah or hubble-bubble, in which smoke is allowed to bubble through water before being inhaled.

The drug was also claimed to focus the mind for the purpose of meditation, and was, of course, also taken to promote pleasurable intoxication and as an aphrodisiac. For all these purposes cannabis was commonly adulterated with a variety of substances designed to enhance its psychotropic effects (e.g. strychnine from *Nux vomica*, alcohol, *Datura*, opium). A particularly interesting and widespread use of cannabis described by Chopra and Chopra is the taking of cannabis to promote endurance:

Cannabis drugs are reputed to alleviate fatigue and also to increase staying power in severe physical stress. In India, fishermen, boatmen, laundrymen and farmers, who daily have to spend long hours in rivers, tanks and waterlogged fields, often resort to cannabis in some form, in the belief that it will give them a certain amount of protection against catching cold. Mendicants who roam about aimlessly in different parts of India and pilgrims who have to do long marches often use cannabis either occasionally or habitually. Sadhus and fakirs visiting religious shrines usually carry some bhang or ganja with them and often take it. It is not unusual to see them sitting in a circle and enjoying a smoke of ganja in the vicinity of a temple or a mosque. Labourers who have to do hard physical work use cannabis in small quantities to alleviate the sense of fatigue, depression and sometimes hunger.

Veterinary uses of cannabis included the promotion of bovine lactation, the disinfection of sheep pens, the treatment of intestinal worms, and a general tonic for bullocks. It was even employed as an aphrodisiac for mares prior to mating.

SOUTH AMERICA

It is clear that the use of cannabis has gradually and progressively spread both east and west from the original indigenous area in central Asia. However, the most intriguing details concerning the historical spread of cannabis usage have come from South America. In 1993, a team of German anthropologists published the results of an analysis of various tissues from seventy-two Peruvian mummies dated 200–1500 AD (Parsche *et al.*, 1993). Bones from twenty of them were shown to contain cannabinoids. In the same study, ten bodies from the German Bell Culture (2500 BC) did not contain cannabinoids. In addition, two African mummies from the Sudan (dated at 5000–4000 BC and 400–1400 AD) also did not contain cannabis.

Cannabis is not native to the Americas, suggesting that there may have been some contact between South America and Asia or Egypt in antiquity. Assuming the results of the Peruvian mummy analysis are correct, it is difficult to explain them without conjecturing that some form of transatlantic communication must have occurred before the arrival of Columbus (Moore, 1993). Until this information came to light, it was generally assumed that hemp was introduced to Chile by the Spanish during their conquest of 1545 onwards. Use of cannabis in South America is known to have been boosted by the arrival of African slaves in the seventeenth and eighteenth centuries.

EUROPE AFTER 500 AD

From earliest times, hemp was the basic commodity used for the manufacture of the ropes and rigging of sailing vessels in Europe. Often the ropes were impregnated with tar to render them waterproof and make them more durable.

Hildegard of Bingen mentions the analgesic properties of cannabis in the *Physica* published in the twelfth century. Peter Schoofer discusses the therapeutic properties of the drug in his herbal printed in Mainz in 1485. He advocated use of the plant in various guises as an analgesic, and to treat gastrointestinal disorders, oedema and as a plaster for boils (Mechoulam, 1986).

Cannabis, opium and alcohol were some of the less well known ingredients used by the European witch cult of the late middle ages in association with the more widely recognised solanaceous "hexing herbs" (Rudgley, 1995).

The cannabis plant has formed part of the culture and folklore of several eastern european countries for centuries. This may be because eastern europeans have been more reliant on the plant than those from the west, and cannabis has also had a longer history of association with the east. In the Ukraine, for example, cannabis harvested on St John's Eve was thought to deter evil actions (Richardson, 1988). Slavs from the south of Europe believed that the appearance of cannabis on a wedding day foretold future marital harmony and contentment. Consequently, cannabis seeds were thrown after the ceremony, and the bride was encouraged to brush the walls of her house with the plant. In Poland, the hemp dance was performed on Shrove Tuesday. Sometimes the seeds were eaten on special occasions. In Poland and Lithuania, cannabis seed soup has traditionally been prepared on Christmas Eve, and in Latvia and the Ukraine the seed is eaten on Three Kings' Day.

ENGLAND BEFORE 1800

The earliest record of the use of hemp in Britain comes from seeds found in a Roman well in York. The finding of large tracts of pollen in parts of East Anglia indicate that hemp was grown there throughout the Anglo-Saxon period (Wild, 1988). The British climate of the period would not have supported even a reasonable harvest of resin, so the plants must have been grown for fibre production. There is one reference to the medicinal use of cannabis during this period – it is mentioned as an ingredient in an Anglo-Saxon herbal, the Lacnunga, where it is called *haenep* (Grattan and Singer, 1952). Usually it is the resin that is used medicinally, rather than other parts of the plant, and this must have been imported.

The playwright, William Shakespeare, does not mention the intoxicating or medicinal properties of cannabis in any of his works, although he does refer to its value for the production of rope on several occasions. For example in Henry V, reflecting a tradition that hemp was used to produce the hangman's noose, Pistol declares:

Let gallows gape for dog, let man go free,
And let not hemp his wind-pipe suffocate.

(Henry V, Act iii, Scene vi, line 45–6)

Other Tudor and Jacobean writers discuss the therapeutic value of cannabis in some detail. William Turner produced "The New Herbal" in 1528. In it, he describes the properties of Cannabis as follows:

Of Hempe: Cannabis named both of the Grecians and latines, is called in English hempe, in Duche hanffe, in French chanure. Hemp sayeth Dioscorides, is profitable for many things in mans lyfe, and especially to make strong cables and roopes of. It hath leaves like an Ashe tree, with a strong sauour [savour], longe stalkes and round sede.

The herbals of three other Englishmen were widely known during the seventeenth century. John Gerard first published "The Herbal or Generall Historie of Plantes" in 1597 but it was reprinted several times; John Parkinson published "The Theater of Plantes, an Universall and Complete Herbal" in 1640, and Nicholas Culpeper produced his "English Physician and Complete Herbal" in 1653. All three publications explain the medicinal value of cannabis. Their accounts are based largely on the works of Greek and Roman authors such as Pliny, Galen and Dioscorides with little new information. They are remarkably similar. Culpeper's herbal is perhaps the most famous, and the complete entry for hemp is given below:

This is so well known to every good housewife in the country, that I shall not need to write any description of it.

Time It is sown in the very end of March, or beginning of April, and is ripe in August or September.

Government and virtues It is a plant of Saturn, and good for something else, you see, than to make halters only. The seed of Hemp consumes wind, and by too much use thereof disperses it so much that it dries up the natural seed for procreation; yet, being boiled in milk and taken, helps such as have a hot dry cough. The Dutch make an emulsion out of the seed, and give it with good success to those that have the jaundice, especially in the beginning of the disease, if there be no ague accompanying it, for it opens obstructions of the gall, and causes digestion of choler.

The emulsion or decoction of the seed stays lasks and continual fluxes, eases the cholic, and allays the troublesome humours in the bowels, and stays bleeding at the mouth, nose, or other places, some of the leaves, being fried with the blood of them that bleed, and so give to them to eat. It is held very good to kill the worms in men or beasts; and the juice dropped into the ears kills worms in them; and draws forth earwigs, or other living creatures gotten into them. The decoction of the root allays inflammations of the head, or any other parts: the herb itself, or the distilled water thereof doth the like. The decoction of the root eases the pains of the gout, the hard humours of the knots in the joints, the pains and shrinking of the sinews, and the pains of the hips. The fresh juice mixed with a little oil or butter, is for any place that hath been burnt with fire, being thereto applied.

(Culpeper, 1653)

Parkinson offers some interesting additional information. He, like Culpeper and Gerard, suggests that hemp is useful to kill worms in "man or beast", but intriguingly he adds that a decoction of hemp "powred into the holes of earthwormes, will draw them forth, and fishermen and anglers have used this feate to get wormes to baite their hookes" (Parkinson, 1640). He also quotes Matthiolus:

...as Matthiolus saith, the women in Germany went a wrong course, to give their children the decoction of Hempe seeds for the falling sickness [ie epilepsy], which it did rather augment, then helpe to take away... Matthiolus saith that Hempe seede, given to Hennes in the winter, when they lay fewest egges, will make them lay more plentifully.

Robert Burton, an English clergyman, mentions cannabis in his study entitled "The Anatomy of Melancholy" which he published in 1621. He proposed cannabis as a treatment for depression. John Quincy in the English Dispensatory of 1728 reported that the seeds of hemp "are claimed to abate venereal desires and jaundice", but this was the first publication to doubt the veracity of these traditional uses: "there is no authority which has justified their being included in prescriptions." The New English Dispensatory of 1764 advocated the application of the root of the cannabis plant to treat inflammation of the skin, while the Edinburgh New Dispensatory of 1794 submitted that cannabis seed had been used to treat coughs and "heat of urine" (presumably cystitis), and might also restrain venereal appetites (Lewis, 1794). The author noted, however, that the value of cannabis in ameliorating the symptoms of these conditions was not borne out by experience and that "other parts of the plant may be considered as deserving further attention". The updated version of this publication from 1804 described hemp succinctly; "smell weak, taste mawkish, effects emmollient, anodyne" (Duncan, 1804). The term 'anodyne' usually indicated an analgesic effect.

FRANCE IN THE NINETEENTH CENTURY

In France, in the 1840s, cannabis was used by Dr Jacques Joseph Moreau ('Moreau de Tours') in an attempt to treat mental illness. He probably first took the drug himself whilst travelling in Arab countries during his youth. However, from the beginning of the 19th century, cannabis was also brought back to France by Napoleonic soldiers returning from the east. French physicians accompanying the army to the east had seen the unusual properties of the resin and had used it to combat pain amongst the soldiers. Moreau's colleague, Louis Albert-Roche, had produced a book in 1840 in which he claimed that cannabis was an effective treatment for plague. Moreau himself published several papers describing the successful use of cannabis to alleviate symptoms of mental illness.

His most famous publication was *Du Hachisch et de l'Alienation Mentale, Etudes Psychologiques* (Hashish and Mental Illness; Psychological Studies). This was published in 1845. He introduces the second chapter as follows:

At first, curiosity led me to experiment upon myself with hashish. Later, I readily admit, it was difficult to repress the nagging memory of some of the sensations it revealed to me. But from the very outset I was motivated by another reason.

I saw in hashish, or rather in its effect upon the mental faculties, a significant means of exploring the genesis of mental illness. I was convinced that it could solve the enigma of mental illness and lead to the hidden source of the mysterious disorder that we call "madness".

(Moreau, 1973 edn [a])

Moreau obtained cannabis as imports from French possessions in the east, but he also grew the plant in Paris (Siegel and Hirschman, 1991). In himself he reported that cannabis produced symptoms akin to an acute psychotic state and, in Egypt, Moreau had observed that chronic, excessive use could cause symptoms similar to the mental illnesses that he later studied in institutions in France.

Anyone who has visited the Orient knows how widely used hashish is, especially among the Arabs, who have developed no less pressing a need for it than the Turks and Chinese for opium or the Europeans for alcoholic beverages.

(Moreau, 1973 edn [b])

Moreau describes the effect of cannabis intoxication upon himself at great length, including detailed accounts of a series of bizarre hallucinations and illusions. At one point he writes:

Twenty times I was on the verge of indiscretion, but I stopped myself, saying, "I was going to say something, but I must remain quiet." I cannot describe the thousand fantastic ideas that passed through my brain during the three hours that I was under the influence of the hashish. They seemed too bizarre to be credible. The people present questioned me from time to time and asked me if I wasn't making fun of them, since I possessed my reason in the midst of all that madness.

(Moreau, 1973 edn [b])

He is keen to suggest that cannabis may have therapeutic value, but does not exaggerate the significance of his published observations:

Unfortunately, I have only a few cases to present, and I am not ready to assert that these cases can justify any opinion concerning the effectiveness of extract of Indian hemp upon a specific mental illness.

(Moreau, 1973 edn [c])

Edmond DeCourtive, a pharmacist and student of Moreau, began to experiment with hashish in 1847. He formulated cannabis resin into highly concentrated pills and supplied them to Moreau and colleagues. He is also credited with writing the first thesis on hashish, a complete copy of which has recently come to light (Siegel and Hirschman, 1991). In it he describes, in a charmingly youthful style, the effects of cannabis administration on animals:

Jays, magpies, sparrows, canaries are relaxed and sad and do not eat. Music awakens and animates them a great deal. Geese are anxious and dazed. ...Dogs, besides showing greater agility and extreme good humour, are greatly impressed by music, much more so than in the normal state.

(Siegel and Hirschman, 1991)

He also recounted the story of a friend whom he found under the influence of the drug: *"he was laughing so hard that he made me laugh and he laughed even harder"* The friend cried *"I am too happy; I want some hashish; give me enough to die!"*.

Shortly after the publication of his most renowned work Moreau, together with the writer Théophile Gautier, established *Le Club des Haschischins* in about 1846. It was heavily inspired by Marco Polo's tale of Alaodin and the Assassins from six centuries before (Rudgley, 1993). *Le Club* met initially amidst the decaying splendours of the Hôtel Pimodan in Paris, and later in the apartments of Roger de Beauvoir, a wealthy Parisian socialite. At various times, members of *Le Club* included literary figures such as Gérard de Nerval, Honoré de Balzac, Hector Horeau and Alexandre Dumas, as well as Boissard de Boisdenier the painter and Charles Baudelaire the poet.

Dumas evokes beautifully the spirit and atmosphere of *Le Club des Haschischins* in his classic novel 'The Count of Monte-cristo', published in 1844–5. In it, the Count tempts a young visitor to try the cannabis that he holds before him in a silver cup:

Are you a man of imagination – a poet? taste this, and the boundaries of possibility disappear; the fields of infinite space open to you, you advance free in heart, free in mind, into the boundless realms of unfettered reverie. Are you ambitious, and do you seek after the greatness of the earth? taste this, and in an hour you will be a king, not a king of a petty kingdom hidden in some corner of Europe like France, Spain or England, but king of the world, king of the universe, king of creation; without bowing at the feet of Satan, you will be king and master of all the kingdoms of the earth...when you return to this mundane sphere from your visionary world, you would seem to leave a Neapolitan spring for a Lapland winter – to quit paradise for earth – heaven for hell! taste the hashish, guest of mine, – taste the hashish!

(Dumas, 1936 edn)

The visitor is persuaded to try the drug and then:

with his eyes closed upon all nature his senses awoke to impassable impressions, and he was under the painful yet delicious enthralment produced by the hashish...

In 1858 Baudelaire published 'The Poem of Hashish'. In it he briefly describes the history of cannabis use, the effects of the drug upon himself and his friends, and discusses the moral, psychological and philosophical impact of cannabis use. He explains that *"hashish reveals to the individual nothing except himself"* (Baudelaire, 1950 edn) and that although the drug stimulates imagination and creative thinking it is a spur to deep thought and philosophy, not a substitute for it: *"what hashish gives with one hand it takes away with the other: that is to say, it gives power of imagination and takes away the ability to profit by it.*" He warns that:

"It is said, and is almost true, that hashish has no evil physical effects; or, at worst, no serious ones. But can it be said that a man incapable of action, good only for dreaming, is truly well, even though all his members may be in their normal condition? ... hashish, like all other solitary delights, makes the individual useless to mankind, and also makes society unnecessary to the individual."

(Baudelaire, 1950 edn).

The taking of cannabis seems to have been uncommon in France outside the circles of the Parisian artistic elite. In the rest of Europe and the US, cannabis abuse was similarly rare. Opium, tobacco and alcohol were much more popular. Cannabis was a drug with interesting pharmacological properties, but no clear-cut medical uses. Throughout the nineteenth century it was never regarded as a 'drug of abuse'.

VICTORIAN BRITAIN

Samuel Gray's *Supplement to the Pharmacopoeia and Treatise on Pharmacology*, published in 1833, noted that the juice of the cannabis plant could be made into an "agreeable inebriating drink" which would "produce fatuity". The dried leaves, Gray noted, could be used as tobacco. This is the first hint of cannabis use for intoxication in the British literature.

Cannabis was brought to the attention of the western medical profession formally by O'Shaughnessy's paper of 1838 in the Bengal Medical Journal. The opening paragraph of his paper highlights the lack of interest in cannabis amongst Europeans of the time:

But in Western Europe, its use either as a stimulant or as a remedy, is equally unknown. With the exception of the trial, as a frolic, of the Egyptian 'Hasheesh', by a few youths in Marseilles, and of the clinical use of the wine of Hemp by Mahneman... I have been unable to trace any notice of the employment of this drug in Europe.

(O'Shaughnessy, 1838)

O'Shaughnessy administered cannabis to animals and described the effects produced. His conclusion from this experimentation was that the drug did not appear to be harmful, but that not all animals were affected equally:

It seems needless to dwell on the details of each experiment; suffice it to say that they led to one remarkable result – That while carnivorous animals, and fish, dogs, cats, swine, vultures, crows, and adjutants [=storks], invariably and speedily exhibited the intoxicating influence of the drug, the graminivorous, such as the horse, deer, monkey, goat, sheep, and cow, experienced but trivial effects from any dose we administered.

O'Shaughnessy administered cannabis to several human subjects and claimed success in alleviating symptoms in a small number of cases of rheumatism, cholera, tetanus and infantile convulsions. He also used it as a sedative in an incurable case of rabies. He described the intoxication ("delirium") arising from "incautious use of hemp prepara-tions... especially among young men first commencing the practice." His observations are reproduced below:

The state is at once recognized by the strange balancing gait of the patient's; a constant rubbing of the hands; perpetual giggling; and a propensity to caress and chafe the feet of all bystanders of whatever rank. The eye wears an expression of cunning and merriment which can scarcely be mistaken. In a few cases, the patients are violent; in many highly aphrodisiac; in all that I have seen, voraciously hungry.

A blister to the nape of the neck, leeches to the temples, and nauseating doses of tartar emetic with saline purgatives have rapidly dispelled the symptoms in all the cases I have met with, and have restored the patient to perfect health.

However, it was not only in the colonies of the British Empire that the intoxicating powers of cannabis came to the attention of doctors. In 1845, DeQuincey described the case of a Scottish farmer who found the taking of cannabis much to his liking:

One farmer in Midlothian was mentioned to me eight months ago as having taken it, and ever since annoyed his neighbours by immoderate fits of laughter; so, that in January it was agreed to present him to the sheriff as a nuisance. But for some reason the plan was laid aside; and now, eight months later, I hear that the farmer is laughing more rapturously than ever, continues in the happiest frame of mind, the kindest of creatures and the general torment of the neighbourhood.

(Walton, 1938c)

In 1850, David Urquhart, an English Member of Parliament described his experience of cannabis intoxication whilst travelling through Morocco:

Images came floating before me – not the figures of a dream, but those that seem to play before the eye when it is closed... The music of the wretched performance was heavenly, and seemed to proceed from a full orchestra... I was following a new train of reasoning; new points would occur, and concurrently there was a figure before me throwing out corresponding shoots like a zinc tree...

(Walton, 1938c)

Edward Birch, a British doctor based in Calcutta in 1887, described the use of cannabis to cure chloral or opium addiction. He reported two cases where cannabis was successful and was surprised by the "_immediate action of the drug in appeasing the appetite for the chloral or opium_". He advocated further research, but concluded with a warning:

Upon one point I would insist – the necessity of concealing the name of the remedial drug from the patient, lest in his endeavour to escape from one form of vice he should fall into another, which can be indulged in any Indian bazaar...

(Birch, 1889)

In the Lancet of 15th March 1890, a correspondent, identified only as "W.W.", reported the ill effects of the drug on himself after taking it to treat neuralgic pains. The experience was extremely unpleasant, consisting of delusions of sensory deprivation, partial paralysis, anxiety, hysteria and suicidal ideation. Not surprisingly the correspondent was "_determined never again to take cannabis_". He concluded: "_I may state that I experienced no pleasurable intoxication or feeling of happiness, but the very reverse._" (W.W., 1890). Awareness by British doctors that cannabis could be taken to provoke 'pleasurable intoxication', suggests that the drug was already taken for this purpose in Victorian England by a minority. Indeed, Sir Richard Burton, in a footnote to The Arabian Nights, written in 1885, notes that:

I heard of a "Hashish-orgie" in London which ended in half the experimentalists being on their sofas for a week. The drug is useful for stokers, having the curious property of making men insensible to heat.

(Burton, 1894 edn [c])

I have smoked it and eaten it for months without other effect than a greatly increased appetite and a little drowsiness.

(Burton, 1894 edn [f]).

The letter from "W.W." in the Lancet prompted a famous reply from Dr J. Russell Reynolds, Queen Victoria's physician (Reynolds, 1890). He explained that contrary to W.W.'s experience, "*Indian hemp, when pure and administered carefully, is one of the most valuable medicines we possess.*" Reynolds gave details of medical conditions which he had treated with cannabis including senile dementia, neuralgia, migraine, cramps, asthma, non-classical epilepsy and dysmenorrhoea. Conditions which he found were not amenable to cannabis included mania, sciatica, tinnitus, dystonias and chronic epilepsy. He concluded his letter as follows:

...the object of this communication will be attained if, by giving my experience of the great value of Indian hemp, my brethren may be deterred from abandoning its use by any dread of its causing "toxic effects", unless it be given in a "toxic" dose.

One year later, a letter in the British Medical Journal reported the value of cannabis for treating "*insanity... brought on usually by mental worry*", depression, mania, migraine, gastric ulceration, insomnia and "*chorea when arsenic fails*" (Anon., 1891). Cannabis, like so many medicaments in Victorian England, was used to treat a wide variety of disparate conditions, often in a most inappropriate and illogical manner. One writer, for example explains that:

The most popular and in many respects the most effectual remedies for corns are those containing salicylic acid. In these, extract of Indian Hemp [Cannabis] is generally found – why, it is difficult to say, but it gives a nice colour and acts faintly as a sedative. The original formula... came from Russia in 1882.

(MacEwan, 1901)

The benefits of sedation in the treatment of corns is not generally recognised today.

The first account in the British medical literature of the adverse consequences of cannabis taken deliberately for abuse are described by Foulis in the Edinburgh Medical Journal of 1900. He relates the story of two brothers who took a large dose of cannabis to investigate its intoxicating effects. One brother (referred to as 'A') became very depressed as a result of taking the preparation, the other ('B') became exhilarated, boisterous and almost manic ("*it was impossible to get B to do anything else than dance and sing and talk*"). In a description of their admission to the local infirmary in a highly intoxicated state, brother A later recounted:

We were then handed over to the nurse. I am afraid that she did not look upon us in a very favourable light, although I tried to impress upon her that it was an extremely interesting experience. Indeed, it was with difficulty that I could persuade anyone that we had not been taking the common drink.

In 1925, attempts were made via the League of Nations to limit the widescale use of cannabis largely because of an alleged association with insanity in Egypt. Consequently trade in cannabis was restricted by the International Opium Convention. However, cannabis abuse in the UK continued to be rare until after World War II. It is known that cannabis was imported into the country for coloured seamen and entertainers in London in the late 1940s (Zacune and Hensman, 1973). The spread of cannabis use was fuelled because of its association with the increasingly popular jazz scene, and because many of the immigrants entering Britain at this time had used the drug in their countries of origin. Usage was encouraged by *avant garde* literary figures of the era. As in the USA, it was not until the 1960s that cannabis smoking became fashionable with young white people, and it has remained popular ever since. Cannabis abuse was made illegal in the UK in 1964 and at the time of writing is subject to the Misuse of Drugs Act (1971). Around 15–30% of teenagers have taken cannabis and it accounts for 80–90% of drug-related offences dealt with by the police. In 1990, there was estimated to be one million cannabis users in the UK, a figure which does not differ significantly from mid-1980s estimates.

CANNABIS AND NORTH AMERICA

The British navy used hemp extensively for the manufacture of ropes and rigging in the seventeenth century, but UK farmers were unwilling to grow the plant because it was believed to deprive the soil of nutrient. In addition, hemp demanded time-consuming preparation, which was also unpleasant due to the smell produced by the retting process. Consequently, the British Government tried to persuade settlers in Virginia and Maryland to grow cannabis to meet the navy's demand, but this measure met with only limited success because tobacco was a much more profitable crop. The hemp industry became more important during the American Revolution, when fibre was difficult to obtain from Europe, but production declined rapidly after the Civil War due to spiralling labour costs and the preferential use of alternative types of fibre such as cotton. In the Midwest, particularly, hemp was still grown for a while as a source of birdseed. Throughout this period of cannabis farming, the plant gradually escaped the confines of farmland and began to grow in the wild. In Nebraska it was estimated in 1969 that there was approximately 150,000 acres of wild hemp growing (Richardson, 1988).

In 1854, Bayard Taylor recounted his experiences under the influence of cannabis whilst taking a holiday in Damascus. He had on a previous occasion taken a small dose whilst in Egypt which produced a pleasurable experience "*so peculiar in its character, that my curiosity instead of being satisfied, only prompted me the more to throw myself, for once wholly under its influence*" (Walton, 1938c). Unfortunately, he inadvertently took a rather large overdose and suffered an experience which although initially pleasurable and controlled, rapidly became very frightening.

The spirit (demon, shall I not rather say?) of Hasheesh, had entire possession of me... My perceptions now became more dim and confused. I felt that I was in the grasp of some giant force; and in the glimmering of my fading reason, grew earnestly alarmed, for the terrible stress under which my frame labored increased every moment... By this time it was nearly midnight. I had passed through the Paradise of Hasheesh, and was plunged at once into its fiercest hell... Every effort to preserve my reason was accompanied with a pang of mortal fear, lest what I now experienced was insanity, and would hold mastery over me forever..."

It took several days for Bayard Taylor's alarming reaction to completely resolve. He later wrote of his subsequent dependence upon cannabis. It became a daily habit which lasted four or five years. He was only released from his compulsion when overtaken by a fever in Syria.

While the sickness continued, I could not take the hasheesh; and when I recovered, I had so far gained my self-control, that I resolved to fling the habit aside forever.

In 1857, another American, Ludlow, published accounts of his experiences of cannabis intoxication. He reported effects which were frightening, pleasant or enlightening. In one magazine article he described the manner in which cannabis intoxication could act as a form of revelation:

How is it that the million drops of memory preserve their insulation, and do not run together in the brain into one fluid chaos of impression?... How does spirit communicate with matter, and where is their point of tangency?... Problems like these, which have been the perplexity of all my previous life, have I seen unravelled by hasheesh, as in one breathless moment the rationale of inexplicable phenomena has burst upon me in a torrent of light.

(Walton, 1938c)

In many respects the rise in cannabis use in the USA followed a similar course to that seen in the UK. The first reference to cannabis abuse in a major medical publication appeared in the Boston Medical and Surgical Journal of 1857. John Bell described the effects of the drug upon himself and suggested as Moreau did before him, that cannabis, by producing symptoms similar to a manic state, *may in a degree serve as a key to unlock some at least of the mysteries of mental pathology*". Throughout the late 19th century the drug was used in the US to treat a variety of disparate medical conditions.

In 1860, a committee of the Ohio State Medical Society published a report on the medical uses of cannabis. Information was provided on its use to treat puerperal psychosis, various pains (inflammatory, neuralgic, abdominal), gonorrhoea, cough and so forth. One of the contributors, Dr Fronmueller, compared cannabis favourably to opium as an analgesic. He suggested that although cannabis was a less reliable and less potent analgesic, the side effect profile was better. Unlike opium, cannabis did not reduce appetite, produce constipation, cause nausea or vomiting, affect lung function and "the nervous system is also not so much affected" (McMeens, 1860).

Cannabis smoking for pleasure was probably introduced to the US on a wide scale by immigrant Mexican agricultural labourers. However, in an "anthropological footnote" to his translation of the Arabian Nights, written in 1885, Burton discusses the taking of cannabis to produce intoxication:

I found the drug well known to the negroes of the Southern United States and of Brazil, although few of their owners had ever heard of it.

(Burton, 1894 edn [b])

Whether the Afro-Caribbeans' knowledge was based on experience from their native lands, or whether they acquired it subsequent to arrival (presumably from Mexican labourers) is unclear. The word 'marihuana' is the term predominantly used in the US today to describe cannabis. It is derived from the Mexican-Spanish *marijuana*, the original meaning of which is now obscure. Initially, it may have been used to describe rough grade tobacco (Bowman and Rand, 1981). It has also been suggested that the word is derived from the Portuguese *maran guango* meaning "intoxication" (Brunner, 1973). By the time that cannabis began to be abused more widely for its intoxicating effects, the medical usage had declined considerably. By 1942, cannabis was no longer listed in the United States Pharmacopoeia.

From the immigrant agricultural community, cannabis use soon spread to the jazz scene (Musto, 1991). Throughout the 1930s there were almost frenetic anti-cannabis crusades, public education drives, newspaper campaigns and police actions against suppliers and users alike. Possession, sale and administration of the pharmacologically active parts of the cannabis plant was eventually made illegal by Public Law Number 238 on 2nd August 1937. This is commonly referred to as the Marihuana Transfer Tax Law or Marihuana Tax Act. In 1938, Robert Walton, Professor at the Mississippi School of Medicine, wrote of the growing popularity of indulging in cannabis intoxication which "*is currently expanding among the idle and irresponsible classes of America and Russia and which has maintained itself for centuries among the more dissolute populations of the East*" (Walton, 1938a). He further explained that:

Marihuana smoking in the United States is a development which has taken place almost entirely within the last 20 years. In fact, the epidemic proportions of this vice have only become manifest in the last 5 or 10 years.

As in Europe, it was not until the 1960s that cannabis use became very widespread. Legal restrictions were extended to the isolated active principle (tetrahydrocannabinol) in 1968. In 1970, the Comprehensive Drug Abuse Prevention and Control Act was introduced. However, by the early 1970s, US tolerance of the drug was such that a Presidential Commission in 1972 recommended 'decriminalisation' of cannabis. In 1973, Oregon was the first state to rescind legislation which prohibited the personal use of cannabis. Subsequently many states issued similar laws. These State Laws do not allow users to grow cannabis. In 1977, President Carter's government supported legalised possession of small amounts of the drug (Musto, 1991). Since then the popularity of cannabis in the US has waxed and waned, but not all new laws have been in favour of liberalisation – in 1990, for example, the people of Alaska voted to re-criminalise the possession of cannabis.

Cannabis is used extensively in the West Indies. It is thought to have been introduced there by workers arriving from India, and elsewhere in Asia, during the mid-1800s. The drug is particular popular with the Rastafarian culture in Jamaica, who attribute divine power to it, although many other West Indians also use the drug (Edwards *et al.*, 1982).

THE ANALYSIS OF CANNABIS

Throughout the nineteenth century various attempts were made to isolate and purify the psychoactive constituents of cannabis. The first chemical analysis of the resin was probably undertaken by Tscheepe in 1821 and this was followed by a large number of other studies (Mechoulam, 1973). None of these was successful in isolating a single active principle, probably because investigators usually assumed that the substance would be an alkaloid like most other plant-derived drugs known at the time. Knowledge of analytical chemistry and the available technology was also primitive – most investigations relied on fractional distillation of resin dissolved in numerous different solvents. Various alkaloids were isolated, and nicotine was even proposed as the active constituent (Hay, 1883). This was quickly refuted by others who pointed to the common practice of mixing cannabis with tobacco for smoking.

In 1899 Wood *et al.* isolated a crude liquid from which they later separated cannabinol. This was the first cannabinoid to be isolated in pure form. It was initially claimed to be the major pharmacologically active component, despite an apparently low yield. Subsequent research became confused by the inability to repeat this work and, in many cases, a failure to realise that extracts assumed to be pure were, in reality, often mixtures of several chemicals with different pharmacological properties, but similar chemical properties and structure.

In 1965 Mechoulam and Gaoni isolated delta-1-tetrahydrocannabinol ("delta-1-THC") amongst a group of other constituents using chromatography. The chemical structure was elucidated and, in 1970, it was shown to be psychoactive in rhesus monkeys (Mechoulam *et al.*, 1970).

Since 1965, a revision of nomenclature has meant that delta-1-THC is now known as delta-9-THC. It is known to be the main psychoactive constituent of cannabis resin, and is usually referred to simply as "THC".

REFERENCES

Anon. (1891), On the therapeutic value of Indian hemp, Br. Med. J. (4th July), **II**, 12.

Baudelaire, C. (1950), The Poem of Hashish, In *My Heart Laid Bare and Other Prose Writings*, Weidenfeld & Nicolson, London, pp. 75–123.

Bell, J. (1857), On the haschisch or Cannabis indica, *Boston Medical and Surgical Journal*, **56**(11), reprinted in *J. Substance Abuse Treatment* **2** (1985), pp. 239–243.

Bellonci, M. (1984 edn), tr. Waugh, T., *The Travels of Marco Polo*, Book Club Associates, London, Ch. XLI-XLIII, pp. 38–39.

Birch, E.A. (1889), The use of Indian hemp in the treatment of chronic chloral and chronic opium poisoning, *Lancet* (30th March), **I**, 625.

Bowman, W.C. and Rand, M.J. (1980), Cannabis, In *Textbook of Pharmacology*, Blackwell, Oxford, pp. 42.43–42.61.

Boyle, J.A. (1968), *The Cambridge History of Iran*, Cambridge University Press, London, p. 443 and pp. 453–4.

Burton, R.F. (1894 edn [a]), *The Book of the Thousand Nights and a Night*, Nichols & Co, London, Vol. I, p. 65 (footnote).

Burton, R.F. (1894 edn [b]), *Ibid.*, Vol. II, p. 25–27.

Burton, R.F. (1894 edn [c]), *Ibid.*, Vol. II, p. 315–6.

Burton, R.F. (1894 edn [d]), *Ibid.*, Vol. III, p. 159 (footnote).

Burton, R.F. (1894 edn [e]), *Ibid.*, Vol. III, p. 196–7.

Burton, R.F. (1894 edn [f]), *Ibid.*, Vol. XI, p. 14–30.

Chopra, I.C., and Chopra, R.N. (1957), The use of the cannabis drugs in India, *Bull. Narcotics*, **9**, 4–29.

Columella, L.J.M. (1960 edn [a] tr. Boyd), *On Agriculture*, Heinemann, London, Vol. I, Ch. 7, v. 1, p. 139.

Columella, L.J.M. (1960 edn [b] tr. Boyd), *Ibid.*, Vol. I, Ch. 10, v. 21, p. 169–70.

Cooper, P. (1995), Herbs of Homer, *Pharm. J.*, **255**, 898–900.

Creighton, C. (1903), On indications of the haschish vice in the Old Testament, *Janus*, **8**, 241 and 297.

Culpeper, N. (1653), *The Complete Herbal*, Petert Cole, London, p. 128–9.

Cunliffe, B. (1994), *Oxford Illustrated Prehistory of Europe*, OUP, Oxford, p. 405.

Dodwell, H.H. (1974) (ed.), *The Cambridge History of India*, Cambridge University Press, London, Vol. IV, p. 230.

Dumas, A. (1936 edn), *The Count of Monte-Cristo*, Odhams Press, London, Vol. 1, Ch. 31, pp. 301–306.

Duncan, A. (1804), *Edinburgh New Dispensatory*, Bell and Bradfute, Edinburgh, p. 363.

Edwards, G., Arif, A. and Jaffe, J. (1982), Jamaica: Contrasting patterns of cannabis use, In *Drug Use and Misuse – Cultural Perspectives*, Croom Helm, London, pp. 70–73.

Foulis, J. (1900), Two cases of poisoning by Cannabis indica, *Edinburgh Medical Journal*, **8**, 202–210.

Grattan, J.H.G. and Singer, C. (1952), *Anglo-Saxon Magic and Medicine*, Oxford University Press, London, p. 84 and p. 123.

Gray, S.F. (1833), *Supplement to the Pharmacopoeia and Treatise on Pharmacology*, Underwood, London, p. 36.

Guthrie, D. (1946), *A History of Medicine*, Thomas Nelson and Sons, London p. 36.

Hay, M. (1883), A new alkaloid in cannabis indica, *Pharm. Journal*, June 2nd, pp. 998–999.

Herodotus tr. Rawlinson, G. (1949) *The History of Herodotus*, Everyman's Library, J.M. Dent & Sons, London, Vol. 1, Book IV, Ch. 73–75, pp. 315–316.

Homer (1992 edn), *The Odyssey*, tr. Cowper, W., J.M. Dent & Sons, Everyman's Library, London, Book IV, Lines 275–284, p. 53.

Lewis, Dr. (1794), *Edinburgh New Dispensatory*, Edinburgh Press, Bell and Bradfute, Edinburgh, p. 126.

MacEwan, P. (1911), *Pharmaceutical Formulas, Being "The Chemist and Druggist's" Book of Useful Recipes for the Drug Trade*, Chemist and Druggist, London, p. 824.

Mechoulam, R. and Gaoni, Y. (1965), *J. Amer. Chem. Soc.*, **8**, 3273.

Mechoulam, R., Shani, A., Edery, H. and Grunfield, Y. (1970), *Science*, **169**, 611.

Mechoulam, R. (1973), Cannabinoid Chemistry, In *Marijuana – Chemistry, Pharmacology, Metabolism and Clinical Effects*, Mechoulam, R. (ed.), Academic Press, London, pp. 1–29.

Mechoulam, R. (1986), The Pharmacohistory of Cannabis sativa. In Mechoulam, R. (ed.), *Cannabinoids as Therapeutic Agents*, CRC Press, Boca Raton, Florida, pp. 1–19.

Mechoulam, R., Devane, W.A., Breuer, A., Zahalka, J. (1991), A random walk through a cannabis field, *Pharmacol. Biochem. Behav.*, **40**, 461–464.

McMeens, R.R. (1860), Report of the committee on Cannabis indica, *Transactions of Fifteenth Annual Meeting of the Ohio State Medical Society*, Follet, Foster and Co., Columbus, USA, pp. 75–100.

Moore, N. (1993), Drugs in ancient populations (letter), *Lancet*, **341**, 1157.

Moreau, J.J. (1973 edn [a]), *Hashish and Mental Illness*, Peters, H. and Nahas, G.G. (eds.), tr. Barnett, G.J., Raven, New York, Ch. 2, p. 15.

Moreau, J.J. (1973 edn [b]), *Ibid.*, Ch. 1, pp. 1–14.

Moreau, J.J. (1973 edn [c]), *Ibid.*, Ch. 7, pp. 205–226.

Musto, D.F. (1991), Opium, cocaine and marijuana in American history, *Scientific American*, July, 20–27.

Nunn, J.F. (1996), *Ancient Egyptian Medicine*, British Museum Press, London, p. 156.

O'Shaughnessy, W.B. (1838), On the preparations of the Indian hemp or Gunjah, *Transactions of the Medical and Physical Society of Bengal* (1838–40), pp. 421–461.

Parkinson, J. (1640), *The Theater of Plantes – an Universal and Compleate Herbal*, Cotes, London, p. 42.

Parsche, F., Balabanova, S. and Pirsig, W. (1993), Drugs in ancient populations (letter), *Lancet*, **341**, 503.

Pliny (1950 edn [a] tr. Rackham), *Natural History*, Heinemann, London, Vol. V, Book XIX, Ch. XXII, p. 461.

Pliny (1950 edn [b] tr. Rackham), *Ibid.*, Ch. LVI, p. 531–3.

Pliny (1950 edn [c] tr. Rackham), *Ibid.*, Book XX, Ch. XCVII, p. 153.

Prioreschi, P. and Babin, D. (1993) Ancient use of cannabis (letter), *Nature*, **364**, 680.

Quincy, J. (1728), *Pharmacopoeia Officinalis et Extemporanea or A Compleat English Dispensatory*, A. Bell, London, p. 226.

Reynolds, J.R. (1890), Therapeutic uses and toxic effects of Cannabis indica, *Lancet* (22nd March) **I**, 637–8.

Richardson, P.M. (1988) *Flowering Plants – Magic in Bloom*, Burke Publishing Co., London, pp. 32–50.

Rudgley, R. (1993), *The Alchemy of Culture*, British Museum Press, London, pp. 28–37.

Rudgley, R. (1995), The archaic use of hallucinogens in Europe: an archaeology of altered states (editorial), *Addiction*, **90**, 163–4.

Siegel, R.K. and Hirschman, A.E. (1991), Edmond Decourtive and the first thesis on hashish: a historical note and translation, *J. Psychoactive Drugs*, **23**, 85–86.

Singer, C. and Underwood, E.A. (1962), *A Short History of Medicine*, Clarendon Press, Oxford, p. 341.

Turner, W. (1568), *New Herball*, Arnold Birckman, Cologne, p. 105.

Twitchett, D. and Loewe, M. (1986), *Cambridge History of China* (Volume 1), Cambridge University Press, Cambridge, p. 589.

Walker, K. (1954), *The Story of Medicine*, Hutchinson, London, p. 207.

Walton, R.P. (1938a), Chapter 1. In Walton, R.P., *Marihuana America's New Drug Problem*, J.P. Lippincott Company, Philadelphia, pp. 1–18.

Walton, R.P. (1938b), *Ibid.*, p. 142.

Walton, R.P. (1938c), *Ibid.*, p. 56–114.

Wild, J.P. (1988), *Textiles in Archaeology*, Shire Publications, Aylesbury, UK, p. 22.

Wilson, E.H. (1913), *A Naturalist in Western China with Vasculum, Camera and Gun* (Volume II), Methuen, London, p. 81.

Wood, T.B., Spivey, W.T.N., Easterfield, T.H. (1899), *J. Chem. Soc.*, **75**, 20.

W.W. (1890), Toxic effects of Cannabis indica, *Lancet*, (15th March), **I**.

Zacune, J. and Hensman, C. (1971), *Drugs Alcohol and Tobacco in Britain*, Heinemann, London, pp. 95–101.

Zlas, J. *et al.* (1993), Early medical use of cannabis (letter), *Nature*, **363**, 215.

2. THE CANNABIS PLANT: BOTANY, CULTIVATION AND PROCESSING FOR USE

AMALA RAMAN

Department of Pharmacy, King's College London, UK

Cannabis plants have been cultivated in Europe, Asia, Africa and the Americas for hundreds, perhaps even thousands of years as a source of three main products – hemp fibre, cannabis seeds and medicinal or narcotic preparations (Fairbairn, 1976). Hemp fibre is obtained from cannabis stems, and has been used over the centuries for the production of textiles, rope and sacking. It is strong and durable, composed of about 70% cellulose and reaches lengths of 3–15 feet (Schultes, 1970). The fibre has been used in the past to make paper, and has been proposed as a replacement for wood pulp in modern paper production (Kovacs, 1992). However, there are many technological limitations to be overcome before this becomes a commercially viable proposition (Judt, 1995). The "seeds" (which technically are the fruit or achene) may be roasted and consumed by man, used as birdseed or anglers' bait or pressed to yield a greenish yellow, fixed oil which has been used in foodstuffs and in varnishes, paints and soap (Schultes, 1970; Fairbairn, 1976). Cannabis leaves and flowering tops and preparations derived from them have many pharmacological effects in man, including narcotic properties; the latter is the most widely known use of cannabis in the present day.

GEOGRAPHICAL ORIGIN OF CANNABIS

The history of cannabis is complex and it has not been possible to ascribe a precise geographical origin to the plant. The picture is complicated by a long history of human use of cannabis in many parts of the world, doubts over the distinction between wild, cultivated and escaped types, extensive transfer of its use and cultivation between cultures and the existence of much variety in the physical and chemical properties of the plant (Schultes, 1970). There is general agreement however, that cannabis is Asiatic in origin, but locations ranging from the Caspian Sea and Central and Southern Russia to Northern India and the Himalayas have been proposed as its native habitat (Schultes, 1970). Of these, an area of Central Asia just North of Afganisthan is favoured by most experts (Schultes and Hoffman, 1980).

TAXONOMY AND NOMENCLATURE OF CANNABIS

The botanical classification of cannabis plants has been the subject of much debate and repeated alteration since the time of Linnaeus (late 18th Century). Controversy has surrounded both the family in which the genus *Cannabis* is placed, and the question of whether the genus is mono- or polytypic (i.e. consisting of one or many species). A number of authors have dealt extensively with this subject (Schultes, 1970; Small and Cronquist, 1976; Schultes and Hoffman, 1980). The history of cannabis taxonomy and nomenclature is briefly reviewed in this section.

In Europe cannabis had been grown from ancient times for the production of hemp fibre. Gerarde in his 1597 Herball lists the following common European names for the plant: *hempe* (English), *kemp* (Brabanders), *zamer hanff* (Dutch), *canape* (Italian), *cananio* (Spanish), *chanvre* (French), *kannabis* (Greek) and *cannabis* (Latin). The first recorded use of the Latin binomial *Cannabis sativa* was by Caspar Bauhin in 1623 (Schultes, 1970), but the official publication of this name was in Linnaeus's *Species Plantarum* of 1753, the internationally acknowledged starting point for modern botanical nomenclature. The species name *Cannabis* is stated by Bloomquist (1971) to mean "canelike" whilst the genus name "*sativa*" has the meaning "planted or sown" and signifies that the plant is propagated from seed and not from perennial roots.

In 1785, Lamark assigned the Latin binomial *Cannabis indica* to cannabis grown in India, classifying it as a unique species on the basis of its different growth habit, morphological characteristics and stronger narcotic properties than the European plant. However many taxonomists regarded the plant as a variety of *Cannabis sativa* using the nomenclature *Cannabis sativa* var. *indica* (Schultes and Hoffman, 1980) to distinguish it from the fibre hemp. In colloquial language the term "Indian hemp" was coined and still persists as a name for narcotic cannabis.

In 1924, Janischevsky described a wild form of the fibre type of cannabis found in Western Siberia and Central Asia, which could be distinguished from the cultivated form on the morphological characteristics of the seed (achene). He assigned the domes-ticated or cultivated form to the species *C. sativa* and named the wild type *C. ruderalis* or *C. sativa* var. *ruderalis*. Similarly, Vavilov and Bukinich (1929) described a wild form of narcotic cannabis growing in eastern Afganistan, and described this as *C. indica* var. *kafiristanica*. Thus distinctions were made based on both narcotic potential and morphological variations observed at different levels of domestication.

There are divergent views as to whether the numerous forms of *Cannabis* observed are variations of a single species or distinct species in their own right. An examination of entries in *Index Kewensis* from 1893–1990 reveals that many species, subspecies and varieties of *Cannabis* have been proposed during this period (*C. chinensis, erratica, foetens, lupulus, macrosperma, americana, generalis, gigantea, ruderalis, intersita* and *kafiristanica*), but that many of these have subsequently come to be regarded as equivalent to *C. sativa*. However many Russian botanists have held the view that several species do exist within the genus (Schultes and Hoffman, 1980). Morphological and microscopical differences between *C. sativa, C. indica* and *C. ruderalis* plants have been observed, and following extensive literature, herbarium and field studies, Schultes and co-workers at the University of Mississippi declared in 1974 their support for the Russians' view that the three should be delimited as distinct species (Schultes and Hoffman, 1980). The need to resolve the issue assumed some importance in the 1970s since the identity of cannabis samples had to be clarified for criminal proceedings against those involved in its abuse as an intoxicant. National and state laws in the United States of America defined marijuana specifically as *Cannabis sativa*, and consequently any other *Cannabis* species could not technically be considered illegal.

The most extensive consideration of cannabis taxonomy to date is that published by Small and Cronquist in 1976. Based on an examination of the literature relating to chemical factors, plant and achene morphology, as well as the influence of selection of characteristics through cultivation over centuries, they concluded that *Cannabis sativa* is a single but highly variable species. Two subspecies, namely *C. sativa* subsp. *sativa* and *C. sativa* subsp. *indica*, of low and high narcotic potential respectively were proposed and each subspecies was further classified into cultivated and spontaneous (wild) varieties as follows:

C. sativa subsp. *sativa* var. *sativa* (cultivated)
C. sativa subsp. *sativa* var. *spontanea* (spontaneous)
C. sativa subsp. *indica* var. *indica* (cultivated)
C. sativa subsp. *indica* var. *kafiristanica* (spontaneous)

A taxonomic key was given to identify the four varieties, and synonyms of each listed to show relationships to previously published species names (Table 1). Further chemotaxonomic evidence for a single *Cannabis* species was provided by Lawi-Berger and Kapetanidis (1983) and their co-workers (Lawi-Berger *et al.*, 1983a, b) who showed that the type and proportion of fatty acids, as well as the proteins and enzymes present in cannabis seeds from 14 different geographical locations were virtually identical.

At a higher level of classification, the genus *Cannabis* belongs to the family Cannabaceae (often erroneously rendered Cannabinaceae, Cannabidaceae or even Cannaboidaceae) of the order Urticales; this too has been the subject of debate. Taxonomists had initially placed the plant in Urticaceae (the nettle family), but in the early part of this Century there was some support for classifying it under Moraceae (the fig family) (Schultes, 1970). Both are families in Urticales and *Cannabis* has some features of each,

Table 1 Synonyms of *Cannabis sativa* subspecies and varieties according to Small and Cronquist (1976)

C. sativa Linn. subsp. sativa var. sativa Small et Cronq.
C. sativa var. *vulgaris* Alefield; *C. chinensis* Delile; *C. sativa* δ *chinensis* A. DC.;
C. gigantea Delile, *C. sativa* var. *gigantea* Alefield; *C. gigantea* Crevost, *C. sativa*
β *vulgaris* A. DC; *C. sativa* γ *pedemontana* A. DC; *C. sativa var culta* Czern.;
C. sativa subsp. *culta* Serebr.; *C. sativa* var. *praecox* Serebr.; *C. sativa* var. *monoica* Hol.;
C. generalis Krause; *C. americana* Houghton et Hamilton.

C. sativa subsp. sativa var. spontanea Vavilov
C. sativa var. spontanea Czernj.; *C. sativa* subsp. *spontanea* Serebr.; *C. ruderalis* Janisch.;
C. sativa var. *ruderalis* Janisch.

C. sativa subsp. indica var. indica (Lam.) Wehmer
C. sativa var. *indica*; *C. macrosperma* Stokes; *C. sativa* α *kif* A. DC.; *C. sativa*
forma *afghanica* Vav.; *C. indica* var. *kafiristanica* forma *afghanica* Vav.

C. sativa subsp. indica var. kafiristanica (Vavilov) Small et Cronq.
C. indica Lam. var. *kafiristanica* Vav.

---

but also sufficiently significant differences to prevent it being placed with confidence in either. Morphological and chemical studies led to the creation, in the 1960s, of the distinct family Cannabaceae which contains two genera only – *Cannabis* and *Humulus* (hop plants). Table 2 shows the larger view of the taxonomical position of *Cannabis* within the plant kingdom, with classifications above the level of species as described by Quimby (1974) and subspecies as classified by Small and Cronquist (1976).

The establishment of the family Cannabaceae has obtained widespread support, but the debate over whether or not *Cannabis* is a monotypic genus has not been satisfactorily resolved. Small and Cronquist's classification is not universally accepted and support for a polytypic view of the genus still exists as evidenced by the comments of Schultes and Hoffman (1980) and subsequent entries in *Index Kewensis* (1976–1990). Nevertheless, the tendency in recent literature is to refer to all types of cannabis as *Cannabis sativa* L. with an indication of the fibre or narcotic characteristics of the plant. As Schultes and Hoffman (1980) point out, divergent definitions of what constitutes a distinct species are largely to blame for the controversies that have arisen in the classification of *Cannabis*. A second important factor is the extensive cultivation and selection of high yielding strains of the plant (fibre or narcotic) over many centuries giving rise to a wide range of phenotypes. The development of modern analytical techniques and easier access to material from different parts of the world has led to a substantial body of work on the chemical variation of cannabis plants, particularly in relation to whether the plants are predominantly fibre type or drug type, and the factors influencing this property. Chemical variation in *Cannabis* is discussed in more detail in Chapter 3.

Table 2   Taxonomic classification of Cannabis. (based on the descriptions of Quimby (1974) and Small and Cronquist (1976))

Kingdom:	Plant
Division:	Tracheophyta
Subdivision:	Pteropsida
Class:	Angiospermae
Subclass:	Dicotyledoneae
Superorder:	Dilleniidae
Order:	Urticales
Family:	Cannabaceae
Genus:	*Cannabis*
Species:	*sativa*
Subspecies:	*C. sativa* L. subsp. *sativa* (L.) Small *et* Cronquist.
	C. sativa subsp. *indica* (Lam.) Small *et* Cronquist.
Varieties:	*C. sativa* L. subsp. *sativa* (L.) Small *et* Cronquist var. *sativa* (L.) Small *et* Cronquist, Taxon **25** (1976) 421.
	C. sativa L. subsp. *sativa* (L.) Small *et* Cronquist var. *spontanea* Vavilov, Taxon **25** (1976) 423.
	C. sativa L. subsp. *indica* (Lam.) Small *et* Cronquist var. *indica* (Lam.) Wehmer, Die Pflanzenstoffe (1911) 248.
	C. sativa L. subsp. *indica* (Lam.) Small *et* Cronquist var. *kafiristanica* (Vavilov) Small *et* Cronquist, Taxon **25** (1976) 429.

BOTANICAL FEATURES OF CANNABIS

Despite the debate over the mono or polytypic status of the genus *Cannabis*, all cannabis plants are easily recognised by certain distinct common botanical characteristics. Early herbals (Dodonaeus, Gerarde) speak of two types of cannabis – the seed bearing type and the barren type, which reflect the fact that the plant is dioecious i.e. that it bears male and female flowers on separate plants (Figure 1). The male plant bears staminate flowers and the female plant pistillate flowers which eventually develop into the fruit and seeds. In the early herbals the sexes were in fact erroneously assigned, with the plant bearing seeds referred to as seede hempe, male hempe or *Cannabis mas* and the plant bearing flowers alone referred to as barren hempe, female hempe or *Cannabis femina*.

Occasionally monoecious plants are encountered bearing both male and female flowers; these may arise as a result of special breeding (Small and Cronquist, 1976). They are particularly frequent in varieties developed for hemp production (Clarke, 1981). Feminisation of male plants using ethephon (Ram and Sett, 1982a) and masculinisation of female plants with silver nitrate and silver thiosulphate complex (Ram and Sett, 1982b) have been reported, and irradiation (Nigam *et al.*, 1981a) or treatment with streptovaricin (Nigam *et al.*, 1981b) can also induce changes in flower formation. An interesting observation was that in a collection of wild cannabis plants growing along

Figure 1   Male (left) and female (right) plants of *Cannabis sativa* L. Photograph courtesy of Professor R. Brenneisen, University of Bern

streets and highways in the United States, only 41% were male as compared to 55% of all plants collected from varying sites for the study (Haney and Bazzaz, 1970). It has been demonstrated (Heslop-Harrison, 1957) that exposure to low levels of carbon monoxide for short periods of time can cause a shift of sex expression from male to female.

Information published elsewhere (Stearn, 1970; Schultes and Hoffman, 1980; Bloomquist, 1971; Clarke, 1981: pp. 1–10), gives detailed technical descriptions of cannabis morphology; the information has been simplified in the present text. Cannabis plants have tap roots, about one-tenth the length of the stalk, and with small branches diversifying out from the main root. Above ground, the plants vary in height from 1–20 feet (1–6 m). Up to the flowering season, plants of both sexes have a similar appearance, although the male plants may be more slender than the females which tend to be somewhat stocky. The stems are angular and furrowed, sometimes hollow and can be either branched or unbranched, depending on the proximity of neighbouring plants. Both sexes have compound, green leaves composed of 3–15 leaflets or blades with toothed margins. The leaflets are arranged in a palmate fashion, i.e. radiating from a single point at the end of a stalk. At the base of the stem, the leaves are arranged in pairs, but this changes to an alternate, spiral arrangement, generally with an increasing number of leaflets in the upper parts. The leaflets are 6–11 cm long and between 2 to 15 mm wide. Variations in leaf shape within this broad description have been recorded (Clarke, 1981: p. 89).

Often the sex of the plant is only determinable at the onset of flowering, when the distinct male and female flowers emerge. The two types of inflorescence are easily distinguished. The male inflorescence is composed of many individual flowers borne on flowering branches up to 18 cm long and stands out from the leaves. The individual flowers are small, consisting of 5 whitish or greenish sepals less than 5 mm in length and containing 5 pendulous stamens. By contrast the female inflorescences, do not project beyond the surrounding leaves and are formed in the axillae or terminals of branches. They are compact, short and contain only a few flowers grouped in pairs. Each flower consists of an ovary surrounded by a green bract (the calyx) which forms a tubular sheath about 2 mm in length around the ovary. Two stigmata project out of this sheath. Following fertilisation, the ovary (containing a single ovule) develops into a thin wall surrounding a single seed with a hard shell. This sort of fruit is technically termed an achene. In practice the whole fruit is regarded as the "seed" known as hempseed or cannabis seed. The achene is 2.5–5 mm long and slightly less in width.

Virtually every aerial part of the cannabis plant is covered in minute hairs or trichomes. These are either simple hairs (covering trichomes) or glandular trichomes (Figure 2) containing a resin. Five main types of trichomes have been identified (Fairbairn, 1976; Turner et al., 1981a) and described (Clarke, 1981: p. 97). These are:

(a)   long, unicellular, smooth, curved, covering trichomes;
(b)   more squat, unicellular, cystolith covering trichomes, containing calcium carbonate;
(c)   bulbous, glandular trichomes;
(d)   capitate-sessile (i.e. without a stalk), glandular trichomes, and
(e)   capitate-stalked, glandular trichomes.

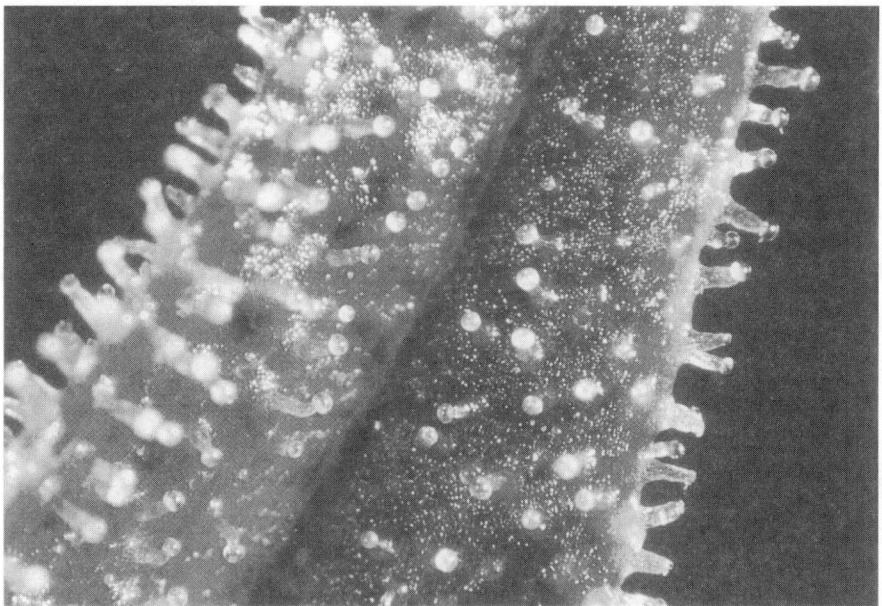

Figure 2 Resin-containing capitate-stalked trichomes on a pistillate bract of *Cannabis sativa* L. Photograph courtesy of Professor R. Brenneisen, University of Bern

Types (a)–(d) are found on the vegetative leaves and pistillate bracts, while type (e) is found on the bracts and floral leaves only (Hammond and Mahlberg, 1977; Turner *et al.*, 1981a). The capitate glandular trichomes have been shown to contain cannabinoids, the unique phytochemicals found in cannabis, some of which are responsible for the intoxicant properties of the plant (Fairbairn, 1972; Turner *et al.*, 1977, 1978). Within the trichomes, the cannabinoids form a resinous substance. This is present in secretory sacs which consist of a distended area bounded by a sheath, formed between secretory cells of the trichome (Lanyon *et al.*, 1981). Cannabinoids are not found in the non-glandular (covering) trichomes (Malingré *et al.*, 1975).

It has been shown that the density of capitate sessile and bulbous glands on the pistillate bract decreases as the bract matures, whereas that of capitate stalked glands increases (Turner *et al.*, 1981b). In leaf development however, bulbous and sessile capitate gland density remains virtually unchanged (Turner *et al.*, 1981a). In both leaves and bracts, the density of non-glandular trichomes show an overall decrease in density over time (Turner *et al.*, 1981a, b). A comparison of the cannabinoid content and total numbers of glands in the two organs reveals differences. During bract development, the total number of capitate glands, total cannabinoid content and cannabinoid content per unit dry weight increases (Turner *et al.*, 1981b). However, in the leaf, although total numbers of capitate glands and total cannabinoid content per leaflet increase, the concentration of cannabinoids decreases during leaf maturation (Turner *et al.*, 1981a).

These studies further strengthen the association between the glandular trichomes and cannabinoid content.

Glandular trichomes of cannabis have been used as a diagnostic agent in identifying the plant material by microscopy when more gross morphological characteristics are not discernable (e.g. in a fragmented leaf). Corrigan and Lynch (1980) have carried out extensive experiments to find a suitable staining agent which will allow cannabis to be distinguished from other plant materials that may have trichomes of a similar appearance. Fast Blue B was found to be a highly selective stain for the cannabinoids in the trichomes of *Cannabis sativa*, having no effect on the trichomes of over two hundred other plant species tested. The stalked capitate trichomes of the bract have a characteristic appearance, being 130–250 μm in length, and having either a multicellular or unicellular stalk and 8–16 cells in the head (Evans, 1989; Bruneton, 1995). The cystolithic covering trichomes of cannabis, although unusual, are not sufficiently unique to the species to be of diagnostic value (Thornton and Nakamura, 1972).

GROWTH CYCLE OF CANNABIS

Cannabis is an annual herb, which grows during the warm season, and then dies down, with new generations springing up from seed the following year. It can grow in any sort of soil, even when nutrition is poor (Bloomquist, 1971), although for commercial purposes good soils are required since cannabis is a heavy feeder and can deplete the soil of nutrients (Schultes, 1970). It has been observed that mature seeds from domesticated varieties of cannabis are larger and germinate more readily than those of wild plants (Janischevsky, 1924). Seeds usually germinate within 3–7 days of sowing (Clarke, 1981: p. 1), and under favourable conditions, the height of the plant can increase by as much as 7 cm per day (Clarke, 1981: p. 2).

Flowering is usually initiated at a critical daylength (photoperiod), which varies depending on the strain of the plant (Clarke, 1981: p. 3). As cannabis is usually dioecious, male and female flowers are produced on separate plants and pollination is reported to occur mainly by the agency of wind (Bloomquist, 1971). The male plants die down soon after pollination whilst the females survive until the onset of inclement weather (frost in temperate areas and drought in the tropics). However, female plants kept indoors are reported to survive for many years (Anon., 1972). Seeds mature towards the end of the warm season and both man and birds are important in their dispersal (Haney and Bazzaz, 1970).

The maturation time for cannabis varies from two to ten months (Anon., 1972). Typically in the Northern hemisphere, cannabis seeds would be sown in May and the plants harvested in September (Bloomquist, 1971). Brenneisen and Kessler (1987), studying cannabis cultivation in Switzerland, have described early and late maturing ecotypes of cannabis. Early maturing forms ("Bolivia" and "Italia") generally originate in temperate zones or at higher altitudes in the subtropics e.g. from the Andes, Atlas, and Himalaya mountain ranges where vegetation time is limited. They have thick and feather-like leaves. Female flowers are produced in mid August, with male flowers appearing four to eight weeks before this. Maturation of the fruit occurs from mid-September to mid-October. The late maturing ecotypes ("Hellas", "Africa I" and "Africa II") originate from tropical or subtropical zones such the Caribbean, Central America, California, Africa and Asia. They have thinner leaves, and produce flowers between mid-September and the end of October. They can only be made to bear mature fruit in European countries if grown under artificial heat and lighting conditions. The pattern of production of Δ^9-tetrahydrocannabinol (THC), the main narcotic cannabinoid, also differs in the two ecotypes. In the early maturing forms, THC concentrations in the plant are relatively low initially and peak at the time of seed ripening. By contrast, levels of THC in vegetative parts of the plant are generally higher in the late maturing types, peak towards the end of the vegetative growth and begin to decrease in the reproductive phase. This latter type is favoured by growers seeking to produce cannabis for narcotic purposes. Rosenthal (1984) has reviewed the maturation period and general characteristics of narcotic cannabis varieties from around the world.

Cannabis is reported to have few natural enemies other than man (Bloomquist, 1971). It is generally resistant to weather change, although heavy frosts may destroy it. Established cannabis plants are able to control the growth of competing weeds, possibly through the agency of volatile terpenes and sesquiterpenes produced by the plant (Haney and Bazzaz, 1970). Young cannabis plants are unable to produce terpenes and may become smothered by surrounding weeds if not controlled. Pests attacking cannabis include the root parasite, branched broom rape, which has been known to cause some damage in European plants (Haney and Bazzaz, 1970). In India, young plants of the drug crop are reported to be prone to wilt disease caused by *Sclerotium rolfsii* Sacc. whilst the hemp crop may suffer leafspot disease caused by *Phomopsis cannabina* Curzi, and infestation by cut worms (Anon., 1992).

CULTIVATION OF CANNABIS

History of Cannabis Cultivation

The histories of the main uses of cannabis can be traced separately but are inevitably linked as a result of the intermingling of knowledge from diverse cultural streams. Archaeological evidence suggests that the use of cannabis can be traced back at least 6000 years (Anon., 1972) and specimens have been found in a 3000–4000 year old Egyptian excavation site (Schultes, 1970).

Schultes (1970) states that cannabis as a source of hemp fibre was probably introduced into Western Europe from the East at different times and by different invaders. Hempen cloth, estimated to have survived from about 6000 years ago, has been excavated from sites in Europe and samples of hemp seeds, leaves, textiles and rope dating to between 500 BC and 300 AD have been found in Germany and England (Schultes, 1970). Pollen evidence shows increasing cultivation of cannabis in England from the early Anglo-Saxon era to the Norman period (Godwin, 1967). The cultivation and processing of hemp in eastern England during medieval times is also suggested by pollen and fossil evidence (Bradshaw *et al.*, 1981). The fibres were apparently obtained by a process known as "water-retting" in which bundles of mature stems were placed in deep ponds of standing water (retting pits) to allow rotting of the tissue surrounding the fibres (Bradshaw *et al.*, 1981). There is evidence for similar cultivation and proces-sing of cannabis in Central Wales during the Tudor period (French and Moore, 1986).

The cultivation of hemp spread to the Americas following the influx of European colonists. It was introduced to South America by the Spanish in 1545, to Canada in 1606 by Louis Hébert, apothecary to the French explorer Samuel de Champlain (Anon., 1972), and to New England by the pilgrims from England. Hemp cultivation in North America was actively encouraged by both France and England in order to supply both European and local American demand. By 1630, cannabis had become a staple crop on the East Coast of North America. Thus until relatively recently, the predominant use of cannabis in the West was as a source of hemp, with the narcotic and medicinal uses only being recognised following contact with Asian and Northern African cultures to whom these properties were well known (Kalant, 1968).

In Asia, the pharmacological effects of cannabis had been discovered in ancient times and the plant was used for medicinal, narcotic and ceremonial purposes. Early writings (Anon., 1972) on the effects of the herb include the medicinal treatise of the legendary Chinese Emperor, Shen Nung (ca 2700 BC, although there is some dispute about the date and authorship of the text), the ancient *Atharvaveda* of India (pre 1400 BC) and the *Zend-Avesta* of Northern Iran (ca 600 BC). Herodotus (ca 450 BC) described the use of cannabis in funeral rites of the Scythians, who occupied an area near the Black Sea. The herb (Emboden, 1972) or seeds (Schultes, 1970) were thrown onto heated stones and the vapour inhaled as a post-funeral purification rite. A similar method of use has been reported in the pre-Portuguese era among people living in the Zambezi valley of Africa, where vapours from a smouldering pile of cannabis would be inhaled either directly or through reeds. The incorporation of cannabis-based rituals into important social ceremonies is believed by some anthropologists to indicate a long period of contact with the herb. The use of cannabis as an euphoriant is thought to have spread from India to the Middle East and then to North Africa (Kalant, 1968).

Although Galen (130–200 AD) recorded the intoxicating effects of excessive con-sumption of hemp cakes (Schultes and Hoffman, 1980), in Western Europe the narcotic properties of cannabis do not appear to have been widely known until relatively recently. The German Schöffer in the "Latin Herbarius" of 1474 lists numerous medicinal uses such as clearing scales from the head, growing hair, aiding digestion, analgesia and drying up the sperm, but makes no mention of any psychoactive effects (Nigg and Seigler, 1992). The herb does not appear at all in Herba Plantes by Sanitatis (1491), but by the next Century can be found listed either as "hempe" or "cannabis" or "cannabina herba" in herbals by Dodonaeus (1578) and Gerarde (1597, 1633). Neither of these authors makes any mention of the narcotic properties of the herb, but list similar uses to Ortus Sanitatis as well as the intriguing use of the herb to increase egg laying in hens! It is thought (Kalant, 1968) that it was only in the 19th Century that the euphoriant properties of cannabis were learnt of by the British in India, and the French in North Africa. The infamous "Club des Hachischins" founded in Paris in 1844 gained a small European following who experimented with the narcotic effects of the herb (Anon., 1972; Bergel and Davis, 1970) but the widespread use of cannabis for hedonistic purposes in Western Europe is believed to be more recent (Anon., 1972), following the popularisation of American practices of the early twentieth Century.

It is believed that cannabis (origin unknown) was already being used as a narcotic in Central and South America in the 16th Century (Anon., 1972), when it was introduced for hemp production by the Spanish. Knowledge of its narcotic properties in Brazil however, was said to have spread only following the arrival of slaves from Africa who were already familiar with its use (Anon., 1972). The Mexicans are credited with the dissemination of this knowledge to English-speaking North America between 1920 and 1930 (Kalant, 1968). The hedonistic use of the herb spread so rapidly in North America, both in the criminal underworld and in fashionable circles, that even by the mid-1930s there was considerable official concern both about the dangers to the user and its connection with criminality (Bergel and Davis, 1970). At present, in the majority of countries of the world, cannabis cultivation and use for narcotic purposes is a criminal offence. Nevertheless there is widespread illicit cultivation of the plant and an international trade in narcotic preparations derived from it.

The history of the medicinal use of cannabis mirrors its narcotic use, although the herb possesses numerous other pharmacological properties. Cannabis appears in ancient Chinese and Indian works on medicine and features in the 15th and 16th century Western herbals. In Medieval Europe, preparations made from the root or seed of hemp cannabis were used for gout, cystitis, gynaecological problems and various other conditions, some of which have been listed above (Le Strange, 1977). However, the superior medicinal properties of the Indian variety of cannabis were recognised by W.B. O'Shaughnessy, a British physician working in Calcutta, who is believed to have introduced the herb to Western medicine. His report in 1842 on the analgesic, anticonvulsant and muscle relaxant properties of the drug generated much interest and led to its widespread use in the 19th Century. It became a recognised official drug, featuring in a number of Pharmacopoeias. Despite extensive cultivation of cannabis in Europe and America, the principal variety used in official medicines came from India (Le Strange, 1977). However, from the beginning of the 20th Century, the popularity of cannabis preparations declined due not only to the variable properties and erratic availability of the plant, but also as a consequence of the emergence of more reliable synthetic medicaments without narcotic effects. Cannabis was removed from the British Pharmacopoeia in 1932, The United States Pharmacopoeia in 1942 and the Indian Pharmacopoeia in 1966.

Although medicinal use of cannabis has declined, its popularity as an euphoric narcotic has continued to increase to the present day, especially among the younger generation. In recent years, there has been much social debate about the legal position of cannabis use, with many calling for its decriminalisation on the grounds of its low addictive potential and its non-narcotic pharmacological properties which could still make a valuable contribution to modern medicine (Gray, 1995).

Present Day Cultivation of Cannabis

Cannabis is now widely distributed throughout the world, both in cultivated forms and as wild plants escaped over the years from cultivation sites. Large scale commercial production takes place in relatively few areas and a distinction can be drawn between its legitimate cultivation as a source of hemp fibre and seeds and the usually illicit cultivation of cannabis as a source of narcotic materials. Whether a cannabis plant predominantly produces fibre (hemp) or narcotic resin is governed by both genetic and climatic factors (see Chapter 3). However, in general terms it can be said that the two properties seem inversely related and individual varieties can be classified as either drug type or fibre type (Bruneton, 1995) depending on the concentrations of the psychoactive compound THC and the non-narcotic cannabinoid, cannabidiol (CBD). The "drug" or "resin" type has a high THC concentration ($>1\%$) and virtually no CBD. This property is observed amongst plants growing in warm climates and producing abundant narcotic resin. The "fibre" or "hemp" type, grown in northern temperate zones has very low THC levels ($<0.3\%$, or even $<0.03\%$ for most cannabis varieties cultivated for fibre) and high CBD concentrations. However "intermediate" varieties are also found, with high levels of both THC and CBD. The general growth cycle of plants from the different varieties is similar, except for variations in maturation period (Clarke, 1981: p. 124) although the forms selected, harvesting methods and further processing depend on the purpose for which the plant is cultivated. The breeding and cultivation of cannabis plants with different characterisitics have been described in detail by Clarke (1981).

Cultivation of Cannabis for Hemp Fibre or Cannabis Seed

Despite its widespread cultivation in Europe and North America in the late 19th and early 20th centuries, the large scale production of hemp for fibre or seed is now restricted to a few areas of Eastern Europe (Hungary, Romania, Ukraine, Russia, former Czechoslovakia, Serbia, Croatia) and China (De Meijer, 1995). This is partly due to a great reduction in the demand for hemp fibre following the advent of more attractive alternatives (synthetic or natural), and secondly due to concerns and restrictions throughout the world on the cultivation of cannabis for narcotic purposes. Since 1961, the cultivation, trade and consumption of cannabis have been placed under restrictions worldwide following the United Nations Organisation's "Single Convention on Narcotic Drugs" (Brenneisen, 1983).

Whilst cultivation of drug type cannabis is universally prohibited or legally regulated, rules concerning cultivation of fibre (hemp) forms vary. In Canada, for example, the cultivation of any cannabis without special authorization has been prohibited since 1938, and all hemp fibre used in the country has to be imported (Anon., 1972). De Meijer (1995) has reviewed the availability and registration status of hemp cultivars in Europe. A number have been registered by the European Union (EU), which implies that their cultivation should be permitted by any member state. In practice, individual states may obstruct seed distribution and cultivation on a number of grounds including national drug legislation. In Italy, for instance, hemp cultivation is prohibited as long as a cultivar cannot be identified with a morphological marker linked to low THC-content (De Meijer, 1995). However, in France, a 1990 statute specifies 12 fibre varieties with not more than 0.3% THC which may be cultivated for the manufacture of speciality papers, non-woven products, furniture particle board, animal litter and so on. The producers must hold a contract specifying a buyer and use certified seeds from the authorised varieties (Bruneton, 1995). Cannabis is also legally cultivated for fibre and seed in Switzerland (Brenneisen, 1983).

There has been considerable recent interest in the development of hemp as an industrial crop in the Netherlands (De Meijer and Van Soest, 1992). Under the National Hemp Programme, the Centre for Plant Breeding and Reproduction Research (CPRO) have carried out a number of studies surveying variations in cannabis cultivars in terms of stem yield and quality, psychoactive potency, resistance to root-knot nematodes and plant morphology (De Meijer and Keizer, 1996), principally to evaluate their suitability as an arable source of paper pulp. A germplasm collection has been established at CPRO (De Meijer and Van Soest, 1992).

Fibre type cannabis is best grown in cold or temperate regions where the subsoil is moist and rain is abundant, since fibres produced in hot, dry climates are too brittle to be of commercial value (Bloomquist, 1971). The use of well-manured soil is recommended since this improves the quality of the fibres (Schultes, 1970). Commercially produced, high fibre varieties of even maturation time are selected to facilitate efficient harvesting. These are often monoecious strains which tend to mature more evenly than dioecious ones (Clarke, 1981: p. 15). Hemp bast fibre is produced in the phloem tissue of the stem. Consequently, the plants are grown close to each other so that branching is limited and long slender stems are produced (Fairbairn, 1976). The properties of hemp fibre, its harvesting and processing have been described by Judt (1995) and Clarke (1981: p. 150). Two types of fibres are present in cannabis stems – bark or bast fibres (23–28% by weight) and core fibres (75–70%). The two vary in their physical characteristics and chemical composition (Judt, 1995). Bark fibres are 20–22 mm long and contain nearly 70% cellulose and small amounts of hemicellulose (10%) and lignin (5%). Compared to these, core fibres are considerably smaller in length (0.55 mm) and have a lower proportion of cellulose (35%) and greater amounts of hemicellulose and lignin (20% of each).

Cannabis stalks are harvested at a point in the plant's growth (usually a prefloral stage) most appropriate to the best yield of fibre, before extensive lignification sets in (Clarke, 1981: p. 150). This is often a critical matter of a few days (Judt, 1995). A portion of the crop may be left to develop mature seed which can be used the following year. The harvested stalks are stripped of leaves, dried and stored in bales before further processing. Whole stalks may be pulped by chemical or mechanical means to obtain a heterogenous mix of fibres, whilst bast fibres may be separated by a process known as retting – partial rotting of the stems in water to destroy the other parts of the plant (Clarke, 1981: p. 150). Natural retting takes from a week to a month. The fibres are then dried, wrapped in bundles and stored in a cool, dry area.

Hemp fibre is relatively expensive to produce; in a study carried out in 1994 (Judt, 1995) to determine the viability of using hemp in the paper industry, the suggested prices of whole hemp stalks and the more valuable bast fibre were US $200 and US $630 respectively per air-dry ton as compared with US $78–199 for hardwood pulp.

When the seeds are required as the commercial product, plants are cut only after seed maturation. Seeds fall easily from the floral clusters when mature and may be collected by hand or machine. The remainder of the plant may be used as pulp material (Clarke, 1981: p. 150).

Cultivation of Cannabis for Narcotic Use in India and Surrounding Areas

In Asia, cannabis grows wild throughout the Himalayas from Kashmir to Eastern Assam, up to altitudes of 10,000 ft (Chopra *et al.*, 1958). It extends down into parts of Pakistan, Bangladesh and India (Punjab, Bengal, Uttar Pradesh and Bihar) but in many of these areas, it is possible that wild growth of cannabis is supplemented by human factors arising from local use of the herb (Chopra *et al.*, 1958; Evans, 1989). There are a few licensed growers of cannabis in Madhya Pradesh and Orissa (Anon., 1992), but illicit cultivation of cannabis is widespread not just in India, but throughout the tropical and sub-tropical areas of the world.

Due to its long historical association with the medicinal and narcotic use of cannabis, cultivation and harvesting practice in India is well documented (Chopra *et al.*, 1958; Anon. 1992). Seed is sown in rows about 1.3 m apart in rich, well-manured, weed-free soils; light or loamy soils are preferred (Evans, 1989; Chopra *et al.*, 1958). Sowing takes place in June or July and harvesting in December or January. When the plants reach a height of about 20 cm, they are thinned out and the lower branches removed to stimulate growth of the flowering branches. The narcotic components of cannabis (mainly THC) are found in a resin secreted by glandular trichomes on the leaves and flower bracts, particularly on pistillate flowers (Clarke, 1981).

In some areas, the tradition is that, as soon as flowering begins, the male plants are identified and systematically removed by the roots (Bloomquist, 1971). A common belief that the male plants are pharmacologically inactive is not true since similar amounts of cannabinoids may be produced in plants of either sex (Chiesa *et al.*, 1973; Valle *et al.*, 1968). However, male plants often yield less plant material, and the staminate bracts have fewer glandular trichomes than the pistillate ones (Clarke, 1981). It has further been shown that buds from unfertilized flowering tops of female plants are more potent (i.e. contain higher THC levels) than fertilised buds, and may even exceed some resin samples in cannabinoid content (El Sohly *et al.*, 1984). When unpollinated, the pistillate plants start to produce more capitate glandular (resin producing) trichomes probably as a protection for the unfertilised ovule (Clarke, 1981). The product consisting of unfertilised flowering tops is referred to as "sinsemilla", derived from the Spanish words "sin" (without) and "semilla" (seed) (Rosenthal, 1984). It is highly valued not only for its greater potency, but also for its more intense aroma and enhanced appearance (Rosenthal, 1984).

To obtain sinsemilla, it is imperative that the male plants are removed meticulously before any large flower clusters appear, since even a single male flower is capable of yielding sufficient pollen to fertilise a large number of females. Male flowers growing in plant internodes can be used to to distinguish them from females at an early stage of their development (Rosenthal, 1984; Clarke, 1981). However, this practice requires much care and attention and the removal of male plants to produce sinsemilla buds is generally carried out only in small cultivation sites (El Sohly *et al.*, 1984). Consequently seeds are commonly encountered in many of the commercial cannabis products (Baker *et al.*, 1980b).

When grown for narcotic or medicinal use in India, the main parts of the cannabis plant harvested are the leaves, female flowering tops or the resin itself. Different harvesting and processing methods are used depending on the final product required (Chopra *et al.*, 1958; Anon., 1992). *Bhang* consists of larger leaves and twigs of the plant and is prepared by simply cutting the plants (wild or cultivated), drying them and beating them against a hard surface to separate the leaves. Both male and female plants may be used (Evans, 1989) and flowering parts are frequently present. *Ganja* or *gunja* consists of the dried flowering and fruiting tops of the female plant from which the resin has not been removed. Harvesting for *ganja* begins when the lower leaves begin to turn yellow. Spikes bearing the inflorescences are cut off and taken to the manufacturer's yard. For Bombay *ganja*, the plant material is piled into ridges and furrows and the material subjected to repeated treading by foot, turning over, drying and retreading. This results in the formation of compact sheaves which are made into piles and kept under pressure for a few days. The heaps are turned over, spread again and the treading repeated. The material is sifted to separate out dust, stones, seeds and leaves and then packed into a flat cake. For Bengal *ganja*, the withered flowering tops are not trampled on, but rolled by hand or foot to form rounded or sausage shaped masses.

According to Clarke (1981: p. 152), flowering tops or floral clusters are best dried by hanging the plants or clusters upside down, a method practised by some growers. This has the effect of allowing the leaves to hang next to the clusters and protect them from mechanical damage which may cause loss of the resin. The method also serves to enhance the appearance of the clusters when dry, since they appear larger than if they are compressed by laying flat to dry. During the drying process, the characteristics of the leaves and flowers change in that the unpleasant "green" taste of the cannabis is gradually lost in a process known as "curing". This does not happen if drying occurs too rapidly. However, too slow or incomplete drying may lead to deterioration of the plant material by the agency of micro-organisms (Clarke, 1981: p. 153). Removal of the outer leaves from the dried floral clusters known as "manicuring". Manicuring before drying may result in loss of resin potency due to greater breakdown of THC (Clarke, 1981: p. 153).

Another Indian product, *charas* (or *churrus*) is the actual resin, in crude form, from the leaves and flowering tops. Men dressed in leather suits, jackets or aprons walk through the fields rubbing and crushing against the plants in the morning shortly after sunrise. The resin exuding from the leaf and flower trichomes sticks to the leather and can be scraped off (Samuelsson, 1992). Other methods include rubbing the flowering tops with the hands, from which resin is later scraped off, beating the flowering tops over a piece of cloth on which the resin collects as a greyish powder, or thrashing the tops against smooth concrete walls and collecting the powder and resin that stick to the wall (Samuelsson, 1992; Chopra *et al.*, 1958).

Bhang, ganja and *charas* have been in use for many centuries in India. The dried and crushed flower heads and small leaves (*ganja*) from any geographical source, are commonly referred to as marijuana and the resin (*charas*) is referred to as hashish. Other names encountered for the different types of cannabis products described above are discussed further on in this chapter. Relatively recently, a further product of cannabis has entered the illicit market and is a concentrated liquid extract or oil produced by hot-solvent extraction or distillation of the resin (Brenneisen, 1983), or occasionally similar treatment of the herb or flowering tops (Baker *et al.*, 1980b). Three to six kilograms of resin are needed to produce one kilogram of oil (Stamler *et al.*, 1985), which may subsequently be dissolved in a vegetable oil (Anon., 1992). The product contains high levels of THC and is commonly referred to as hash oil. This is not to be confused with cannabis seed oil which is a fixed oil devoid of narcotic properties.

Worldwide Cultivation Sites of Narcotic Cannabis

The international trafficking of cannabis products is mainly supplied from a few major source countries in tropical or sub-tropical areas of the world. Products originating in India or Pakistan are often seized by Customs officials, but other equally important producers are Colombia, Mexico, Jamaica, Morocco, Lebanon and Thailand (Bruneton, 1995; Stamler *et al.*, 1985; Brenneisen and El Sohly, 1988). However, since this illicit trade is a lucrative one, smaller seizures of cannabis products originating from diverse parts of the world are encountered. These include other South American and Caribbean countries, Southern parts of North America, Egypt, Turkey and Nepal as well as various non-Mediterranean African countries *viz.* Ghana, Nigeria, Sierra Leone, Kenya, Zaire, Malawi, Zambia, Zimbabwe and South Africa (Baker *et al.*, 1980a, b; 1982). Different strains of cannabis are grown in these locations, and consequently, the gross phenotype of the cannabis plants can vary from the short, broad strains of the Hindu Kush to tall meandering varieties found in Thailand (Clarke, 1981; pp. 102–118). Chemical characteristics of plants from different geographical origins are also known to vary (see Chapter 3).

Indoor Cultivation of Cannabis

As well as large scale cultivation in the aforementioned tropical or sub-tropical regions, a proportion of the cannabis used for hedonistic purposes is cultivated by individuals in temperate end-user countries either for personal use or supply. Indoor or greenhouse cultivation, which has been described extensively by Rosenthal (1984), reduces the problem of poor resin potency due to low outdoor temperatures and minimises the risk of detection by law-enforcement agencies (Stamler *et al.*, 1985). Photoperiod can be controlled with the use of, for example, blackout screens in order to force flowering (Clarke, 1981, p. 148). Growth from seed can be successful, but in many of these illicit operations, vegetative propagation is carried out from stem cuttings of female plants, firstly in order to speed up propagation and secondly to ensure the sex of the plant. The cuttings may be grown either in soil or hydroponically, often without roots (Rosenthal, 1984). Hydroponic propagation of stem cuttings has the legal advantage, in the United States of America for instance, that the cutting is not classed as a "plant" if it lacks roots (Taylor *et al.*, 1994). This assumes significance where the severity of sentencing is based on the total count of "plants" in the defendant's possession. As well as indoor cultivation of cannabis plants, clandestine small scale operations in Canada for producing hash oil have also been reported (Stamler *et al.*, 1985).

ILLICIT CANNABIS PRODUCTS

Synonyms for Cannabis Preparations

The main illicit products of cannabis plants are female flowering tops with or without leaves, the resin from the flowering tops and an oil extracted or distilled from the resin, or more rarely, directly from the leaves and flowering tops. Many names are in use for these products in different parts of the world.

Preparation of the leaves and flowering tops are generally referred to in English speaking countries as *cannabis, Indian hemp* or very often *marijuana* (sometimes rendered *marihuana*). Various etymological sources suggested for the latter name include the Mexican-Spanish *mariguana* or Portuguese *mariguango* meaning "intoxicant", the Mexican-Spanish slang *Marijuana* (Mary-Jane) or *Maria y Juana* (Mary and Jane), or an earlier Aztec word *milan-a-huan* (Bloomquist, 1971). Names in other parts of the world include *bhang* (India, leaves) *ganja* or *gunja* (India, Jamaica, flowering tops), *kif* or *kief* (Morocco), *dagga* (Southern Africa), *maconha* (Brazil), *kabak* (Turkey), and rarely *hashish* (Egypt) although this latter name usually refers to the resin. Common slang names in the West include *grass, pot, dope, weed, Mary Jane, hash* and less often, *shit, bush, tea, Texas tea, locoweed, griefo, hay, hemp, jive, mor-a-griefa, rope, boo, wacky backy,* or *black* (Bloomquist, 1971; Anon., 1972; Bergel and Davies, 1970; Gosden, 1987).

Cannabis resin, known as *charras* or *churrus* in India is almost universally referred to as *hashish*. The word has in many texts been linked to the terms "ashashin" or "hashashi" (hashashan = herb eaters), fanatical religious followers of an 11th Century Persian leader, known to the Crusaders as the Old Man of the Mountains (Bergel and Davis, 1970; Anon., 1972). It is said that their political and military activities led to their name forming the root of the English word "assassin", whilst their connection with the use of cannabis formed the basis of the derivation of "hashish". This linguistic derivation is under some dispute, and it has been said that the Arabic word "hashish" which means "dry herb", "grass" or even "hemp" is a far more likely origin (Anon., 1972). Slang names for the resin include *hash, shit* and *stuff* (Brenneisen, 1983). The oil obtained from cannabis resin is known as *hashish oil* or more commonly *hash oil* (Gosden, 1987). Slang names include *oil, red oil,* and *Indian oil* (Brenneisen, 1983).

Cigarettes containing marijuana, hashish or hash oil, are known as *reefers* or *joints* (Wills, 1993), and plastic bags containing leafy plant material, usually with seeds present are known as a *stash* (Bloomquist, 1971).

International Trafficking of Illicit Cannabis Products

Stamler *et al.* (1985) have reviewed the local geographical sites of production within the major source countries, as well as local cultivation, storage and trafficking practices. For example, two annual harvests are obtained in Colombia, and there is evidence that crops are staggered in order to ensure a continuous supply. Generally, the crop is cultivated by individual farmers and sold to drug traffickers who may also cultivate the plant. The cut and dried plants are packaged in 20–35 kg bundles and transported by mule to clandestine airstrips or more commonly beaches. Maritime rather than airplane transport is common in the illicit supply of Colombian marijuana which is more frequently encountered in North America than in Europe. Sea routes are often used by smugglers operating from Mediterranean sources and Thailand, whereas hashish smuggling from India and Pakistan is usually done on a small scale by couriers travelling on commercial airlines. Small quantities of cannabis products may also be sent between countries via the normal postal service.

Drug trafficking patterns are subject to change over time (Stamler *et al.*, 1985), influenced no doubt by political factors in source countries as well as the work of customs and police officials worldwide. There is also evidence that cannabis products from one source country may be shipped to their eventual destination via an intermediate country, itself perhaps involved in cannabis cultivation. For instance, Colombian or Thai cannabis may be transported through Mexico to the United States (Brenneisen and El Sohly, 1988), whilst the geographical location of Jamaica makes it an important storage and forwarding site for the drugs from other sources (Stamler et al., 1985).

The predominant type of product supplied may differ from country to country. Countries such as Colombia, Mexico, Bolivia, Thailand and the non-Mediterranean African countries generally supply only marijuana, whereas in addition to marijuana, hashish and hash oil are produced in Morocco, the Middle East, India and Pakistan (Brenneisen, 1983), and Jamaica (Stamler et al., 1985).

Cannabis products from different parts of the world vary in appearance not only due to variations in plant characteristics, but also processing methods and packaging. A range of these products is shown in Figures 3–5. El Sohly et al. (1984), Baker et al. (1980a, b) and Brenneisen (1983) have described some visual characteristics indicative of the geographical origin of cannabis products. Marijuana for instance can sometimes be obtained in the form of loose plant material consisting of various combinations of dried leaves, stems and seeds, with the colour varying from shades of green to brown

Figure 3 Cannabis products from around the world. Top row, left to right: Indian, Lebanese, Turkish and Pakistani hashish. Bottom row, left to right: Swiss hashish, Zairean marijuana, Swiss marijuana, Morrocan hash oil. Photograph courtesy of Professor R. Brenneisen, University of Bern

Figure 4 Thai sticks of cannabis. Photograph courtesy of Professor R. Brenneisen, University of Bern

Figure 5 Moroccan hashish with characteristic imprint. Photograph courtesy of Professor R. Brenneisen, University of Bern

depending on the source country (Baker *et al.*, 1980b). However, a classical Mexican packaging method is to compress the plant material into a block or kilobrick, whereas marijuana from Thailand is obtained as "Thai sticks" (Figure 4) consisting of leafy material tied around stems (El Sohly *et al.*, 1984). Sinsemilla is obtained as loose flowering tops characterised by the absence of seeds (El Sohly *et al.*, 1984).

Cannabis resins (hashish) from Mediterranean countries are characteristically powdery and pale green or brown whereas those from India, Nepal and Pakistan are much darker, varying from brown to almost black (Baker *et al.*, 1980a, b). The resin is often moulded into characteristic shapes (Figure 3), indicative of the country of origin (Baker *et al.*, 1980b) such as sticks from India, or rectangular slabs from Morocco with a characteristic imprint (Figure 5).

Variations in the chemical profiles of cannabis products of different origins have been examined, particularly as a means of identifying the source country of seizures; this is reviewed more fully in Chapter 3. Crosby *et al.* (1984) have suggested that a study of the insects present in a sample of cannabis can allow an exact indication of its country of origin.

THC Content and Stability of Cannabis Preparations

The content of the main psychoactive constituent, THC, in cannabis products varies greatly depending on the type of preparation, geographical source, plant strain, quality and age of the preparation. Estimates vary, but according to Fairbairn (1976) marijuana contains up to 8% THC, hashish up to 14% THC, and hash oil up to 60% THC. Sinsemilla buds may contain up to 10% THC (Wills, 1993). In a study comparing THC content of a number of cannabis products confiscated over a ten year period in the United States of America (El Sohly *et al.*, 1984), sinsemilla buds were found to be more potent than the average hashish sample.

The figures given above are approximate upper limits, and values commonly encountered may be considerably lower and occasionally even higher. For example, the average THC content in hash oil samples collected in the American study (El Sohly *et al.*, 1984) was about 18%. A popular cannabis preparation often seized from users is reefers or joints consisting of cigarettes containing marijuana, hashish or hash oil (Wills, 1993). These have been shown to vary greatly (Fairbairn *et al.*, 1974; Humphreys and Joyce, 1982) in the amount of cannabis plant material (11–1090 mg) or cannabis resin (6–838 mg) and consequently in the THC content (0.14–41 mg) per reefer.

Baker *et al.* (1980a) have examined the variation in THC content of Cannabis products illicitly imported into the United Kingdom over the period 1975–1978. Geographical variations in potency were seen but these were found to be changeable. For instance, in the early part of the study, herbal cannabis from South East Asia was found to be of a higher quality (i.e. THC content) than from other parts of the world, but its quality had declined by 1978. Jamaican cannabis, however, improved in quality over the same period. Hash oil from India, Kenya and Pakistan had very similar mean THC contents (33, 34 and 30%) in 1977 whereas the values were very different (40, 16 and 18%) in 1978. However, in both this and a further study (Baker *et al.*, 1982) it was noted that there was considerable variation in THC content even within samples from a given country of origin, and so a single sample could not be considered indicative of the quality of all similar products from that source.

In both the above studies, THC was estimated as the total of THC and delta-9-tetrahydrocannabinolic acid (THCA) since the latter is converted to THC both during the analytical process and during the smoking of cannabis preparations. In a different study however (Baker et al., 1981), the two compounds were estimated individually in a range of marijuana and hashish samples. THC values ranged from 1–10.6% in cannabis herb and 6.0–12.5% in the resin. The THCA:THC ratios in the resins also varied from 0.5:1 to 6.1:1. Higher ratios were encountered in resins from the Mediterranean area, whereas the lower ones were measured in samples from the Indian sub-continent. Thus despite some general trends, there appears to be considerable variation in the THC content of products derived from a single source country both at any given time, and over longer time scales.

The age and storage conditions of cannabis products can affect their potency due to changes in THC content. A 1931 herbal (Grieve, 1974) indicates that two-year old *ganja* is almost inert. However, according to Fairbairn et al. (1976) the stability of a cannabis preparation depends on its preparation and storage. In one experiment, about 90% of the THC content of marijuana herb was still present after storage for a year at room temperature in the dark (Fairbairn, 1976). Exposure to air and daylight but not air alone had a deleterious effect, particularly on solutions of cannabinoids (Fairbairn et al., 1976). Temperatures of up to 20°C had little effect on stability although higher values caused breakdown of THC. However, another important factor was the integrity of the resin glands (Fairbairn et al., 1976). Damage to these by rubbing or scraping leads to rupture of the glands and exposure of the cannabinoids to oxidation even in the dark. However, the authors concluded that herbal or resin cannabis is reasonably stable for one to two years if stored in the dark at room temperature.

Light alone produces polymerisation of THC whereas oxidation converts THC mainly to CBN, the non-narcotic compound cannabinol (Fairbairn, 1976). Turner et al. (1973) had shown in an earlier study that hexahydrocannabinol and other minor products were produced in addition to CBN. A decomposition pathway of THC to CBN involving a number of hydroxylated intermediates has been suggested by Turner and El Sohly (1979). Harvey et al. (1985) found that in a 140 year old ethanolic cannabis extract, most of the THC had decomposed to CBN and that the intermediates postulated by Turner and El Sohly (1979) could be detected. However, these intermediates were not detected in samples of dried material containing high levels of CBN and dating back to the beginning of this century (Harvey, 1990). This result suggests that possible alternative or additional mechanisms of breakdown may occur in the dried material. It was also observed that although CBN was the main cannabinoid found in the samples, significant amounts of THC had survived, indicating that breakdown was not as rapid as would have been predicted from the work of previous authors. The acid analogues of THC, CBN and other cannabinoids were also present, indicating their relative stability under the storage conditions (dry material at room temperature with possible exposure to light).

In a separate study in which hashish resin samples were exposed to temperatures of 80°C in the dark, the acid forms CBNA, THCA and CBDA (cannabidiolic acid) underwent decarboxylation to CBN, THC and CBD respectively. However the total amounts of acid and decarboxylated forms decreased with time indicating that other reactions such as polymerisation and disproportionation were also occuring under these conditions (Kovar and Linder, 1991). Thus light, air and heat may all contribute to the breakdown of THC in cannabis products.

It has been found (Smith and Vaughan, 1977) that there are considerable differences in the relative amounts of different cannabinoids in the inner and outer layers of resin samples, although the pattern is somewhat erratic. THCA concentrations are higher in inner layers, indicating that exposed areas are prone to decarboxylation.

Although the narcotic properties of cannabis have generally been ascribed to THC, *Cannabis sativa* in fact contains a very large number of phytochemicals, whose pharmacological activities have not been fully investigated. It is more than likely that the overall pharmacological profile of cannabis preparations is due to a number of substances present. The phytochemical characteristics of *Cannabis sativa* are described in the next Chapter.

ACKNOWLEDGEMENT

The services of the Herbarium Library at the Royal Botanic Gardens, Kew, the libraries at the Chelsea Physic Garden and the Royal Pharmaceutical Society of Great Britain and the Cannabis Bibliographic Library at the University of Mississippi are gratefully acknowledged.

REFERENCES

Anonymous (1972) *A report of the commision of inquiry into the non-medical use of drugs*, Ministry of National Health and Welfare, Information Canada, Ottawa.

Anonymous (1992). *The wealth of India; a dictionary of Indian raw materials and industrial products*. Raw materials volume 3: Ca-Ci, Publications and Information Directorate, Council of Scientific and Industrial Research, New Delhi, India, pp. 195–205.

Baker, P.B., Bagon, K.R. and Gough, T.A. (1980a) Variation in the THC content in illicitly imported *Cannabis* products. *Bull. Narc.*, **32**(4), 47–54.

Baker, P.B., Gough, T.A., Johncock, S.I.M., Taylor, B.J. and Wyles, L.T. (1982) Variation in THC content in illicitly imported *Cannabis* products – Part II. *Bull. Narc.*, **34**(3–4), 101–108.

Baker, P.B., Gough, T.A. and Taylor, B.J. (1980b) Illicitly imported *Cannabis* products: some physical and chemical features indicative of their origin. *Bull. Narc.*, **32**(2), 31–40.

Baker, P.B., Taylor, B.J. and Gough, T.A. (1981) The tetrahydrocannabinol and tetrahydro-cannabinolic acid content of cannabis products. *J. Pharm. Pharmacol.*, **33**, 369–372.

Bergel, F. and Davies, D.R.A. (1970) *All about drugs*, Thomas Nelson and Sons Ltd, London, pp. 42–47.

Bloomquist, E.R. (1971). *Marijuana the second trip* (Revised Edn), Glencoe Press, Collier-Macmillan Canada Ltd, pp. 1–11.

Bradshaw, R.H.W., Coxon, P., Greig, J.R.A. and Hall, A.R. (1981) New fossil evidence for the past cultivation and processing of hemp (*Cannabis sativa* L.) in Eastern England. *New Phytol.*, **89**(3), 503–510.

Brenneisen, R. (1983) Psychotrope Drogen. I. *Cannabis sativa* L. (Cannabinaceae). *Pharm. Acta Helv.*, **58**(11), 314–320.

Brenneisen, R. and Kessler, T. (1987) Psychotrope Drogen 1. *Pharm. Acta Helv.*, **62**(5–6), 134–139.

Brenneisen, R. and El Sohly, M.A. (1988) Chromatographic and spectroscopic profiles of cannabis of different origins: Part I. *J. Forensic Sci.*, **33**(6), 1385–1404.

Bruneton, J. (1995) Orcinols and Phloroglucinols. In *Pharmacognosy, Phytochemistry, Medicinal Plants*. Intercept Ltd, Hampshire, UK, pp. 371–379.

Chiesa, E.P., Rondina, R.V.D. and Coussio, J.D. (1973) Chemical composition and potential activity of Argentine marihuana. *J. Pharm. Pharmacol.*, **25**, 953–956.

Chopra, R.N., Chopra, I.C., Handa, K.L. and Kapoor, L.D. (1958) In *Chopra's Indigenous Drugs of India*, U.N. Dhur and Sons Pvte Ltd, Calcutta, India, pp. 84–92.

Clarke, R.C. (1981) *Marijuana botany. An advanced study: the propagation and breeding of distinctive cannabis*, And/Or Press, Berkeley, California, USA.

Corrigan, D. and Lynch, J.J. (1980) An investigation of potential staining reagents for the glandular trichomes of *Cannabis sativa*. *Planta Med.*, Suppl., 163–169.

Crosby, T.K., Watt, J.C., Kistemaker, A.C. and Nelson, P.E. (1984) Entomological identification of the origin of imported cannabis plants. *J. Forens. Sci. Soc.*, **24**(4), 290.

De Meijer, E. (1995) Fibre hemp cultivars: A survey of origin, ancestry, availability and brief agronomic characteristics. *J. Int. Hemp Assoc.*, **2**(2), 66–73.

De Meijer, E.P.M. and Keizer, L.C.P. (1996) Patterns of diversity in *Cannabis*. *Genetic Resources and Crop Evolution*, **43**, 41–52.

De Meijer, E.P.M. and Van Soest, L.J.M. (1992) The CPRO *Cannabis* germplasm collection. *Euphytica*, **62**, 201–211.

Dodonaeus seu Dodoens, R. (1578). *Niewe Herball or Historie of Plantes*. Translated out of French by Henry Lyte, p. 71.

El Sohly, M.A., Holley, J.H., Lewis, G.S., Russel, M.H. and Turner, C.E. (1984) Constituents of *Cannabis sativa* L. XXIV: The potency of confiscated marijuana, hashish and hash oil over a ten-year period. *J. Forensic Sci.*, **29**(2), 500–514.

Emboden, W. (1972) *Narcotic plants*, Studio Vista, London, UK, pp. 14–16.

Evans, W.C. (1989) Hallucinogenic, allergenic, teratogenic and other toxic plants. In W.C. Evans, (ed.), *Trease and Evans' Pharmacognosy* 13th Edn, Ballière Tindall, London, UK, pp. 743–748.

Fairbairn, J. (1976) The Pharmacognosy of Cannabis. In J.D.P. Graham (ed.), *Cannabis and Health*. Academic Press Inc., London, UK, pp. 3–19.

Fairbairn, J. (1972) The trichomes and glands of *Cannabis sativa* L. *Bull. Narc.*, **24**(4), 29–33.

Fairbairn, J.W., Hindmarch, I., Simic, S. and Tylden, E. (1974) Cannabinoid content of some English reefers. *Nature*, **249**, 276–278.

Fairbairn, J.W., Liebman, J.A. and Rowan, M.G. (1976) The stability of cannabis and its preparations on storage. *J. Pharm. Pharmacol.*, **28**, 1–7.

French, C.N. and Moore, P.D. (1986) Deforestation, *Cannabis* cultivation and schwingmoor formation at Cors Llyn (Llyn mire) Central Wales. *New Phytol.*, **102**, 469–482.

Gerarde, J. (1597) *Herball: or Generall Historie of Plantes*, London, p. 572.

Gerarde, J. (1633) *The Herball or Generall Historie of Plantes*. Enlarged and amended by Thomas Johnson, London, p. 708.

Godwin, H. (1967) Pollen analytical evidence for the cultivation of Cannabis in England. *Rev. Palaeobot. Palynol.*, **4**, 71–80.

Gosden, T.B. (1987) *Drug abuse: the facts about today's drug scene – (Take control)*, David and Charles Publishers plc, Devon, UK, pp. 116–118.

Gray, C. (1995) Cannabis – the therapeutic potential. *Pharm. J.*, **254**, 771–773.

Grieve, M. (1974) *A modern herbal*, Jonathan Cape Ltd, London, UK. First published 1931, reprinted 1974. C.F. Leyel (ed.), p. 396.

Hammond, C. and Mahlberg, P. (1977) Morphogenesis of capitate glandular hairs of *Cannabis sativa* L. (Cannabaceae). *Am. J. Bot.*, **65**, 1023–1031.

Haney, A. and Bazzaz, F.A. (1970) Some ecological implications of the distribution of hemp (*Cannabis sativa* L.) in the United States of America. In C.R.B. Joyce and S.H. Curry (eds.), *The Botany and Chemistry of Cannabis*, Churchill, London, UK, pp. 39–48.

Harvey, D.J. (1985) Examination of a 140 year old ethanolic extract of Cannabis: identification of new cannabitriol homologues and the ethyl homologue of cannabinol. In D.J. Harvey (ed.), *Marihuana '84: Proceedings of the Oxford Symposium on Cannabis*, IRL Press, Oxford, UK, pp. 23–30.

Harvey, D.J. (1990) Stability of cannabinoids in dried samples of cannabis dating from around 1896–1905. *J. Ethnopharmacol.*, **28**, 117–128.

Heslop-Harrison, J. (1957) The experimental modification of sex expression in flowering plants. *Biol. Rev.*, **32**, 1–51.

Humphreys, I.J. and Joyce, J.R. (1982) A survey of the cannabis content of unsmoked reefer cigarettes. *J. Forens. Sci. Soc.*, **22**, 291–292.

Index Kewensis – an Enumeration of the Genera and Species of Flowering plants. (1893 and supplements to 1990), Clarendon Press, Oxford, UK.

Janischevsky, D.E. (1924) Forma konopli na sornykh mestack v Yugovostochnoi Rossii. Ucen. Zap. Saratovsk. gosud. *Černysevskogo Univ.* **2**(2), 3–17.

Judt, M. (1995) Hemp (*Cannabis sativa* L.) – salvation for the earth and for the papermakers. *Agro Food Ind. Hi-tech.*, **6**(4), 35–37.

Kalant, O.J. (1968) *An interim guide to the cannabis (marihuana) literature.* Addiction Research Foundation, Bibliographic series 2, p. 1.

Kovacs, I., Rab, A., Rusznak, I. and Annus, S. (1992) Hemp (*Cannabis sativa*) as a possible raw material for the paper industry. *Cellulose Chem. Tech.*, **26**(5), 627–635.

Kovar, K.A. and Linder, H. (1991) Investigation of Hashish: Content uniformity of different samples by coupled HPLC/PC-analysis. *Arch. Pharm.* (*Weinheim*), **324**, 329–333.

Lamark, J.B. de (1785). *Encyclopédique de Botanique* I (part 2), 694–695.

Lanyon, V.S., Turner, J.C. and Mahlberg, P.G. (1981) Quantitative analysis of cannabinoids in the secretory product from capitate-stalked glands of *Cannabis sativa* L. (Cannabaceae). *Bot. Gazette*, **142**(3), 316–319.

Lawi-Berger, C. and Kapetanidis, I. (1983) Contribution à l'étude chimiotaxonomique due genre *Cannabis* (Cannabaceae). 2e partie: Analyse quantitative des acides gras dans les akènes de *Cannabis sativa* L. *Pharm. Acta Helv.*, **58**(3), 79–81.

Lawi-Berger, C., Miège, J. and Miège, M.N. (1983a) Contribution à l'étude chimiotaxonomique due genre *Cannabis* (Cannabaceae). 3e partie: Analyse des protèines et des enzymes contenus dans les akènes. *Pharm. Acta Helv.*, **58**(5–6), 165–171.

Lawi-Berger, C., Stephanou, E., Buchs, A. and Kapetanidis, I. (1983b) Contribution à l'étude chimiotaxonomique due genre *Cannabis* (Cannabaceae). 1re partie: Analyse qualitative des acides gras dans les akènes de *Cannabis sativa* L. *Pharm. Acta Helv.*, **58**(2), 48–51.

Le Strange, R. (1977) *A history of herbal plants*, Arco Publishing House Inc., New York, USA, pp. 64–65.

Malingré, T., Hendricks, H., Batterman, S., Bos, R. and Visser, J. (1975) The essential oil of *Cannabis sativa*. *Planta Med.*, **28**, 56–61.

Nigam, R.K., Varkey, M. and Reuben, D.E. (1981a) Irradiation induced changes in flower formation in *Cannabis sativa* L. *Biologica plantarum*, **23**(5), 389–391.

Nigam, R.K., Varkey, M. and Reuben, D.E. (1981b) Streptovaricin induced sex expression in male and female plants of *Cannabis sativa* L. *Ann. Bot.*, **47**(1), 169–172.

Nigg, H.N. and Siegler, D. (1992) *Phytochemical Resources for Medicine and Agriculture*, Plenum Press, New York, USA, pp. 39–41.

O'Shaughnessy, W.B. (1842) On the preparation of the Indian hemp, or gunjah (*Cannabis indica*): the effects of the animal system in health, and their utility in the treatment of tetanus and other convulsive diseases. *Trans. Med. Phys. Soc. Bombay*, 1842, **8**, 421.

Quimby, M.W. (1974) Botany of *Cannabis sativa. Arch. Inv. Med.*, **5**, Suppl. 1, 127.

Ram, H.Y.M. and Sett, R. (1982a) Reversal of ethephon-induced feminization in male plants of cannabis by ethylene antagonists. *Zeitschrift fur pflanzenphysiologie*, **107**(1), 85–89.

Ram, H.Y.M. and Sett, R. (1982b) Induction of fertile male flowers in genetically female *Cannabis sativa* plants by silver nitrate and silver thiosulphate complex. *Theor. Appl. Gen.*, **62**(4), 369–375.

Rosenthal, E. (1984) *Marijuana growers handbook; indoor/greenhouse edition*, Quick American publishing company, San Fransisco, California, USA.

Samuelsson, G. (1992) *Drugs of Natural Origin*, Swedish Pharmaceutical Press, Stockholm, Sweden, pp. 155–160.

Sanitatis, O, (1491) *Herba plantes.*

Schultes, R.E. (1970) Random thoughts and queries on the botany of cannabis. In C.R.B. Joyce and S.H. Curry (eds.), *The Botany and Chemistry of Cannabis*, J.A. Churchill, London, UK, pp. 11–38.

Schultes, R.E. and Hoffman, A. (1980) *The botany and chemistry of hallucinogens*, Charles C. Thomas, Springfield, Illinois, USA, pp. 82–116.

Small, E. and Cronquist, A. (1976) A practical and natural taxonomy for *Cannabis. Taxon.*, **25**, 405–435.

Smith, R.N. and Vaughan, C.G. (1977) The decomposition of acidic and neutral cannabinoids in organic solvents. *J. Pharm. Pharmacol.*, **29**, 286–290.

Stamler, R.T., Fahlman, R.C. and Vigeant, H. (1985) Illicit traffic and abuse of cannabis in Canada. *Bull. Narc.*, **37**(4), 37–49.

Stearn, W.T. (1970) The Cannabis Plant: Botanical Characteristics. In C.R.B. Joyce and S.H. Curry (eds.), *The Botany and Chemistry of Cannabis*, Churchill, London, UK, pp. 1–10.

Taylor, R., Lydon, J. and Anderson, J.D. (1994) Anatomy and viability of *Cannabis sativa* stem cuttings with and without adventitious roots. *J. Forensic Sci.*, **39**(3), 769–777.

Thornton, J.I. and Nakamura, J. (1972) Identification of marijuana. *J. Forens. Sci. Soc.*, **12**, 461–519.

Turner, C.E. and El Sohly, M.A. (1979) Constituents of *Cannabis sativa* L., XVI. A possible decomposition pathway of delta-9-tetrahydrocannabinol to cannabinol. *J. Heterocyclic Chem.*, **16**, 1667–1668.

Turner, C.E., Hadley, K.W., Fetterman, P.S., Doorenbos, N.J., Quimby, M.W. and Waller, C. (1973) Constituents of *Cannabis sativa* L. IV. Stability of cannabinoids in stored plant material. *J. Pharm. Sci.*, **62**(10), 1601–1605.

Turner, J., Hemphill, J. and Mahlberg, P. (1977) Gland distribution and cannabinoid content in clones of *Cannabis sativa* L. *Am. J. Bot.*, **64**, 687–693.

Turner, J.C., Hemphill, J.K. and Mahlberg, P.G. (1978) Quantitative determination of cannabinoids in individual glandular trichomes of *Cannabis sativa* L. (Cannabaceae). *Am. J. Bot.*, **65**, 1103–1106.

Turner, J.C., Hemphill, J.K. and Mahlberg, P.G. (1981a) Interrelationships of glandular trichomes and cannabinoid content. II: Developing vegetative leaves of *Cannabis sativa* L. (Cannabaceae). *Bull. Narc.*, **33**(3), 63–71.

Turner, J.C., Hemphill, J.K. and Mahlberg, P.G. (1981b) Interrelationships of glandular trichomes and cannabinoid content. I: Developing pistillate bracts of *Cannabis sativa* L. (Cannabaceae). *Bull. Narc.*, **33**(2), 59–69.

Valle, J.R., Lapa, A.J. and Barros, G.G. (1968) Pharmacological activity of cannabis according to sex of the plant. *J. Pharm. Pharmacol.*, **20**, 798–799.

Vavilov, N.I. and Bukinich, D.D. (1929) Zemledel'cheskii Afghanistan. *Trudy Prikl. Bot.*, Suppl. 33, 380–382. Reissued in 1959 by Izdatel'stuo Akademi Nauk SSSR, Moskva-Leningrad.

Wills, S. (1993) Drugs and substance misuse (5). Cannabis and Cocaine. *Pharm. J.*, **251**, 483–485.

3. THE CHEMISTRY OF CANNABIS

AMALA RAMAN[1] and ALPANA JOSHI[2]

[1]*Department of Pharmacy, King's College London, UK*
[2]*National Center for the Development of Natural Products, University of Mississippi, USA*

The phytochemistry of *Cannabis sativa* has been extensively researched and more than four hundred compounds belonging to a variety of phytochemical groups have been reported to occur in the plant. According to one estimate, over 7000 scientific papers had been published on cannabis, its constituents and their pharmacological activities by 1980 (Turner *et al.*, 1980). Many detailed descriptions of the chemistry of cannabis have been published over the years, such as those of Mechoulam (1973), Razdan (1973), Crombie and Crombie (1976), Schultes and Hoffman (1980), Harvey (1984) and a major review article dealing exhaustively with the phytochemistry of cannabis by Turner *et al.* (1980). In the present text, only the most important features of cannabis phytochemistry will be described; the interested reader is referred to one of the more extensive treatments listed above for greater detail. A further source of information is the annotated bibliography of cannabis covering the literature from 1964 published by Waller *et al.* in 1976 (Volume I) and 1982 (Volume II), updated with regular supplements from 1980 onwards (Waller *et al.*, 1980–1993/4).

The psychoactive effects of cannabis and its preparations have been ascribed in the main to the presence of tetrahydrocannabinols (THCs), in particular the compound Δ^9-tetrahydrocannabinol (Δ^9-THC), which was first isolated and identified in 1964 (Gaoni and Mechoulam, 1964a). Δ^9-THC is one of a group of mostly C_{21} compounds known as cannabinoids, which appear to be unique to *Cannabis sativa*. More recent studies have demonstrated that cannabinoids other than Δ^9-THC also exhibit a range of pharmacological activities (Formukong *et al.*, 1989). Cannabis also contains non-cannabinoid compounds whose effects have not been so widely investigated. An important point regarding *Cannabis sativa* is that it shows considerable variation in its chemistry, as described later in this chapter.

CANNABINOID CONSTITUENTS OF CANNABIS

Numbering Systems for Cannabinoids

Over the years, at least 5 numbering systems have been used for cannabinoids (Eddy, 1965). Only two of these, however, are in widespread use (Figure 1). One is based on the formal chemical rules for numbering dibenzopyran type compounds, and is the system used by Chemical Abstracts. This system will be adopted in the present text. In the second system the cannabinoids are numbered as substituted monoterpenoids (based on p-cymene) due to their biogenetic origin. A reader scanning the literature on cannabis may therefore encounter a number of ways of referring to the same compound. The major

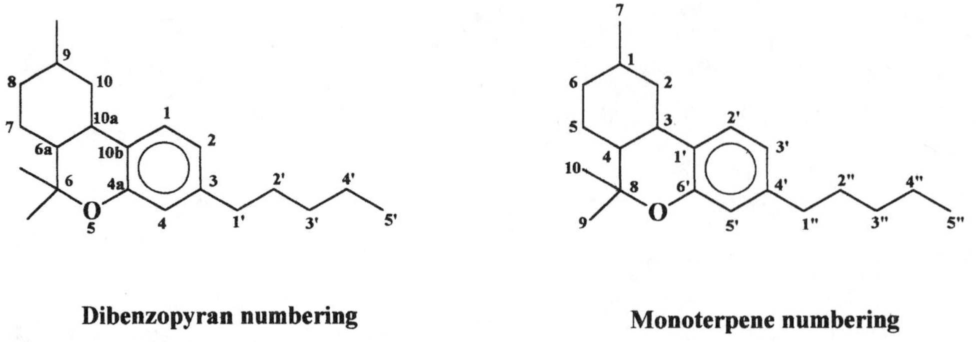

Dibenzopyran numbering   **Monoterpene numbering**

Figure 1   Two common numbering systems used for cannabinoids (Eddy, 1965)

psychoactive component Δ^9-THC, for instance, may be described as either Δ^9-tetrahydrocannabinol (dibenzopyran system) or Δ^1-tetrahydrocannabinol (mono-terpenoid system). Similarly its minor structural isomer, Δ^8-tetrahydrocannabinol (dibenzopyran system), may be referred to as $\Delta^{1(6)}$-tetrahydrocannabinol (mono-terpenoid system).

Structural Groups of Cannabinoids

The very large number of cannabinoids (over 60) known to occur in cannabis (Turner *et al.*, 1980) can be divided into a few main structural types as illustrated in Figure 2. These are the cannabigerol (CBG), cannabichromene (CBC), cannabidiol (CBD), Δ^9-tetrahydrocannabinol (Δ^9-THC), Δ^8-tetrahydrocannabinol (Δ^8-THC), cannabicyclol (CBL), cannabielsoin (CBE), cannabinol (CBN), cannabinodiol (CBND) and cannabit-riol (CBO) types. Variations on these basic types are fairly standard: presence or absence of a carboxyl group on the phenolic ring (at R^2 or R^4), a methyl, propyl or butyl side chain replacing the pentyl one (at R^3), or a methoxy group in place of one of the hydroxyl moieties. Some of the known compounds in each group are listed in Table 1 (from Turner *et al.*, 1980). For each type, the neutral compound with the pentyl side chain is normally referred to by the name and abbreviation listed above. In general, acid analogues have the letter A suffixed to the abbreviation, methyl ethers the letter M and methyl, propyl and butyl side chain analogues the suffix -C_n where n equals the number of carbons in the side chain. However, propyl analogues often have an abbreviation incorporating the letter V as their complete name usually includes the term "varin" e.g. cannabivarin, cannabi chromevarin (C_3 analogues of cannabinol and cannabichromene respectively).

Most natural cannabinoids have at least two chiral centres at carbons 10a and 6a (Figure 1). The absolute configuration at these centres was determined by Mechoulam and Gaoni (1967) for THC (10a R, 6a R) and CBD (10a S, 6a R). Further details regarding the isolation and absolute stereochemical configuration of the various cannabinoids in Figure 2 and Table 1 can be found in Turner *et al.* (1980).

In addition to the main cannabinoid groups described above, some usually very minor constituents belonging to related structural types have been shown to be pre-sent in cannabis. They include dehydrocannabifuran (DCBF), cannabifuran (CBF),

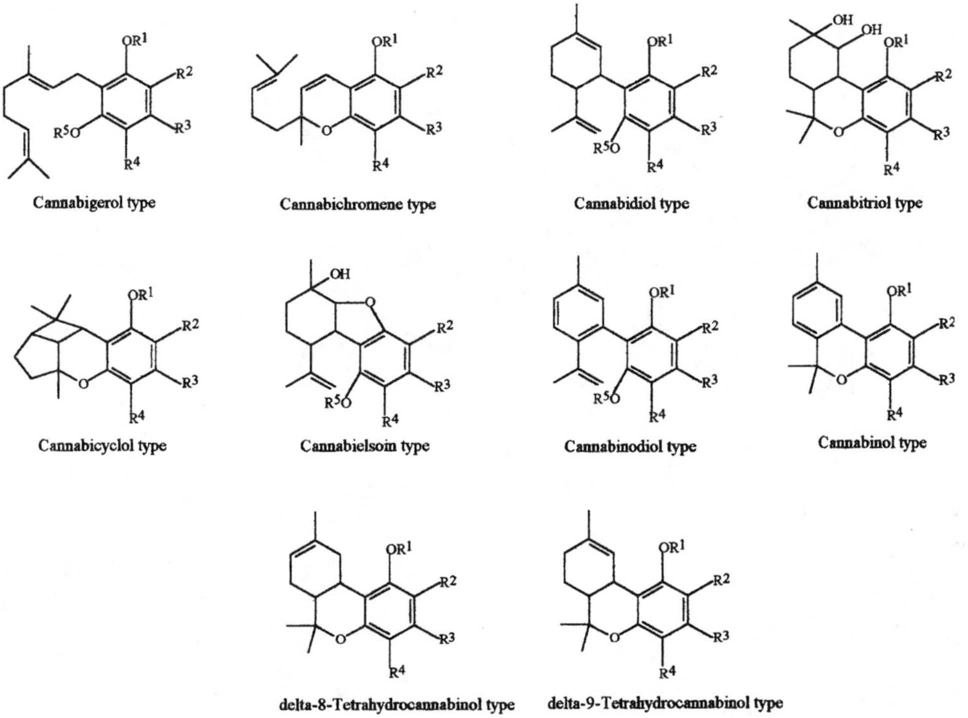

Cannabigerol type

Cannabichromene type

Cannabidiol type

Cannabitriol type

Cannabicyclol type

Cannabielsoin type

Cannabinodiol type

Cannabinol type

delta-8-Tetrahydrocannabinol type

delta-9-Tetrahydrocannabinol type

Figure 2 Main structural types of cannabinoids; see Table 1 for examples of compounds

cannabicitran (CBT), cannabichromanon (CBCN) and a dimeric cannabinoid formed by esterification of cannabidiolic acid with tetrahydrocannabitriol (Turner *et al.*, 1980). One of the most recent cannabinoids isolated from cannabis is cannabinerolic acid – the *trans* isomer of CBG (Taura *et al.*, 1995).

Chemical alteration of cannabinoids may occur during harvesting, storage or processing of cannabis preparations. CBN type compounds isolated from cannabis preparations are degradation products of the corresponding THC derivatives (Garret and Tsau, 1974; Turner and El Sohly, 1979; Harvey, 1985), and are not formed biosynthetically. The acid forms of THC are decarboxylated during storage probably by the agency of heat or light; this reaction occurs during smoking of cannabis preparations and in some analytical processes (Baker *et al.*, 1981). Δ^9-THC may isomerise to Δ^8-THC in the presence of strong acids (Mechoulam, 1973).

Biogenesis of Cannabinoids

Despite the interest in this group of compounds, surprisingly few actual experimental investigations have been conducted into the biogenesis of cannabinoids. Existing reports have variously involved either neutral compounds or the carboxylated forms. A general outline of the biogenetic origin of the cannabinoids, based on these studies as well as postulates, is depicted in Figure 3 (adapted from Harvey, 1984; Clarke, 1981;

Table 1 Examples of cannabinoids belonging to each of the main structural types shown in Figure 2

Cannabinoid type	Name of cannabinoid	R^1	R^2	R^3	R^4	R^5	Abbreviation
Cannabigerol	Cannabigerol	H	H	C_5H_{11}	H	H	CBG
	Cannabigerolic acid	H	COOH	C_5H_{11}	H	H	CBGA
	Cannabigerol monomethylether	H	H	C_5H_{11}	H	CH_3	CBGM
	Cannabigerolic acid monomethylether	H	COOH	C_5H_{11}	H	CH_3	CBGAM
	Cannabigerovarin	H	H	C_3H_7	H	H	CBG-C_3
	Cannabigerovarinic acid	H	COOH	C_3H_7	H	H	CBGA-C_3
Cannabichromene	Cannabichromene	H	H	C_5H_{11}	H	—	CBC
	Cannabichromenic acid	H	COOH	C_5H_{11}	H	—	CBCA
	Cannabichromevarin	H	H	C_3H_7	H	—	CBC-C_3
	Cannabichromevarinic acid	H	COOH	C_3H_7	H	—	CBCA-C_3
Cannabidiol	Cannabidiol	H	H	C_5H_{11}	H	H	CBD
	Cannabidiolic acid	H	COOH	C_5H_{11}	H	H	CBDA
	Cannabidiol monomethylether	H	H	C_5H_{11}	H	CH_3	CBDM
	Cannabidiorcol	H	H	CH_3	H	H	CBD-C_1
	Cannabidivarin	H	H	C_3H_7	H	H	CBDV, CBD-C_3
	Cannabidivarinic acid	H	COOH	C_3H_7	H	H	CBDVA, CBDA-C_3
	Cannabidiol-C_4	H	H	C_4H_9	H	H	CBD-C_4
Cannabitriol	Cannabitriol	H	H	C_5H_{11}	H	—	CBO
Cannabicyclol	Cannabicyclol	H	H	C_5H_{11}	H	—	CBL
	Cannabicyclolic acid	H	COOH	C_5H_{11}	H	—	CBLA
	Cannabicylovarin	H	H	C_3H_7	H	—	CBLV, CBL-C_3

Compound						Abbreviation
Cannabielsoin						
Cannabielsoin	—	H	C_5H_{11}	H	H	CBE
Cannabielsoic acid A	—	COOH	C_5H_{11}	H	H	CBE acid A
Cannabielsoic acid B	—	H	C_5H_{11}	COOH	H	CBE acid B
Cannabinodiol						
Cannabinodiol	H	H	C_5H_{11}	H	H	CBND
Cannabinodivarin	H	H	C_3H_7	H	H	CBVD, CBND-C3
Cannabinol						
Cannabinol	H	H	C_5H_{11}	H	—	CBN
Cannabinolic acid	H	COOH	C_5H_{11}	H	—	CBNA
Cannabinol monomethylether	CH_3	H	C_5H_{11}	H	—	CBNM
Cannabiorcol	H	H	CH_3	H	—	CBN-C1
Cannabivarin	H	H	C_3H_7	H	—	CBV, CBN-C3
Cannabinol-C4	H	H	C_4H_9	H	—	CBN-C4
Δ^8-Tetrahydrocannabinol						
Δ^8-Tetrahydrocannabinol	H	H	C_5H_{11}	H	—	Δ^8-THC
Δ^8-Tetrahydrocannabinolic acid	H	COOH	C_5H_{11}	H	—	Δ^8-THCA
Δ^9-Tetrahydrocannabinol						
Δ^9-Tetrahydrocannabinol	H	H	C_5H_{11}	H	—	Δ^9-THC
Δ^9-Tetrahydrocannabinolic acid A	H	COOH	C_5H_{11}	H	—	Δ^9-THC acid A, Δ^9-THCA
Δ^9-Tetrahydrocannabinolic acid B	H	H	C_5H_{11}	COOH	—	Δ^9-THC acid B, Δ^9-THCA
Δ^9-Tetrahydrocannabiorcol	H	H	CH_3	H	—	Δ^9-THC-C1
Δ^9-Tetrahydrocannabiorcolic acid	H	H/COOH	CH_3	COOH/H	—	Δ^9-THCA-C1
Δ^9-Tetrahydrocannabivarin	H	H	C_3H_7	H	—	Δ^9-THCV, Δ^9-THC-C3
Δ^9-Tetrahydrocannabivarinic acid	H	H/COOH	C_3H_7	COOH/H	—	Δ^9-THCVA, Δ^9-THCA-C3
Δ^9-Tetrahydrocannabinol-C4	H	H	C_4H_9	H	—	Δ^9-THC-C4
Δ^9-Tetrahydrocannabinolic acid-C4	H	H/COOH	C_4H_9	COOH/H	—	Δ^9-THCA-C4

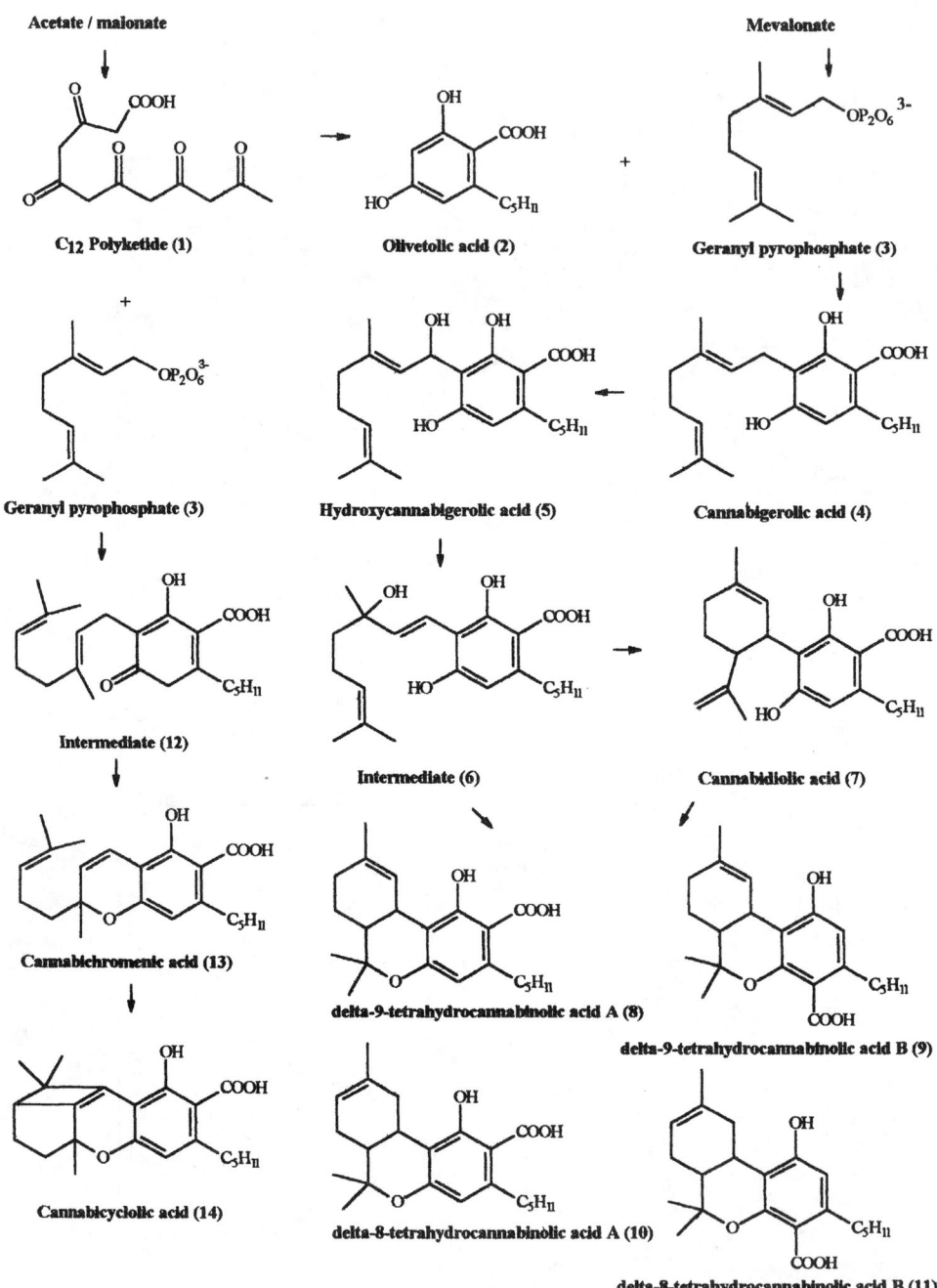

Figure 3 Proposed biogenetic pathway for the main cannabinoids

Schultes and Hoffman, 1980; Turner and Mahlberg, 1985). Numbers in parentheses in this section refer to structures shown in Figure 3. For simplicity, only the acid forms are shown; the neutral cannabinoids commonly encountered in cannabis products may arise either by decarboxylation of the corresponding acids during harvesting and storage (Shoyama *et al.*, 1975) or by a biosynthetic pathway analogous to that shown, but involving the equivalent neutral precursors (Kajima and Piraux, 1982). In support of an independent pathway for neutral compounds, it has been observed that radiolabelled neutral precursors (olivetol and cannibigerol) are incorporated into THC and other neutral cannabinoids but not into THCA (Kajima and Piraux, 1982).

Some of the earliest articles on the biosynthesis of cannabinoids were published by Simonsen and Todd (1942), Farmilo *et al.* (1962) and Ni (1963) who proposed menthatriene, limonene and p-mentha-3,8-diene-5-one repectively as terpene compounds which condensed with olivetolic acid, the precursor for the aromatic ring of the cannabinoids. However, it was Mechoulam and colleagues (Gaoni and Mechoulam, 1964b; Mechoulam and Gaoni, 1965; Mechoulam, 1970, 1973), who suggested the presently accepted route involving initial condensation of a phenolic compound, either olivetolic acid **(2)** or its decarboxylated analogue, olivetol with the terpene derivative geranyl pyrophosphate **(3)**. This has since been supported by experimental studies (Shoyama *et al.*, 1975) in which malonate, mevalonate (precursors of olivetolic acid and geranyl pyrophosphate) and also geraniol and nerol were incorporated into THCA. CBC, however, appears to be formed by a different pathway; Turner and Mahlberg (1985) have shown that labelled olivetol administered to cannabis seedlings is incorporated only into CBG and THC, but not into CBC. This, and their finding that the developing plant first produces CBC and only later CBG and THC (Vogelmann *et al.*, 1988), implies the possible existence of two divergent pathways.

In the first route, CBCA **(13)** arises from combination of geranyl phosphate with a precursor of olivetolic acid (Turner and Mahlberg, 1985), possibly a C_{12} polyketide **(1)** derived from acetate/malonate (Shoyama *et al.*, 1975). However, there is also evidence that CBC can arise from CBG in some variants (Shoyama *et al.*, 1975). CBC and its acid form **(13)** are believed to be the precursors for CBL and CBLA **(14)** respectively.

In the second pathway, geranyl phosphate and olivetolic acid condense to form CBGA **(4)**. Hydroxylation to hydroxycannabigerolic acid **(5)** is followed by rearrangement to an intermediate **(6)** which can then cyclise to form CBDA **(7)**. Further cyclisation involving one or other of the phenolic hydroxyl groups leads to the potential (only three have actually been isolated from cannabis) formation of four isomeric THCAs **(8–11)** which vary in the position of the double bond and carboxylic acid group. However, Kajima and Piraux (1982) showed experimentally that CBD is not necessarily involved in THC biosynthesis . They suggest, in agreement with Turner and Hadley (1973), that a common intermediate **(6)** may give rise to either CBD or rearrange directly to THC. Variation in the levels of enzymes controlling these pathways may account for the chemical variation seen in different varieties of cannabis.

It is of interest to note that despite support for its involvement in cannabinoid biosynthesis, olivetol itself has not been reported to occur in cannabis. On the contrary, a prominent phenolic component of the glandular trichomes was found by Hammond and Mahlberg (1994) to be phloroglucinol (1,3,5-trihydroxybenzene), which they suggest may have some significance in cannabinoid biosynthesis.

Related components of cannabis, such as CBNA and its neutral analogue CBN, are not thought to be biogenetic products, but artefacts arising from the degradation of THCA and THC respectively (Harvey, 1984; Turner and El Sohly, 1979). Radiotracer studies show that the propyl side chain analogues of the cannabinoids do not arise by degradation of the pentyl side chain of the more common cannabinoids (Kajima and Piraux, 1982) and may involve a parallel biogenetic pathway.

Chemical Methods for Cannabinoid Synthesis

Interest in their pharmacological activity, as well as the need for reference materials for analytical purposes, has prompted the development of stereospecific synthetic methods for the production of cannabinoids in high yields. Synthetic processes for cannabinoids generally mirror the proposed biosynthetic sequence, involving the condensation of an optically active monoterpene with olivetol (5-pentylresorcinol). The monoterpene, reaction conditions and subsequent treatment of intermediates can be varied to obtain the desired cannabinoid product. Monoterpenes used by different researchers include p-mentha-2,8-dien-1-ol (Petrzilka et al., 1969), carene oxides (Razdan and Handrick, 1970), chrysanthenol (Razdan et al., 1975), citral (El Sohly et al., 1978) and p-menth-2-ene-1,8-diol (Handrick et al., 1979). Methods for the synthesis of Δ^9-THC and other cannabinoids have been reviewed in detail by Mechoulam et al. (1976), Crombie and Crombie (1976), and Razdan (1984).

NON-CANNABINOID CONSTITUENTS OF CANNABIS

Non-cannabinoid constituents isolated from various parts of the cannabis plant include a range of nitrogenous compounds (including alkaloids), sugars, sugar polymers, cyclitols, fatty acids, amino acids, proteins, glycoproteins, enzymes, hydrocarbons, simple alcohols, acids, aldehydes and ketones, steroids, terpenes, non-cannabinoid phenolic compounds, flavonoid glycosides, vitamins and pigments (Turner et al., 1980). The majority of these compounds are found in many other species and are not unique to cannabis.

Some of the more unusual constituents of cannabis include an amide formed between p-hydroxy-(trans)-cinnamic acid and 2-(p-hydroxyphenyl)-ethylamine, which was isolated from the roots of Mexican cannabis (Slatkin et al., 1971) and the spermidine alkaloids cannabisativine and anhydrocannabisativine isolated from the roots and aerial parts of various cannabis strains (Turner et al., 1980). Non-cannabinoid phenolic compounds found in cannabis include spiro-indans (e.g. cannabispiran, cannabispirenone), dihydro-stilbenes or bibenzyl compounds (e.g. canniprene) and cannabidihydrophenanthrene (Turner et al., 1980). Additional non-cannabinoids isolated from cannabis since the publication of the review by Turner et al. (1980) include three new dihydro-stilbenes (El-Feraly, 1984; El Sohly et al., 1984)) and three new spiro-indans (El-Feraly et al., 1986) either from hashish or leaves of cannabis, four phenyldihydronaphthalene lignanamides from cannabis fruits (Sakakibara et al., 1991, 1992) and phloroglucinol glucoside from shoot latex (Hammond and Mahlberg, 1994). The volatile oil of indoor-grown cannabis has been analysed and found to contain 68 components of which 57 were found to be known monoterpenes and sesquiterpenes (Ross and El Sohly, 1996).

Tris malonate acetylations and decarboxylations involving *p*-hydroxycinnamic acid have been reported to be involved in the biosynthesis of the dihydrostilbene (bibenzyl) compounds and flavones found in cannabis (Crombie *et al.*, 1988). The dihydrostilbenes are believed to be natural precursors of the spiro-indan compounds (El Sohly and Turner, 1982).

CHEMICAL VARIATION IN CANNABIS

Studies on a large number of cannabis plants originating from different parts of the world have led to the acceptance that a number of chemical races or "chemovars" of *Cannabis sativa* exist. These vary widely in their Δ^9-THC content and therefore psychoactive potency. The types cultivated for fibre production have very low levels of this compound, but show enhanced levels of its non-narcotic, biosynthetic precursor CBD. It has not been possible to correlate the chemovars directly with the different species or subspecies of *Cannabis* (e.g. *sativa, indica, ruderalis*) proposed by various authors (see Chapter 2), as these were primarily distinguished on morphological grounds. It is generally believed that the chemovars do not represent individual species, but owe their existence to centuries of cultivation and breeding for one of the two main products i.e. the intoxicant resin or the stem fibre.

A number of classification systems have been proposed to distinguish psychoactive and fibre strains of cannabis based on their cannabinoid composition (reviewed by Turner *et al.*, 1980). The first classification system, proposed by Grlic (1968), involved the use of a selection of chemical, spectroscopic, microbiological and pharmacological tests whose results were dependent on the levels of CBDA, CBD, Δ^9-THC and CBN in the sample. These markers were regarded as indicative of successive stages of "ripening" or subsequent decomposition of the resin. The more "ripe" samples (with higher levels of Δ^9-THC) were found to originate in tropical areas, commonly associated with production of intoxicant cannabis.

A few years later, a method based on quantitative analysis of specific cannabinoids was suggested by Waller and his colleagues (Waller and Scigliano, 1970; Fetterman *et al.*, 1971), in which the ratio of Δ^9-THC and its breakdown product CBN to the non-narcotic CBD was measured:

$$\text{Phenotype} = (\Delta^9\text{-THC} + \text{CBN}) / \text{CBD}.$$

Samples with ratios greater than 1 were classified as "drug type" and those with ratios below 1 as "fibre type" cannabis. Based on an examination of a large number of samples, Small and Beckstead (1973) further expanded the classification to four phenotypes:

Phenotype I: high ($>0.3\%$) THC and low ($<0.5\%$) CBD,
Phenotype II: at least 0.3% THC and high ($>0.5\%$) CBD,
Phenotype III: relatively little THC and high ($>0.5\%$) CBD,
Phenotype IV: plants consistently showing trace amounts of CBGM.

Turner *et al.* (1980) have outlined some of the limitations of the Waller and Small systems, which essentially only require the measurement of Δ^9-THC, CBD and CBN. These include the inadequate separation of CBD from CBC and CBV in the analytical systems employed at the time, the absence of CBD and CBC from cannabis of certain geographical origins, the presence of homologues (C_3 variants) in some samples which may contribute to psychoactive properties, and the influence of the age of the plant when analysed on its constituents, and consequently the phenotype to which it is assigned. They proposed that other cannabinoids (including C_3 homologues) should also be taken into consideration and derived the formula:

$$\text{Phenotype} = \frac{(\Delta^9\text{-THC} + \Delta^9\text{-THCV} + \text{CBN} + \Delta^8\text{-THC})}{(\text{CBDV} + \text{CBD} + \text{CBC} + \text{CBG} + \text{CBGM})}.$$

They suggest that the drug type (ratio >1) and fibre type (ratio <1) classification could be applied most reliably if the analyses were performed at regular intervals throughout the growing season of the plant, although this would not apply to confiscated samples.

Paris and Nahas (1984) have reviewed these classification systems and point out that the term "phenotype" is somewhat misleading as this generally refers to visible characteristics rather than genetic traits. They suggest classification into three chemical types, similar to phenotypes I–III of Small and Beckstead (1973) based on absolute content of THC and CBD rather than ratios:

(1) Drug type: THC $>1\%$, CBD $= 0$,
(2) Intermediate drug type: THC $>0.5\%$, CBD $>0.5\%$,
(3) Fibre type: THC $<0.25\%$, CBD $>0.5\%$.

This classification into drug, fibre and intermediate types was first suggested by Turner (1980). In addition to the three main groups described above, Fournier *et al.* (1987), have reported a new chemotype of cannabis in which CBG (rather than CBD or Δ^9-THC) is the dominant cannabinoid. These chemotypes, however, cannot be considered as unique species or subspecies as it has been found that the variations in CBD and Δ^9-THC content among the plants is completely continuous, and further that individuals from strains belonging predominantly to one group may show characteristics of another (De Meijer *et al.*, 1992). A germplasm collection in which the predominant chemotype has been assessed has been established at Wageningen, the Netherlands (De Meijer and Van Soest, 1992).

Since the drug type and fibre type of cannabis have historically been associated with tropical and temperate regions of the world respectively, there has been considerable attention focussed on whether genetic or geographical factors govern the chemical nature of individual strains. Much of the work to date favours the primary importance of genetic factors in determining the cannabinoid profile of the plant. Fairbairn (1976), for example, reported that when seeds of specific cannabis strains representing either high Δ^9-THC or high CBD varieties were grown in a range of countries (UK, USA, Norway, Canada, Turkey, Thailand) all the plants from a particular batch showed a consistent CBD/Δ^9-THC pattern, although absolute content varied. Further evidence for genetic influence is that when high Δ^9-THC: low CBD strains are crossed with low

Δ^9-THC: high CBD varieties, the offspring show a cannabinoid content intermediate between the two (Clarke, 1981). That the local climate is not the primary influence on psychoactive potency is indicated by the successful outdoor cultivation of plants with relatively high Δ^9-THC content in Italy (Bertol and Mari, 1980; Avico *et al.*, 1985), Switzerland (Brenneisen and Kessler, 1987) and even the Danish island of Bornholm (Felby and Nielsen, 1985) which lies 55$N of the equator.

It has been suggested that over a number of generations, the chemical characteristics of a plant can alter to match more closely the type common to the area of cultivation. Bouquet (1951) reported that after several generations, plants grown in England and France from Indian seeds were indistinguishable from European (fibre) cultigens, whereas European varieties planted in Egypt as a source of fibre altered to low-fibre psychoactive forms. This may indicate the modifying influence of environmental factors, but the possibility of cross pollination with local strains during open cultivation cannot be ruled out. More recently a group in the United Kingdom has grown cannabis plants from seeds of diverse geographical origin under controlled conditions, and monitored their physical and chemical characteristics over four generations (Baker *et al.*, 1982, 1983; Taylor *et al.*, 1985). Marihuana samples prepared from the plants closely resembled the parent preparation even after four generations, and with a few exceptions within each group, the cannabinoid content was still typical of the profile obtained with the original source sample. A notable change in properties was that the THCA/THC ratios in the offspring were higher than in the source sample. This may be due to environmental factors; according to Mechoulam (1970), neutral cannabinoids are rarely found in cannabis grown in northern countries. However, it may also indicate the occurrence of decarboxylation during the preparation or storage of the original sample.

Genetic control of cannabinoid chemotypes is likely to be mediated via the synthesis of particular enzymes involved in cannabinoid biogenesis. In the proposed biosynthetic sequence (Figure 3), CBG is converted to an intermediate which can form either CBD or THC, and CBD may itself be converted to THC. Thus genetically controlled deficiencies in particular steps of the pathway can lead to CBG, CBD or THC dominant plants.

It is important to note that even though a plant may have the genetic capacity to express a particular enzyme, the environment could still influence the extent to which this occurs and therefore alter the cannabinoid content. In the study by Fairbairn (1976) described earlier, although the dominant cannabinoid remained unchanged in the different growth locations for a particular batch of seeds, variations were noted in the actual cannabinoid levels. In a group of Mexican drug type cannabis plants grown in Mississippi (Turner *et al.*, 1982), the CBC content was found to increase over a two year period. It was also noted that high temperatures and rainfall resulted in higher Δ^9-THC levels. Mahlberg and Hemphill (1983) have shown the importance of daylight in controlling Δ^9-THC and CBC levels. They found that red, blue and green filters had differing effects on the two cannabinoids, suggesting that the effect of light was being mediated via enzymes involved in their separate biosynthetic pathways. Pate (1983) has suggested that enhanced production of Δ^9-THC in regions of higher light intensity may indicate a protective role for the compound against the harmful effects of UVB radiation.

There is considerable evidence that as well as genetic and environmental factors, there is high inherent interplant variability between members of the same chemotype and even the same strain growing under identical conditions (Cortis *et al.*, 1985; De Meijer *et al.*, 1992). Daily and monthly fluctuations in the content of major cannabinoids have also been reported (Phillips *et al.*, 1970; Turner *et al.*, 1975).

Assessment of the chemical profile of a cannabis strain has been important for two main purposes – to distinguish drug and fibre chemotypes and to try to identify the geographical source of illicit samples of cannabis or cannabis products. Taking the first aspect, the recognition that fibre type cannabis generally has low levels of Δ^9-THC has been important in allowing countries to legislate for the cultivation of hemp and against the cultivation of narcotic cannabis. For instance, the maximum permitted Δ^9-THC content in fibre hemp is reported as 0.3% and 0.2% respectively for France (Bruneton, 1995) and the former USSR (De Meijer *et al.*, 1992). A review of the analytical methods that can be used to measure cannabinoid content is beyond the scope of this chapter, but a recent paper by Lehmann and Brenneisen (1995) who report comparative profiles of drug, fibre and intermediate types using high performance liquid chromatography (HPLC) coupled to photodiode array detection may be mentioned here.

A number of studies have examined the possibility of predicting the intoxicant potential of a particular cannabis plant or seed sample without the necessity of growing it to maturity. Independent studies carried out by Barni-Comparini *et al.* (1984) and Cortis *et al.* (1985) show that the cannabinoid profile of vegetative leaves even at an early stage in the plant's development is a good indication of its ultimate chemical characteristics. An attempt has been made to correlate the chemical characteristics of cannabis populations to some non-chemical traits (De Meijer *et al.*, 1992). Morphological features such as achene characteristics, stem width and internode length showed no correlation, but a weak association was found between psychoactive properties, leaflet width and date of anthesis. In another study, although variations were seen in the electrophoretic patterns of seed proteins from different cultivars, these could not be associated with the cannabinoid profile of the plant (De Meijer and Keizer, 1996). The potential use of random amplification of polymorphic DNA (RAPD) in the profiling of cannabis samples has been reported (Gillan *et al.*, 1995), but as yet no correlations to cannabinoid content have been made.

Cannabis strains that can be classified as drug type on the basis of their Δ^9-THC content, nevertheless show considerable variability in their overall phytochemical profile. Brenneisen and El Sohly (1988) have used high resolution gas-chromatography coupled to mass spectrometry (GC–MS), as well as HPLC, to examine the complex profiles of cannabis samples of various known geographical origins. Compounds appearing in the chromatographic profiles included both cannabinoids and non-cannabinoids, and samples from a common source showed similar characteristic peak patterns. Many of the diagnostically important peaks were found in the terpene region rather than amongst the cannabinoids. Certain components were only found in samples from particular sources e.g. allo-aromadendrene and tetrahydrocannbiorcol were characteristic of Mexican and Jamaican cannabis, whereas caryophyllene oxide (the terpene supposedly detected by sniffer dogs) was absent only in USA derived samples. However, only a limited number of samples were analysed from each source and further work is required to confirm these findings. Baker *et al.* (1980) examined samples from various countries (between 5 and 150 samples from each source) by TLC and reported that although more than one type of product originated from a particular country, these could usually be visually and chemically distinguished. THV (Δ^9-THC-C_3) was common in illicit cannabis products from South Africa, Angola, Swaziland and Zimbabwe, sometimes exceeding Δ^9-THC in concentration, whereas samples from Ghana, Jamaica and Nigeria had low THV: Δ^9-THC ratios. CBG and CBC were common in Ghanaian samples. CBD was absent in samples from Kenya, Zambia, South Africa and Thailand (in the latter only THC and THCA were detected), whereas Moroccan, Pakistani and Lebanese hashish had significant levels of CBD. Indian cannabis was found to be highly variable in chemical composition, reflecting either the presence of many chemotypes under cultivation or the large size of the country. However, strict geographical patterns cannot be defined and are unlikely to be consistent over a long period of time due to exchange of seeds between countries, often as part of the illicit products transported.

REFERENCES

Avico, U., Pacifici, R. and Zuccaro, P. (1985) Variations of tetrahydrocannibinol content in cannabis plants to distinguish the fibre-type from drug-type plants. *Bull. Narc.*, **37**(4), 61–65.

Baker, P.B., Gough, L.A. and Taylor, B.J. (1980) Illicitly imported *Cannabis* products: some physical and chemical features indicative of their origin. *Bull. Narc.*, **32**(2), 31–40.

Baker, P.B., Gough, T.A. and Taylor, B.J. (1982) The physical and chemical features of *Cannabis* grown in the United Kingdom of Great Britain and Northern Ireland from seeds of known origin. *Bull. Narc.*, **34**(1), 27–36.

Baker, P.B., Gough, T.A. and Taylor, B.J. (1983) The physical and chemical features of *Cannabis* grown in the United Kingdom of Great Britain and Northern Ireland from seeds of known origin – Part II: second generation studies. *Bull. Narc.*, **35**(1), 51–62.

Baker, P.B., Taylor, B.J. and Gough, T.A. (1981) The tetrahydrocannabinol and tetrahydrocannabinolic acid content of cannabis products. *J. Pharm. Pharmacol.*, **33**, 369–372.

Barni-Comparini, I., Ferri, S. and Centini, F. (1984) Cannabinoid level in the leaves as a tool for the early discrimination of cannabis chemovariants. *Forensic Sci. Int.*, **24**, 37–42.

Bertol, E. and Mari, F. (1980) Observations on cannabinoid content in *Cannabis sativa* L. grown in Tuscany, Italy. *Bull. Narc.*, **32**(4), 55–60.

Bouquet, R.J. (1951) Cannabis. *Bull. Narc.*, **3**, 14–30.

Brenneisen, R. and El Sohly, M.A. (1988) Chromatographic and spectroscopic profiles of *Cannabis* of different origins: Part I. *J. Forensic Sci.*, **33**(6), 1385–1404.

Brenneisen, R. and Kessler, T. (1987). Psychotrope Drogen 1. *Pharm. Acta Helv.*, **62**(5–6), 134–139.

Bruneton, J. (1995). Orcinols and Phloroglucinols. In *Pharmacognosy, Phytochemistry, Medicinal Plants*, Intercept Ltd, Hampshire, pp. 371–379.

Clarke, R.C.C. (1981) *Marijuana Botany*, And/or Press, Berkeley, California. p. 93 (cross-breeding); pp. 169–171 (cannabinoid biosynthesis).

Cortis, G., Luchi, P. and Palmas, M. (1985) Experimental cultivation of cannabis plants in the Mediterranean area. *Bull. Narc.*, **37**(4), 67–73.

Crombie, L. and Crombie, W.M.L. (1976) Chemistry of the cannabinoids. In J.D.P. Graham, (ed.), *Cannabis and Health*, Academic Press, London, pp. 43–76.

Crombie, L.W., Crombie, M.L. and Firth, D.F. (1988) Synthesis of bibenzyl cannabinoids, hybrids of two biogenetic series found in *Cannabis sativa. J. Chem. Soc. Perkin Trans.I*, **5**, 1263–1270.

De Meijer, E.P.M. and Keizer, L.C.P. (1996) Patterns of diversity in *Cannabis. Genetic Resources and Crop Evolution*, **43**, 41–52.

De Meijer, E.P.M. and Van Soest L.J.M. (1992) The CPRO *Cannabis* germplasm collection. *Euphytica*, **62**(3), 201–211.

De Meijer, E.P.M., Van der Kamp, H.J. and Van Eeuwijk, F.A. (1992) Characterisation of *Cannabis* accessions with regard to cannabinoid content in relation to other plant characters. *Euphytica*, **62**(3), 187–200.

Eddy, N.B. (1965) *The Question of Cannabis*, Bibliography, United Nations Commission on Narcotic Drugs, E/CN7/49.

El-Feraly, F.S. (1984) Isolation, characterisation and synthesis of 3,5,4'-trihydroxybibenzyl from *Cannabis sativa. J. Nat. Prod.*, **47**(1), 89–92.

El-Feraly, F.S., El-Sherei, M.M. and Al-Muhtadi, F.J. (1986) Spiro-indans from *Cannabis sativa. Phytochem.*, **25**(8), 1992–1994.

El Sohly, H.N. and Turner, C.E. (1982) Constituents of *Cannabis sativa* L. XXII: isolation of spiro-indan and dihydrostilbene compounds from a Panamanian variant grown in Mississippi, United States of America. *Bull. Narc.*, **34**(2), 51–56.

El Sohly, H.N., Ma, G.E., Turner, C.E. and El Sohly, M.A. (1984). Constituents of *Cannabis sativa* XXV. Isolation of two new dihydrostilbenes from a Panamanian variant. *J. Nat. Prod.*, **47**(3), 445–452.

El Sohly, M.A., Boeren, E.G. and Turner, C.E. (1978) Constituents of *Cannabis sativa* L. An improved method for the synthesis of dl-cannabichromene. *J. Heterocyclic Chem.*, **15**, 699–700.

Fairbairn, J.W. (1976) The Pharmacognosy of Cannabis. In J.D.P. Graham, (ed.), *Cannabis and Health*, Academic Press, London, NY, San Fransisco, pp. 3–19.

Farmilo, C.G., Connell-Davis, T.W.M., Vandenheuval, F.A. and Lane, R. (1962) Studies on the chemical analysis of marihuana, biogenesis, paper chromatography, gas chromatography and country of origin. U.N. Secretariat Document, ST/SOA/Ser.s/7.

Felby, S. and Nielsen, E. (1985) Cannabinoid content of cannabis grown on the Danish island of Bornholm. *Bull. Narc.*, **37**(4), 87–94.

Fetterman, P.S., Keith, E.S., Waller, C.W., Guerrero, O., Doorenbos, N.J. and Quimby, M.W. (1971) Mississippi-grown *Cannabis sativa* L.: preliminary observation on chemical definition of phenotype and variations in tetrahydrocannabinol content versus age, sex and plant part. *J. Pharm. Sci.*, **60**, 1246–1249.

Formukong, E.A., Evans, A.T. and Evans, F.J. (1989) The medicinal uses of cannabis and its constituents. *Phytother. Res.*, **3**(6), 219–231.

Fournier, G., Richez-Dumanois, C., Duvezin, J., Mathieu, J.P. and Paris, M. (1987) Identification of a new chemotype in *Cannabis sativa*: cannabigerol-dominant plants, biogenetic and agronomic prospects. *Planta Med.*, **53**(3), 277–280.

Gaoni, Y. and Mechoulam, R. (1964a) Isolation, structure and partial synthesis of an active constituent of hashish. *J. Am. Chem. Soc.*, **86**, 1646–1647.

Gaoni, Y. and Mechoulam, R. (1964b). The structure and synthesis of cannabigerol a new hashish constituent. *Proc. Chem. Soc.*, March, 82.

Garrett, E.R. and Tsau, J. (1974) Stability of tetrahydrocannabinols I. *J. Pharm. Sci.*, **63**, 1563–1574.

Gillan, R., Cole, M.D., Linacre, A., Thorpe, J.W. and Watson, N.D. (1995) A comparison of *Cannabis sativa* by random amplification of polymorphic DNA (RAPD) and HPLC of cannabinoids: a preliminary study. *Science and Justice*, **35**(3), 169–177.

Grlic, L. (1968) A combined spectrophotometric differentiation of samples of Cannabis. *Bull. Narc.*, **20**(3), 25–29.

Hammond, C.T. and Mahlberg, P.G. (1994) Phloroglucinol glucoside as a natural constituent of *Cannabis sativa*. *Phytochem.*, **37**(3), 755–756.

Handrick, G.R., Uliss, D.B., Dalzell, H.C. and Razdan, R.K. (1979) Hashish: synthesis of (-)-delta-9-tetrahydrocannabinol (THC) and its biologically potent metabolite 3'-hydroxy-delta-9-THC. *Tetrahedron Lett.*, **8**, 681–684.

Harvey, D.J. (1984) Chemistry, metabolism and pharmacokinetics of cannabinoids. In G.H. Nahas, (ed.), *Marihuana in Science and Medicine*, Raven Press, NY, pp. 40–43.

Harvey, D.J. (1985) Examination of a 140 year old ethanolic extract of Cannabis: identification of new cannabitriol homologues and the ethylhomologue of cannabinol. In D. J. Harvey (ed.), *Marihuana '84: Proceedings of the Oxford Symposium on Cannabis*, IRL Press, Oxford, pp. 23–30.

Kajima, M. and Piraux, M. (1982) The biogenesis of cannabinoids in *Cannabis sativa*. *Phytochem.*, **21**(1), 67–69.

Lehmann, T. and Brenneisen, R. (1995) High performance liquid chromatographic profiling of cannabis products, *J. Liquid Chromatog.*, **18**(4), 689–700.

Mahlberg, P.G. and Hemphill, J.K. (1983) Effect of light quality on cannabinoid content of *Cannabis sativa* L. (Cannabaceae). *Bot. Gazette*, **144**, 43–48.

Mechoulam, R. (1970) Marijuana chemistry. *Science*, **168**, 1159–1166.

Mechoulam, R. (1973) *Marihuana: Chemistry, Pharmacology, Metabolism and Clinical Effects*, Academic Press, New York, London, pp. 1–99.

Mechoulam, R. and Gaoni, Y. (1965) A total synthesis of dl-Δ^1-tetrahydrocannabinol, the active constituent of hashish. *J. Am. Chem. Soc.*, **87**, 3273–3275.

Mechoulam, R. and Gaoni, Y. (1967) Recent advances in the chemistry of hashish. In L. Zwxhmeister (ed.), *Fortschritte Chemisch Organischer Naturstoffe*, Vol. 25, Springer, Wien, pp. 175–213.

Mechoulam, R., McCallum, N.K. and Burstein, S. (1976) Recent advances in the chemistry and biochemistry of cannabis. *Chem. Rev.*, **76**(1), 75–112.

Ni, R. (1963) Part II. Studies on the biosynthesis of cannabinol and cannabidiol in *Cannabis sativa*. Thesis, University of Minnesota and University Microfilms Inc., Ann Arbor, Michigan.

Paris, M. and Nahas, G.G. (1984) Botany: The unstabilised species. In G.G. Nahas (ed.), *Marihuana in Science and Medicine*, Raven Press, NY, pp. 3–36.

Pate, D. (1983) Possible role of ultraviolet radiation in evolution of *Cannabis* chemotypes. *Econ. Bot.*, **37**(4), 396–405.

Petrzilka, T., Haefliger, W. and Sikemeier, C. (1969) Synthese von Haschisch-Inhaltsstoffen. *Helv. Chim. Acta*, **52**, 1102–1134.

Phillips, R., Turk, R., Manno, J., Jain, N. and Forney, R. (1970) Seasonal variation in Cannabinolic content of Indiana marihuana. *J. Forensic Sci.*, **15**, 191–200.

Razdan, R.K. (1973) Recent advances in the chemistry of cannabinoids. *Prog. Org. Chem.*, **8**, 78–101.

Razdan, R.K. (1984) Chemistry and structure activity relationships of cannabinoids: an overview. In S. Agurell, W.L. Dewey, and R.E. Willette, (eds.), *The Cannabinoids: Chemical, Pharmacological and Therapeutic Aspects*, Academic Press Inc, Orlando, Florida, pp. 63–78.

Razdan, R.K. and Handrick, G.R. (1970) Hashish: A stereospecific synthesis of (-)-delta-1- and (-)-delta-1(6)-tetrahydrocannabinols. *J. Am. Chem. Soc.*, **92**, 6061–6062.

Razdan, R.K., Woodland, L.R. and Handrick, G.R. (1975) A one-step synthesis of (-)-delta-1-tetrahydrocannabinol from crysanthenol. *Experientia*, **31**(1), 16–17.

Ross, S.A. and El Sohly, M.A. (1996) The volatile oil composition of fresh and air-dried buds of *Cannabis sativa*. *J. Nat. Prod.*, **59**, 49–51.

Sakakibara, I., Ikeya, Y., Hayashi, K. and Mitsuhashi, H. (1992) Three phenyldihydronaphthalene lignanamides from fruits of *Cannabis sativa*. *Phytochem.*, **31**(9), 3219–3223.

Sakakibara, I., Katsuhara, T., Ikeya, Y., Hayashi, K. and Mitsuhashi, H. (1991) Cannabisin A, an arylnaphthalene lignanamide from fruits of *Cannabis sativa*. *Phytochem.*, **30**(9), 3013–3016.

Schultes, R.E. and Hoffman, A. (1980) *Botany and Chemistry of the Hallucinogens*, Charles C Thomas Publishers, Springfield, pp. 100–111.

Shoyama, Y., Yagi, M., Nishioka, I. and Yamauchi, T. (1975) Biosynthesis of cannabinoid acids. *Phytochem.*, **14**, 2189–2192.

Simonsen, J.L. and Todd, A.R. (1942) The essential oil from Egyptian hashish. *J. Chem. Soc.*, 188–191.

Slatkin, D.J., Doorenbos, N.J., Harris, L.S., Masoud, A.N., Quimby, M.W. and Schiff, P.L. (1971) Chemical constituents of *Cannabis sativa* L. root. *J. Pharm. Sci.*, **60**, 1891–1892.

Small, E. and Beckstead, H.D. (1973) Common cannabinoid phenotypes in 350 stocks of *Cannabis*. *Lloydia*, **36**, 144–165.

Taura, F., Morimoto, S. and Shoyama, Y. (1995) Cannabinerolic acid, a cannabinoid from *Cannabis sativa*. *Phytochem.*, **39**(2), 457–458.

Taylor, B.J., Neal, J.D. and Gough, T.A. (1985) The physical and chemical features of *Cannabis* grown in the United Kingdom of Great Britain and Northern Ireland from seeds of known origin – Part III: third and fourth generation studies. *Bull. Narc.*, **37**(4), 75–81.

Turner, C.E. and El Sohly, M.A. (1979) Constituents of *Cannabis sativa* L. XVI. A possible decomposition pathway of Δ^9-tetrahydrocannabinol to cannabinol. *J. Heterocyclic Chem.*, **16**, 1667–1668.

Turner, C.E. and Hadley, K. (1973) Constituents of *Cannabis sativa* L. II Absence of cannabidiol in an African variant. *J. Pharm. Sci.*, **62**(2), 251–255.

Turner, C.E., El Sohly, H.N., Lewis, G.S., Lopez-Santibanez, I. and Carranza, I. (1982) Constituents of *Cannabis sativa* L., XX: the cannabinoid content of Mexican variants grown in Mexico and in Mississippi, United States of America. *Bull. Narc.*, **34**(1), 45–59.

Turner, C.E., El Sohly, M.A. and Boeren, E.G. (1980) Constituents of *Cannabis sativa* L. XVII. A review of the natural constituents. *J. Nat. Prod.*, **43**(2), 169–234.

Turner, C.E., Fetterman, P.S., Hadley, K.W. and Urbanek, J.E. (1975) Constituents of *Cannabis sativa* L. X: Cannabinoid profile of a Mexican variant and its possible correlation to pharmacological activity. *Acta Pharm. Jugoslav.*, **25**, 7–16.

Turner, J.C. and Mahlberg, P.G. (1985) Cannabinoid synthesis in *Cannabis sativa* L. *Am. J. Bot.*, **72**(6), 911.

Turner, C.E. (1980) Marijuana research and problems: an overview. *Pharm. Int.*, **1**, 93–96.

Vogelmann, A.F., Turner, J.C. and Mahlberg, P.G. (1988) Cannabinoid composition in seedlings compared to adult plants of *Cannabis sativa*. *J. Nat. Prod.*, **51**(6), 1075–1079.

Waller, C.W. and Scigliano, J.A. (1970) The national marihuana program. Report to the commission of problems of drug dependence. *Natl. Acad. Sci. NRC*, **4**, 28–32.

Waller, C.W., Johnson, J.J., Buelke, J. and Turner, C.E. (1976) *Marihuana: An Annotated Bibliography*, Vol. I, Macmillan, New York.

Waller, C.W., Baran, K.P., Urbanek, B.S. and Turner, C.E. (1980) *Marihuana: An Annotated Bibliography*, Supplement, Macmillan, New York.

Waller, C.W., Baran, K.P., Urbanek, B.S. and Turner, C.E. (1981) *Marihuana: An Annotated Bibliography*, Supplement, Macmillan, New York.

Waller, C.W., Nair, R.S., McAllister, A.F., Urbanek, B.S. and Turner, C.E. (1982) *Marihuana: An Annotated Bibliography*, Vol. II, Macmillan, New York.

Waller, C.W., Urbanek, B.S. and Wall, G.M. (1985–86; 1987–88; 1989–90; 1991–92; 1993–94) *Marihuana: An Annotated Bibliography*, Supplement, Macmillan, New York.

Waller, C.W., Urbanek, B.S., Wall, G.M., Mack, J.E. and Turner, C.E. (1982) *Marihuana: An Annotated Bibliography*, Supplement, Macmillan, New York.

Waller, C.W., Urbanek, B.S., Wall, G.M., Mack, J.E. and Turner, C.E. (1983–4) *Marihuana: An Annotated Bibliography*, Supplement, Macmillan, New York.

4. ANALYTICAL AND LEGISLATIVE ASPECTS OF CANNABIS

GEOFFREY F. PHILLIPS

1 LEGAL DEFINITIONS

This first section of Chapter 4 is concerned with forensic definitions of cannabis and its products as a controlled drug of abuse. Three following sections address related offences and attitudes, techniques used in forensic analysis, and the pharmaceutical quality of cannabis products.

1.1 International Conventions and National Enactments

The various international conventions and protocols, and the succession of 'Dangerous Drugs' legislation in the UK, are discussed in this section and summarised in Table 1.

There is a long history of medicinal use and social abuse of Indian Hemp (see Chapter 1) but there were strong representations from some delegations to the 1923 Opium Conference, notably reports of Egyptian experience, seeking to ban all non-medicinal uses. In the subsequent 1925 (and 1931) League of Nations Conventions, the description of cannabis restricted the controlled drug to the female plant and named a particular species, *Cannabis sativa*, alias *indica*. In the UK, 'Indian Hemp' was dropped in the 1932 revision of the British Pharmacopoeia 1914 and non-medicinal use was banned in 1928.

This definition was maintained after World War II in the United Nations 'Lake Success' amending protocol, and in the UK in the corresponding Dangerous Drugs Act of 1951, through the wording:

"Indian hemp is the dried flowering or fruiting tops of the pistillate plant known as *Cannabis sativa* [alt.*indica*] from which the resin has not been extracted".

In the United Nations comprehensive 'Single Convention' of 1961, which brought together many classes of narcotic and other drugs of abuse, the definition used in Article 1 removed this gender discrimination by describing Cannabis as –

"the flowering or fruiting tops of the Cannabis plant (excluding the seeds and leaves when not accompanied by the tops) from which the resin has not been extracted, by whatever name they may be designated"; where cannabis plant "means any plant of the genus *Cannabis*".

In a corresponding national enactment, the UK Dangerous Drugs Act (DDA) 1964 recognised this widened scope of control by the crisper wording –

"the flowering or fruiting tops of any plant of the genus cannabis from which the resin has not been extracted, by whatever name they may be designated".

Table 1   Conventions and Enactments

(a) *Relevant International Conventions and Protocols*
   Convention on Narcotic Drugs: Geneva 1931
     amending protocol: Lake Success (USA) 1946
   Single Convention on Narcotic Drugs: New York 1961
     (mostly natural opiates and synthetic opioids)
     various amending protocols, e.g. 1972
   Convention on Psychotropic Substances: Vienna 1971
   Convention Against Illegal Trafficking: Vienna 1988
     (international surveillance and precursors control)
(b) *Post-1945 Dangerous Drugs legislation in Great Britain*
   – there are separate series for Northern Ireland, IoM, Channel Isles
   Dangerous Drugs Act 1951
   Dangerous Drugs Act 1964: redefine Cannabis, prohibit cultivation
   Dangerous Drugs Act 1965: consolidated UN SC/1961
     but continued D.D.Regs. 1964
   Dangerous Drugs Act 1967:
     drug addiction, safe custody, powers of search
   Drugs (Prevention of Misuse) Act 1964:
     primarily 'generic' stimulants but added –
     DPMA, Mod. Order 1966: hallucinogens (semi-generic)
     DPMA, Mod. Order 1970: repealed broad generic stimulants, and
     added cannabinoids and some hallucinogens
   Misuse of Drugs Act 1971: consolidated DDA and DPMA,
     with effect 1973 various amending Orders including new drugs, and
     moving drugs between penalty classes
   Misuse of Drugs Regulations 1973
     various amending Order and consolidated as
   Misuse of Drugs Regulations 1985
     and further amending Orders

These extended definitions had the double advantage of bringing into international control the potentially quite potent male flowering tops and also of sidestepping taxonomic argument as to whether cannabis was a monospecific genus (see below, §1.2, and more detail in Chapter 2). The wider legal definition of cannabis was continued in the UK in the DDA 1965 – which incorporated other narcotic substances newly specified in the UN 1961 Single Convention – and was further sustained in the consolidating enactment of the Misuse of Drugs Act 1971 (MDA).

In France, the production, marketing and use of "Indian hemp, plant and resin, their preparations; and THC and derivatives" (THC is discussed in §1.5) were prohibited except for research or laboratory purposes; there is reference in §2.1 to the French provision for authority for commercial cultivation of hemp fibre. In the UK, the MDA, s. 7(4), gave power for Regulations under the Act (MDR) to make production, supply and possession of specified Controlled Drugs unlawful (except for specified research purposes): Cannabis was placed on such a list in the MDR 1973 and hence became no longer available in clinical practice.

1.2   Distinction and Confusion of Herb and Resin

The UN Multilingual Dictionary (1983) of psychotropic substances under international control lists 194 synonyms for herbal and 54 for resin forms including beverages, confectionery and other preparations containing cannabis. The most frequently encountered terms, according to the country of origin, are marihuana, bhang, dagga, ganja (*) and kif for herbal; and hashish and charas (*) for resin. Some of those terms (e.g. *) may be used interchangeably for either herbal or resin form.

It has become generally accepted that *Cannabis sativa* is a monospecific genus but that there are three genotypes:

1. the resin plant is rich (>1 %) in THC [q.v.], with significant amounts of CBD [q.v.];
2. the hemp fibre plant has low (<0.3%) THC but CBD is essentially absent;
3. an intermediate variety, growing in certain climates.

In practice, the high and low THC/CBD plant types may co-cultivate in the same area, e.g. some Sri Lankan crops.

In those administrations where national legislation for the control of drugs recognises a forensic distinction between herbal specimens and the derived resin (usually obtained by physical separation from the herb and then compacted), the prosecution charge must specify which botanical form has been detected. There have occasionally been problems of mis-identification between cannabis resin and the herbal form, particularly when the seized material is compacted herb (see below), or constitutes a smoking residue (which aspect is discussed later – cf. §1.6).

Small amounts of badly preserved, friable resin may be difficult for the examiner to distinguish from compressed, finely chopped, resin-rich herbal tops. A court may reject a charge which refers to (herbal) "cannabis" if the evidence submitted identifies it as "cannabis resin", e.g. Jersey Royal Court, 19 November 1969. That the separated resin need not be pure was upheld in the UK on Appeal in R *v Janet Thomas* (1981), notwithstanding the presence of herbal debris, including intact oil glands, in a compacted resinous mass. In the criterion of 'separated resin', mechanical separation – e.g. stripping of the lower leaves – should be distinguished from chemical extraction or physical crushing. Thus, in R *v Berriedale-Johnson* (1976) lower leaves were not accepted as constituting 'cannabis resin' [nor would these leaves have qualified – at that time – as herbal cannabis – see following discussion in §1.3].

1.3   Parts of Plant: Fruiting/Flowering Tops;
Aerial Parts, Stalk; Seed (Non)viable

Further forensic uncertainty may arise where there is need for legal discrimination of the flowering tops from the axial buds and lower leaves (which may still be a moderate source of cannabinoid congeners). The forensic difficulty experienced in the UK was not a problem in US Federal Law in the 1970s, when the definition of marihuana subsumed flowering/fruiting tops and lower leaf and viable seed. Moreover, the US Drug Abuse Prevention & Control Act [DAPCA] 1970 also had a very much broader description of synthetic variants of "tetrahydrocannabinols" – see discussion in §1.8 below.

In UK law this anomaly could not be resolved simply by tabling a new statutory 'Modification Order' because the 'Cannabis' definition was contained within the body of the MDA – unlike most other 'dangerous drugs' which were substances and products described in the appended Schedule 2 which is accessible to amendment by an 'Affirmative Resolution' in parliamentary debate. Accordingly, the definition in the principal act, the MDA itself, had to be amended in respect of the cannabis definition. This was conveniently achieved by inserting an additional section [s. 52] among a variety of general amendments to the criminal law that were collated in the Criminal Justice Act of 1977. Herbal cannabis, after excluding the resin, was thereby redefined as –

"all the aerial parts, except the lignified stem and the non-viable seed,
of any plant of the genus *Cannabis*".

Note that this new definition deliberately excluded the roots from control under the MDA.

This revision in the UK in 1977 was forced because previous attempts to control herbal material other than flowering or fruiting parts, such as lower leaf and stalk, could not necessarily rely on demonstrating to the Court the presence within the herbal material of controlled 'cannabinoids' (see below, §1.7 for discussion of controls on congeneric substances constituting the active principles of the herb). After a variety of rulings in lower courts, the link between these other parts of the plant and the 'cannabinoids' controls was finally tested in a key judgment in the House of Lords. Thus, in the case of R *v Goodchild* (1977–78), itself following two Crown Court hearings and two Appeals, their Lordships ultimately ruled that leaves and stalk which had been separated from the harvested plant did not constitute the separated resin and nor were these other parts of the plant legally equivalent to those cannabinoids naturally contained within them (which substances were listed in a controlled drug category attracting a higher level of penalty – see §1.5 and §1.7 below).

1.4   Compendial Definitions

The British Pharmacopoeia 1914 edition was the last to provide a monograph for cannabis; the 'Characters' and 'Tests' are described in §4.1. This monograph defined "Cannabis Indica", alias "Indian Hemp", as the –

"dried flowering or fruiting tops of the pistillate plant *Cannabis sativa* Linn."

and added "grown in India, from which the resin has not been removed".

The British Pharmaceutical Codex 1949 – the last BPC to contain a Cannabis monograph – retained only the first part of the BP 1914 definition (i.e. no reference to origin or to removing resin) but added "(Fam. Cannabinaceae)". This edition of the Codex also prescribed recipes for preparation of (alcoholic) Extract of Cannabis and the appropriate dilution to make Tincture of Cannabis.

The Japanese Herbal Medicines Codex and the Chinese Pharmacopoeia have current monographs for 'hemp fruit' and 'Huo Ma Ren' (i.e. marihuana) respectively: see §4.1 later.

1.5   Cannabis Oils

In the UK, the 'Controlled Drugs' listed in Schedule 2 to the MDA 1971 are subdivided between 'Parts' I, II and III, respectively containing drugs of 'class A', 'B' and 'C', largely following the classification in Schedules I–III in the UN Single Convention 1961; this distribution sought to strike a balance between the perceived degree of social harm and the medicinal value of the substances. In Schedule 4 to the MDA, direction is given for the prosecution and punishment of various relevant offences (cf. §2 below) and a descending scale of penalties reflects the allocation of substances between classes A, B and C. A fourth Part of MDA Schedule 2 provides legal definitions for those entries in Parts I–III of Schedule 2 that do not admit exact chemical composition.

'Cannabinol derivatives' are listed in Part I, i.e. are controlled as 'class A' drugs and explicitly defined in Part IV (see full definition in §1.7). This definition carries the exclusion "except where contained in cannabis or cannabis resin". The latter natural products were already listed in Part II and thus attracted the lower penalties of class B drugs. Following the revised definition of 'Cannabis' in s. 52 of the Criminal Law Act 1977, the herbal form of the drug thereafter legally subsumes "all aerial parts" (except lignified stem and non-viable seeds) of the plant.

In the DDA 1965, following on the UN Single Convention 1961, 'Any extract or tincture of cannabis' (as exemplified by, but not limited to, the alcoholic preparations of the BP 1914 and BP Codex 1949) were listed in a subsequent clause in the schedule containing 'Cannabis' and its resin. These named preparations were deliberately omitted from Schedule 2 of the MDA 1971 and control of 'extract' or 'tincture' thereafter rested on a general reference to "preparations" (of the appropriate class of drug).

The more or less concentrated 'Cannabis oil' is a solvent extract containing 20–40% of the potent principal cannabinoid, tetrahydrocannabinol (THC), and just occasionally, appreciably higher concentrations are detected (cf. Figure 2, later). If judged by UK seizures, so-called hash oil exported from the 'resin belt' countries (i.e. Indian subcontinent, Middle East, Morocco) often contains three or four times the common local level (say, 5–15%) of THC in resin from that region; whereas liquid cannabis preparations from the Caribbean and East Africa may represent tenfold concentration of the much lower level (say, 1–3%) of THC in local herb. Some recent seizures (King, 1997) of the oil appear to derive from extraction of intensively cultivated herbal cannabis. Cannabis oil may be clandestinely prepared, from either herbal or resin forms of cannabis, by extraction with ethyl or methyl alcohol in, for instance, a large drum. The mixture is then filtered and the filtrate concentrated by evaporation of solvent (e.g. in a pressure cooker); it may be purified by treatment with petroleum ether.

Cannabis oil was for many years regarded in UK forensic practice as a preparation of THC, a class A drug (see §1.7). However, if a lower potency oil, only comparable in THC content with the Extract or Tincture of the BPC 1949, is supported by collateral evidence of botanical origin, then 'a preparation of a class B drug' is a more appropriate and equitable prosecution charge. If the solvent were entirely removed, the viscous residual product might then be regarded as 'purified cannabis resin', because the MDA 1971, also places the resin – defined in s. 37(1) as the separated resin, whether crude or purified – in 'class B'. However, where forensic practice treated oily extract of herbal Cannabis as a preparation of THC, then the higher penalties of class A would apply. This situation was recognised as anomalous.

The main constituents of the oil are tetrahydrocannabinol (THC) and cannabinol (CBN); but if made from resin, some cannabidiol (CBD) is also present. Accordingly, identification of the presence of CBD permits assumption of a resin source, whence in UK law the oil can be treated as 'purified cannabis resin' in 'class B'. But in the absence of CBD, or if a herbal source of the oil is otherwise proved or admitted, then 'a preparation of THC', not being Cannabis, predicates assignment as 'class A'. Such a decision was upheld in the case of *R v Carter* in Oxford Crown Court in December 1992, when the oil was not accepted as "a preparation of Cannabis".

This discrimination in Britain was widely considered as unreasonable, especially for low concentration 'cannabirum' extracts from the Caribbean, and various options for changes in UK law were considered:

1. Add a new entity 'Liquid cannabis' to Part II [i.e. with class B] and add a new definition to the list in Part IV [e.g. "a product which has been prepared from cannabis or cannabis resin by solvent extraction".]
2. Transfer cannabinol and cannabinol derivatives from Pt. I to Pt. II of Sch. 2 of MDA.
3. Amend Sch. 2 of the MDA as at 2. but maintain the distinction between non-therapeutic and therapeutic Controlled Drugs through Sch. 1 and 2 of subordinate Misuse of Drugs Regulations.

Option 2 was regarded as administratively tidier but it conflicted with the UN Single Convention placement of cannabinoids in *its* Schedule I and cannabis products in Schedule II. Subsequently, the UN recommended for [potential] medical usage reasons, transferring pure THC {as one potent stereoisomer, under its US and WHO non-proprietary medicinal name 'dronabinol'} from Sch. I to II. This isomer was a clear candidate for the 'medicinal' Schedule, S2, of MD Regulations. Meanwhile, the Home Office Forensic Science laboratories for some years anticipated option 1, by reporting *all* 'hash oil' samples as 'class B' without distinction as to source, conveniently regarding such specimens implicitly or explicitly as 'purified cannabis resin'.

Firm recommendations to resolve this forensic anomaly were presented in 1996 by the Advisory Council on Misuse of Drugs but legal enactment has had to wait on reference to the new parliament in 1997.

1.6   Smoking Residues

The partially pyrolysed residue in a pipe bowl or cigarette butt may still retain some herbal features but is conveniently regarded as a crude specimen of 'separated resin'. Thus, the act of smoking has thermally 'separated' the resin from (herbal) cannabis. The residue may be richer in THC than the original herb through thermal decarboxylation of precursor acids [see next section] but the quantity (and quality!) present will not usually be sufficient to constitute a dose for further smoking. Morphological examination of uncarbonized material may provide forensic evidence of herbal or intact resin but more frequently this distinction is blurred or inconclusive. Nevertheless, the chemical evidence from the nature and proportion of cannabinoid congeners may be capable of three interpretations:

1. The process of distillation – in pipe ('schaum') or cigarette ('joint' or 'reefer') – has produced purified cannabis resin, i.e. supporting a charge of simple possession of a small quantity of a class B substance.
2. This constitutes evidence of having smoked, and therefore having had prior possession of, a limited quantity of cannabis or resin or preparation thereof. As pointed out by Phillips (1973), two lines may be accessible to defence of such a charge: that the smoking implement at the material (prior) time had been in the possession of some other identifiable person; or that the defendant had smoked a preparation (such as Cannabis Tincture BPC) lawfully dispensed for him (cf. Clarke and Robinson, 1970).
3. In the absence of any vegetative matter, the cannabinoid residue might point to prior possession of a small quantity of cannabinol derivatives: but as discussed in the previous Section, this leads to anomalous case law and potentially inequitable penalties.

1.7   Natural Cannabinoids, Including C3 and C1 Analogues

At least 70 terpenoid phenols and acids have been reported to be isolated from Cannabis. The origin and nature of these congeneric substances have been discussed in Chapter 3. For convenience of discussion of structural variation of the natural congeners (discussed in this Section), the synthetic homologues (in §1.8), their esters (§1.9) and their respective stereochemistries (§1.10), their inter-relationships and principal graphic formulae are illustrated in Table 2. Many of the congeneric substances (originally identified as indicated in Table 2), such as cannabichromene, CBCh (with ring-C open), cannabigerol, CBG (both rings B and C open) and cannbicyclol, CBCy (cyclised to a fourth ring), and their various methyl ethers and carboxylic precursors, are of insufficient psychoactivity to warrant international control as potential drugs of abuse.

In the UK the first specific control of the cannabinoid natural constituents cannabinol (CBN; Table 2, I: R = pentyl) and its tetrahydro derivatives, including the most potent stereoisomer, delta^9-*trans*-THC (Table 2, IIa), was their explicit listing as psychotropic substances in 1970 in an addition to the schedule of drugs controlled by the Drugs Prevention of Misuse Act (DPMA) 1964, thereby anticipating UN proposals in the UN Psychotropic Substances Convention 1971, discussed in §1.8.

The distinction in seriousness of criminal offence between cannabinoid substances and herbal specimens of cannabis was heightened by the MDA 1971, which placed 'cannabinol and cannabinol derivatives' in class A (the highest penalty class) but listed cannabis and its resin (whether crude or purified) in class B, in harmony with the UN classification. A range of hydrogenated '*cannabinol derivatives*' was defined in Part IV of Schedule 2 to MDA as –

"the following substances, except where contained in cannabis or cannabis resin,
the tetrahydro derivatives of cannabinol and 3-alkyl homologues of cannabinol or of its tetrahydro derivatives"

This definition comprehensively subsumed not only all tetrahydro (structural) isomers [see below for the US DAPCA 1970 legislation on this point], of which the delta^9

(alias delta^1) and delta^8 (alias delta$^{6(1)}$) (cf. Table 2, IIIa) are the more common, whereas other isomers, such as Adams and Baker's (1940) synthetic delta^{3,4} (Table 2, IIIb), are curiosities; but it also extends to homologous sidechain derivatives. Thus, replacing the 3-pentyl group by alkyl groups with more than 5 carbon atoms thereby brings into control the synthetic 3-alkyl *homologues* of CBN and THC (see §1.8).

Table 2   Cannabinoid congeners and structurally related dibenzopyrans and chromenols

Naturally occuring in cannabis extracts:
CANNABINOL (CBN): R = pentyl
Cannabivarinol (CBV): R = propyl **I**
Cannabiorcinol (CBO): R = methyl

TETRAHYDROCANNABINOL (THC):
 R = pentyl common and most potent
 position isomer is Δ^9 **IIa**
Tetrahydrocannabivarinol (THV): R = propyl
Tetracannabiorcinol (THO): R = methyl

most potent natural stereoisomer
 is (–)-*trans*-6aR, 10aR: **IIb**
 and its unnatural (+)-*SS* enantiomer,

and potential *RS* and *SR cis* pair of
 enantiomers: **IIc**

natural/synthetic position isomer
 Δ^8-THC with two pairs of stereoisomers
 (–)-*trans* is RR: **IIIa**

Synthetic THC products
"Δ^{3,4}" (i.e. Δ^{6a,10a})-THC:
one chiral centre = 2 stereoisomers,
9 R & 9S: **IIIb** and also:
Δ^7-THC: 3 chiral centres = 8 stereoisomers
exocyclic Δ$^{9(11)}$-THC: 4 stereoisomers
(theoretical) Δ^{10}-THC: 2 chiral centres,
 4 stereoisomers
3-alkyl Homologues: 'synhexyl' = 'parahexyl':
IIa, R = 3-n-hexyl 'DMHP': IIa,
 R = 3-(1, 2-dimethylheptyl)

Congeneric related structures
CANNABIDIOL (CBD): R = pentyl [ring-B open] **VI**
 also R = propyl and 3-methyl

CANNABICHROMENE (CBCh) [ring-C open] **VII**

CANNABIGEROL (CBG) [rings-B and C open] **VIII**

CANNABICYCLOL (CBCy): R = pentyl **IX**
 R = propyl

RING NUMBERING:
IV : based on IUPAC
dibenzopyran system

V : based on historic
terpenoid notation

Table 2 *(continued)*

Carboxylic acid precursors:

CBNA is 2-COOH-CBN

THCA(A) (2-COOH-THC); THCA(B) (4-COOH-THC)

CBDA is 2-COOH-CBD

and COOH in the corresponding Varins

CBChA is CBCh-2-COOH precursor

CBGA is CBG-2-COOH precursor:

VIII

Phenolic ester precursors:

e.g. 1-O-acetate of THC and THV (IIa, 'Y' = Ac)

IX

Natural *O*-methyl ethers:

CBNM (I, R = Pen, 'Y' = Me)

CBCM (VII, R = Pen, 'Y' = Me)

CBGM (VIII, 'Y' = Me)

However, note that the naturally occurring congeners of CBN and THC with **shorter** 3-alkyl sidechains, the cannabivarins (CBV and THV), each with a 3-propyl group (Table 2, I and IIa: R = propyl), and cannabiorcinols, each with 3-methyl (Table 2, I and IIa: R = methyl), cannot be construed as 'homologues' of CBN and THC respectively (which have the longer 3-pentyl substituent). Accordingly, the natural cannabivarins and cannabiorcinols are not controlled by the 'homologues' extension in the MDA definition.

The acid precursors of CBN and CBD, such as cannabinolic acid (CBNA) and cannabidiol carboxylic acid, and the tetrahydro derivatives THCA(A) (2-COOH) and THCA(B) (4-COOH), occur naturally in cannabis extracts. These acids are not controlled as such in the UK – but as previously noted [§1.6] thermal decarboxylation (e.g. during smoking) generates additional CBN and THC and thus these acid precursors ultimately contribute to the potency of the cannabis specimen when smoked. Variation in THCA/THC ratios in cannabis specimens of different geographical origin is discussed in §3.6 and some data are summarised in Table 6.

1.8 Synthetic Analogues of Natural Cannabinoids

The DPMA of 1964 was initially designed to control stimulant and anorectic drugs, which in the early 1960s were becoming a distinct public nuisance. These substances were structurally related to amphetamine, but were not subsumed by the UN Single Convention 1961 nor in the UK by the DDA 1965. In the UK in 1966 the DPMA was taken as a convenient vehicle for control of hallucinogens, such as dimethyltryptamine (DMT) and the very potent lysergide (LSD). Concurrent international discussion in WHO and UN working-parties reviewed many classes of psychotropic substances that were not included with the primarily narcotic drugs in the UN Single Convention 1961; and led to the interim Psychotropic Protocol and ultimately to the creation of the Psychotropic Substances Convention 1971. One of the groups of psychedelic drugs which the WHO recommended for control was the series of synthetic analogues of THC, particularly the so-called 'synhexyl' , the 3-hexyl homologue of THC (cf. Table 2, IIa: R = hexyl).

This recommendation was reflected in a definition incorporated in UK legislation in 1970 under the DPMA 1964 and was later consolidated in the S2-Part IV definitions of the MDA 1971 [as set out in §1.7 above]. It should be noted that branched 3-alkyl sidechains (such as 'synhexyl') qualify for control thereby but that derivatives with 3-substitution by alkenyl (ethylenic) or alkynyl (acetylenic) or alkylidene (divalent alkyl radicals) do not qualify for control through this definition.

The US Drug Abuse Prevention & Control Act [DAPCA] 1970 introduced an even broader description of "tetrahydrocannabinols":

"Synthetic equivalents of the substances contained in the plant, or in the resinous extractives of Cannabis sp., and/or synthetic substances, derivatives, and their isomers with similar chemical structure and pharmacological activity"

and then gave as non-exclusive examples –

"such as the following, delta[1] *cis* or *trans* THC and their optical isomers,
delta[6] *cis* or *trans* THC and their optical isomers,
delta[3,4] *cis* or *trans* THC and their optical isomers,
and since nomenclature of these substances is not internationally standardised,
compounds of these structures regardless of numerical designation of atomic positions".

This conflict in nomenclature of the natural cannabinoids stemmed from investigative degradation to, and early synthesis from, monoterpene components. The locants on the non-aromatic ring (of THC and CBD) (cf. ring C in Table 2, IV) were historically prescribed in monoterpene convention (Table 2, V); the ring attachment for CBD was then placed at locant '3' and ring fusion for THC at locants '3,4'. When fully systematic (IUPAC and Chem-Abs approved) nomenclature was applied to substances of the cannabinoid family, they were regarded as derivatives of dibenzo[a,c]pyran: then conventionally orientating the dibenzopyran with ring A top right and using its pre-scribed clockwise numbering system, the ring B oxygen and the dimethyl substituents are given locants '5' and '6' respectively, and the monoterpene carbon bearing the methyl, previously '1', becomes position '9' and erstwhile '6' becomes '8'.

In addition to these early experimental substances, more recently a number of related structures have been synthesised by the pharmaceutical industry and have been screened for potential clinical use. Table 3 lists eight such substances for which WHO non-proprietary names have been assigned, together with their various clinical indications. Seven retain the cannabinoid characteristic oxatricyclic system, five being modified dibenzopyrans and two, Nabitan and Tinabinol, as aza- or thia-analogues respectively. The eighth new drug, Nonabine, at least retains an oxabicyclo (chromenol) moiety. Another commonality resides in their respective 3-alkyl groups: Pirnabine has the simple methyl of the natural orcinol series, whereas the other seven have a homologous nine-carbon branched chain alkyl substituent. Their respective potential clinical uses, as indicated in Table 3, are somewhat varied but closely reflect the different medical applications of natural cannabis – see summary in §2.4 and the extensive treatment in Chapters 7 and 8.

The synthetically prepared selected single stereoisomer (–)-*trans*-THC (cf. discussion of cannabinoid stereochemistry, and legal implications, in §1.10) has been assigned the WHO non-proprietary name 'Dronabinol' and is marketed in USA. It has been included in Table 3 for comparison; quality issues, including the USP monograph, are described in §4.8.

1.9 Esters of Cannabinoids

In the UK, control of naturally occurring esters, such as the *O*-acetates of the phenolic functions of CBN and THC (cf. Table 2, V: OX = OAc, R = pentyl), is achieved through the 'esters' extension clause in Part I of MDA S2, which bites wherever a class A

Table 3 Synthetic dibenzopyran drugs with clinical potential: names, structure and CAS no., control status and indication, 'BAN' = British Approved Names 1994 list and supp.; 'USAN' = United States Adopted Names 1994 list and supp.; 'pINN' and 'rINN' = Proposed and Recommended lists of WHO non-proprietary names

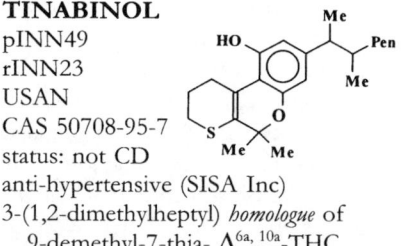

PIRNABIN
pINN41
rINN19
USAN pirnabine
CAS 19825-63-9
status: not CD
anti-glaucoma (SISA Inc)
O-acetyl-$\Delta^{6a, 10a}$-tetrahydrocannabiorcinol

NABILONE
pINN49
BAN
USAN
CAS 51022-71-0
status: not CD
tranquiliser and anti-emetic
(Lilly 'Cesamet' 1983)
(±)-3-(1,1-dimethylheptyl) *homologue* of
  9-demethyl-9-oxohexahydro-cannabinol

CANBISOL
pINN39
'dihydro-Nabilone'
CAS 56689-43-1
status: not CD
anti-hypertensive (Lilly 'Nabidrox')
(±)-3-(1,1-dimethylheptyl) *homologue* of
  9-demethyl-9 *RS*-hydroxyhexahydro-
  cannabinol

NABAZENIL
pINN49
rINN23
USAN
CAS 58019-65-1
status: MDA classA-ester of 3-alkyl
homologue of THC anti-convulsant
(SISA Inc)
'Y' = 4-(perhydroazepin-1-yl)butanoyl
  ester of 3-(1,2-dimethylheptyl)
  homologue of $\Delta^{6a, 10a}$-THC

NONABINE
pINN47
BAN
CAS 16985-03-8
status: not CD
anti-emetic (Beecham)
7-(1,2-dimethylheptyl)-2,2-dimethyl-
4-(4-pyridyl)chromen-5-ol

TINABINOL
pINN49
rINN23
USAN
CAS 50708-95-7
status: not CD
anti-hypertensive (SISA Inc)
3-(1,2-dimethylheptyl) *homologue* of
  9-demethyl-7-thia- $\Delta^{6a, 10a}$-THC

Table 3   *(continued)*

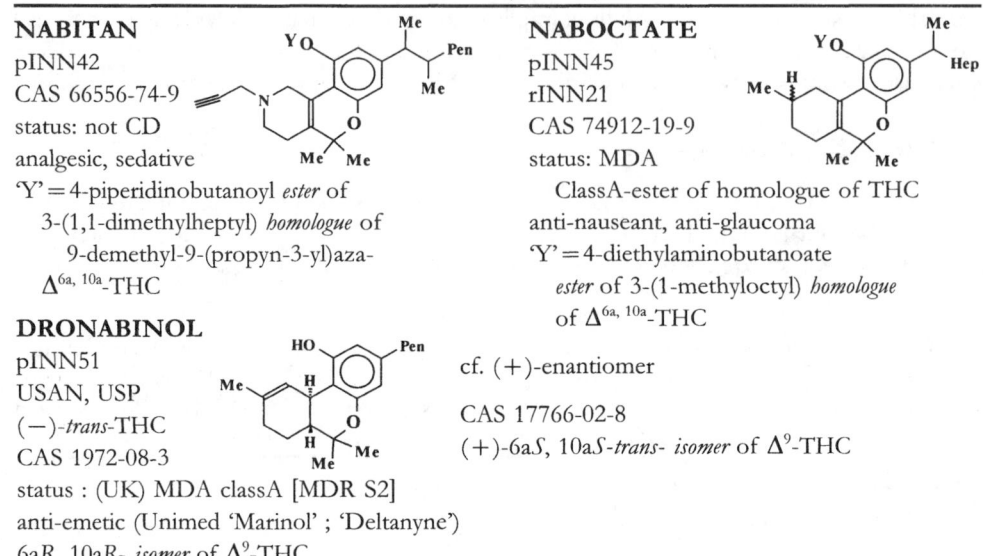

NABITAN
pINN42
CAS 66556-74-9
status: not CD
analgesic, sedative
'Y' = 4-piperidinobutanoyl *ester* of
  3-(1,1-dimethylheptyl) *homologue* of
    9-demethyl-9-(propyn-3-yl)aza-
    $\Delta^{6a, 10a}$-THC

NABOCTATE
pINN45
rINN21
CAS 74912-19-9
status: MDA
  ClassA-ester of homologue of THC
anti-nauseant, anti-glaucoma
'Y' = 4-diethylaminobutanoate
    ester of 3-(1-methyloctyl) *homologue*
    of $\Delta^{6a, 10a}$-THC

DRONABINOL
pINN51
USAN, USP
(−)-*trans*-THC
CAS 1972-08-3
status : (UK) MDA classA [MDR S2]
anti-emetic (Unimed 'Marinol' ; 'Deltanyne')
6aR, 10aR- *isomer* of Δ^9-THC

cf. (+)-enantiomer

CAS 17766-02-8
(+)-6a*S*, 10a*S*-*trans*- *isomer* of Δ^9-THC

drug is capable of forming an ester (or ether). Thereby, all such esters (and ethers) attract similar class A penalties. A more recent uncertainty involved the synthetically prepared acetate and other esters of the cannabinoids which had been detected in an extract of cannabis. Deliberate preparation of the acetate of THC present in an extract of cannabis was ruled in June 1995 as preparation of a class A substance, the product being the ester of the class A substance THC.

In the case before Merthyr Tydfil Crown Court, a significant quantity of alcohol-extracted cannabis, which had been further purified by petroleum ether treatment, was reported by the relevant laboratory of the Forensic Science Service as a 'class B' product; whereas residues of acetic anhydride containing small amounts [ca 150 mg/220 ml] of THC acetate, and also flasks containing solid residues of THC acetate, were both reported as 'an ester of a class A drug', that is to say as an ester of THC. The alternative proposition that the THC acetate present was an ester of an extract of cannabis is not chemically sensible nor indeed is it legally possible – because in the MDA the supplementary clause extending control to 'esters' is only found in Part I of S2 and thus can only bite on esters of class A substances, as such or in admixture. In this case the prosecution submission must be regarded as both morally and legally reasonable.

Thus, 'morally' reasonable because the scale of manufacture did not warrant anticipating forthcoming legislation by unjustifiably assigning the lower penalty class B status to this chemical derivative of hash oil. The defendant's recipe book and his use of acetic anhydride reagent made clear that the acetate was the deliberate target and that the derivative was expected to be at least twice as potent as THC itself. At the very least, the residues were evidence of prior larger scale 'manufacture of a Controlled Drug'.

And 'legally' reasonable because THC acetate should be classified as a class A Controlled Drug in virtue of being an ester of [one or more isomeric] cannabinol derivatives. The fact that the source of the THC from which the THC acetate had been chemically prepared was an extract of cannabis is not relevant and thus whether the starting material had class A or class B status is not in issue. Even if the starting material had been proved to be a class 'B' substance, and then the production of a class A derivative challenged, there are many precedents for retaining their respective classes for derivatives despite chemical conversion – actual or potential – between class B and A drugs; for instance, codeine (class B) is an ether of morphine (class A).

Coincidentally, it was in 1995 that New Zealand police reported their first seizure of THC acetate (confirmed by GC–MS, following various chromatographic separations: Valentine, 1996). Studies in the Institute of Environmental Science and Research in Auckland indicated it to be probable that the seized sample derived from acetic anhydride treatment of cannabis oil.

1.10 Stereochemistry of THC and Legal Implications

It has already been shown (§1.7) that there are several possible double-bond position isomers of THC, of which the 9(10) (i.e. delta^9) and the less potent delta^8 are more usually encountered. From their graphic formulae (see Table 2, IIa and IIIa: R = pentyl) it is evident that two series of stereoisomers, geometrical and optical, are also possible.

At the C:B rings junction, there may be *cis* or *trans* configurations of the two H's (which have IUPAC locant numbers 6a and 10a . It is stated that the delta^9 *trans* isomer (see Table 2, IIb) is the more potent. These same two carbons, 6a and 10a, are chiral centres so that enantiomeric pairs of optical isomers may exist for each centre. In Table 2, *trans* structure (IIb) has a pair of enantiomers, RR and SS; and the corresponding *cis* enantiomeric pair (IIc) have the RS and SR configurations.

In UK law, all possible stereoisomers of a 'Controlled Drug' are automatically controlled in the same way (in virtue of a stereoisomers extension clause in each of Parts I, II and III of S2 of the MDA), unless explicit provision is made for a named stereo-isomer to be treated differently. As an example the useful cough remedy 'Phenyl-propanolamine' is explicitly exempted from the MDA control on its diastereoisomer 'Cathine'. It follows that all four stereoisomers of delta^9 THC have hitherto been controlled equivalently in the UK.

In February 1990, the UN Commission on Narcotic Drugs (UNCND) recommended that the particular (–)-*trans* enantiomer, which is available in restricted clinical use, primarily as an anti-emetic in cancer therapy under the WHO non-proprietary name 'Dronabinol' (cf. Table 3), should be transferred from S1 to S2 of the UN Single Convention. This isomer is marketed in the USA by Unimed as the product 'Marinol' and there was an official monograph introduced in the 2nd (1992) Supplement to USP XXII (cf. §4.8). This isomer is identical with the natural RR-*trans*-delta^9-THC and the UN recommendation had some forensic, analytical and legislative, implications.

The UN Vienna Laboratory and the WHO Expert Committee on Drug Dependence considered that rescheduling all four stereoisomers of delta^9-THC (viz. RR and SS -*trans* and RS and SR -*cis*), would avoid the difficulty of making forensic distinction between an *RR* form placed in a different schedule from the other 3 stereoisomers [and which, as explained above, in UK law would normally be subsumed with the named isomer unless otherwise specified]. In the sequel, the UNCND decided only to reschedule the *RR* isomer but individual states party to the Convention may at their option set or retain a more stringent level of control but must not relax controls below that prescribed in the Convention.

The United States Code of Federal Regulations reproduces the revised schedules of the DAPCA 1970 (cf. §1.8). In the second schedule, used for clinically useful controlled drugs, there is a new Part (f) for hallucinogenic substances, which now includes nabilone and sesame oil preparations of (synthetic) dronabinol.

With similar intent, in 1995 the UK amended their Misuse of Drugs Regulations 1985, specifically naming 'Dronabinol' in S2 of the Regulations and concomitantly removing it and its stereoisomers from the list of drugs designated under s. 7(4) of the MDA 1971 as having no therapeutic value. These two changes enable any stereoisomer of delta^9 THC to be used in medical practice in the UK. It should be noted that although by the same statutory order the entry for 'cannabinol derivatives' in S1 of the Regulations was modified by adding the qualification "not being dronabinol or its stereoisomers", there has been no change in the S2: Part I status of delta^9 THC under the principal Misuse of Drugs Act: unlawful possession, supply and import of THC continue to attract class A penalties.

To summarise the various levels of control on different cannabis products, Table 4 sets out the schedule status or equivalent in the UN Single Convention (as amended), in UK law, and in certain other national administrations.

2   OFFENCES

2.1   Cultivation

In the UK since the DDA 1964 the unauthorised cultivation of any plant of the genus *Cannabis* has been a criminal offence. S.12 of the Misuse of Drugs Regulations 1985 provides authority to grant licences, and set conditions, for the lawful cultivation of plants of the genus *Cannabis*.

In France, Art. 5181 of the Code de la Santé Publique prohibits cultivation of the resin plant but can authorise cultivation of the fibre variety for specified commercial purposes. The product of the fibre plant is defined as less than 0.3% THC as determined by gas chromatography.

In the UK, for a prosecution to succeed, it must be established that the accused *knowingly* cultivated the plant(s). S.28 of the MDA provides some statutory defences. Thus, in s.3(b) there is opportunity for acquittal if it can be shown that the defendant had no reason to suspect that the substance was a controlled drug. For instance, consider a defence submission that cannabis grew adventitiously from viable hemp seed components scattered from bird cage litter: this defence could be refuted if the scale of cultivation, or evidence of involvement in husbandry of the crop, were established. However, a plea by an immigrant of Central European origin that the 'weed' [*sic!*],

Table 4 Controls on Cannabis and Related Products and Substances

	UN Conventions	UK MDA/MDRegs	Singapore	USA DAPCA (see Note 3)
Cannabis				
herb and resin	1961 (S2)	class B/S1	class B	Schedule I(d)
extracts	1961 (S2)	class B/S1	class B	Schedule I(d)
THC concentrate	1971 (S1)	class A/S1	class B	Schedule I(d)
Cannabinol	(no)	class A/S1	class B	(no)
Tetrahydrocannabinol(s)	'all isomers'	'derivatives'	'derivatives'	'isomers and derivatives'
	1971 (S1)	class A/S1		Schedule I(d)
Dronabinol	1993 (S2)	1995 class A/S2	class B	Schedule II(f)
Carboxylic acid derivatives	(no)	(no)	class B	(no)
Homologues, 3-alkyl	Synhexyl and DMHP named in 1971 (S1)	'homologues' class A/S1 –	'homologues' class B	Schedule I(d)

Note 1: Class A of the UK MDA primarily differs from class B in maximum penalty on conviction; but there is also a chemico-forensic difference in that the UN 1961 Convention clause, which generically extends control to esters or ethers of controlled substances, in the MDA only applies to class A substances.

Note 2: Substances listed in the UK Misuse of Drugs Regulations 1985 (MDRegs.) Schedule 1 (i.e. those not in medical use) and in section 1 of Schedule 2 (which are or might have been medically useful) are both drawn from substances in the MDA calss A; while MDRegs. Schedule 2 section 6 substances are mostly drawn from MDA class B.

The same MDRegs. 14, 15, 16, 18, 20, 23, 25 and 26 apply both to S1 and to S2 (s.1); additionally, Reg.21 provides a small derogation in record keeping at sea, on oil rigs and for midwives for S2 substances.

In 1995, Δ^9-THC (as dronabinol and its stereoisomers) was transferred to S2 of MDRegs. but remains in class A of MDA.

Note 3: The US Drug Abuse Prevention & Control Act 1970 (DAPCA) had five schedules and parts within schedules, which are periodically updated. Part (d) of schedule I lists 'hallucinogenic substances' and the entry for 'tetrahydrocannabinols' is cast in very wide terms [the full text is given in Section 1.8 of this Chapter]. Homologues of THC are subsumed where they are substances, which have 'similar chemical structure and pharmacological activity'. Dronabinol, as Δ^9-THC prepared in capsules, and nabilone, are in a new (hallucinogenic) Part (f) of schedule II.

grown from bird seed scattered in the cabbage patch of her South London garden, served as an effective repellant for cabbage pests, in the absence of any evidence of harvesting the 'weed', was accepted without penalty beyond seizure and destruction of the Cannabis crop.

In the UK Cannabis will grow comfortably out-of-doors, in Southern England to a considerable height, although not providing a very high quality product. Even within the Arctic Circle, during UN trials in Northern Norway, successful growth up to about

1 metre was achieved but no resin was formed. As reported in Chapter 2, maximal growth occurs in the Mediterranean area, and in Asian and American sub-tropical regions, as well as in most of the African continent. These then are the likely areas of origin of clandestine trafficking in the drug product, as discussed in the next section; forensic chemotaxonomic differences are reviewed in §3.7.

Greenhouse cultivation in a climatic temperate zone is usually successful, including the official chemotaxonomic trials described in §3.7. Intensive, forced (usually hydroponic) cultivation may give high yields of good quality crop but makes heavy demands on electric power consumption – which is usually the first clue in detecting its unauthorised use. Much of the supply of high quality plants, so-called 'Skunk', derives from the Netherlands. In October 1995, a 'Restricted' publication by the National Criminal Intelligence Service reviewed increasing illegal cultivation in the UK, clandestine growing techniques commonly encountered, horticultural equipment employed and its legitimate sources, and forensic sampling procedures.

The legal situation is complex and raises interesting issues: the supply of cannabis seeds and the sale of cultivation equipment are not, separately, unlawful; the incitement of others to cultivate the plant and to produce cannabis is an offence but it would be a defence to provide written warning that unlicensed cultivation was unlawful. Following extensive police investigation in 1994–95 of clandestine cultivation in Wales and in southern England, three directors of two, linked, companies advertising and separately selling heating lamps and hydroponic growing equipment, and viable cannabis seeds, pleaded guilty at Newport (Gwent) Crown Court to "incitement to produce a controlled drug". The first successful prosecution of the author and publisher of a book offering explicit advice on the domestic cultivation of cannabis was at Worcester Crown Court in February 1996: he was convicted of incitement of others to cultivate cannabis and also of harvesting his own extensive planting (reported in *The Guardian*, 19 March 1996).

2.2 Trafficking Offences

Most of the Cannabis illicitly available in countries outside the main producer regions will therefore be the result of unauthorised importation, whence enforcement will be initially a function of national Customs administrations. In the UK, the Customs & Excise Management Act 1979 distinguishes between "knowingly evading a prohibition on import [or export]" and the incomplete act of "attempting to commit an offence" by some deliberate preparatory action that falls short of the actual evasion. The issue may turn on whether there is a general intention to smuggle drugs or the suspect meant to smuggle a specific substance. It is a test of the "guilty knowledge" of the courier.

Following an Appeal Court decision in 1975, in the case of *Houghton v Smith*, that "to attempt the impossible" was not an offence, intentions to smuggle cannabis (and other drugs) for which substitutions had been made, have been subject to the Criminal Attempts Act 1981. However, this may not apply in charges arising from "handling" goods which are shown subsequently not to have been stolen (cf. the House of Lords ruling in *Anderton v Ryan*, 1985).

The success of enforcement by national Customs administrations will depend on a number of factors. Packages of herbal cannabis, or blocks of the separated resin, occupy substantial volume (and mass) in contrast to the size of consignments of high value potent synthetic drugs, and therefore larger volume concealments tend to be investigated. For countries such as Ireland and the UK, with extensive highly convoluted coastlines, clandestine landings from small craft have vied in frequency with the long favoured land-boundary mechanism of concealed compartments in lorries and caravans. The growth of mixed goods packaging in commercial freight containers, for road or rail movement, has been a popular alternative. Some approaches to detection of cannabis in such concealments are mentioned in §3.1.

Air traffic, once a major route of importation of cannabis, is now more favoured for low volume, high value, drugs. However, the unpleasant problem of internal concealment has been a frequently used mechanism of trafficking from certain disadvantaged regions of the world; see §3.1.

The UK Drug Trafficking Offences Act 1986 introduced some additional offences, such as the unlawful sale of articles for administration or preparation of a Controlled Drug. Of particular interest (and some dispute) is the power for seizure of assets of a convicted smuggler where these could not be shown *not* to have been the proceeds of drug trafficking. Here, the onus falls on the duly convicted smuggler to prove legitimate acquisition of discovered substantial assets.

2.3 Dealing, Handling and Possession

1 Unlawful Possession

Possession without lawful reason, is an 'absolute' offence and thus is the most straightforward offence to prove. Custody by another person may also be deemed possession by the accused where it can be successfully demonstrated that the drugs were held for and on behalf of him. Lawful reasons for possession include production, supply or possession of drugs licensed for use in scientific research or laboratory testing, when being used as such. In those countries where medical use is authorised, the preparation, dispensing and administration of therapeutic presentations of cannabis may be lawful, as well as the corresponding possession by a patient for whom such a preparation has been properly prescribed.

Individual national administrations may establish guidelines, or enact statutory levels, of what may be reasonably claimed to be a "personal supply". As a rough guide, a few grams per week may be taken for recreational purposes, and up to 20 g/week by a habitual or heavy abuser. The charge of unlawful possession constitutes about 85% of cannabis offences in the UK. However, about half of possession cases may be dealt with by formal police "caution" and more by a "suspended" custodial sentence, so that only 10% of such offenders go direct to prison. HM Customs will often impose a "spot fine", and concomitant seizure of the cannabis, when the quantity is small and admitted.

In some other European states, cannabis abuse is so much part of their culture (e.g. in Holland, Italy and Spain) that charges may not be brought for possession of very small quantities. The German Constitutional Court on 28 April 1994 ruled to discontinue prosecution of individual persons arrested for possession of "small amounts" of cannabis stated to be for that individual's personal use [quoted in *Le Monde*, 2 May 1994, p. 8].

Where no lawful reason for possession is established, the nature of the seized substance or product must be unequivocally determined (cf. §3) and continuity of the evidential chain of samples maintained in order to sustain the connection between the accused and the extent of his control over the place where the drugs were found – e.g. on his person or in his clothing, luggage, home or vehicle.

2   Possession With Intent to Supply

Possession with intent to supply, to be substantiated, essentially rests on scale: is the amount of cannabis product discovered consistent with a claim of 'personal use'? Above such levels, some collateral evidence – preferably recorded – is needed of unlawful contact of the suspect with known or putative users of cannabis products.

3   Unlawful Supply

Unlawful supply, or 'dealing', in addition to the facts of quantity and recorded contacts, requires some evidence of the actual transfer of drugs to the control of another person and, usefully, observation of his receipt of payment of some kind for the supply. The proof of supply is crucial because under the MDA the maximum custodial sentence (on indictment in a Crown court) for preparation or trafficking or dealing in cannabis products is 14 years, whereas for simple possession of cannabis the maximum penalty is 5 years (7 years for cannabinoids).

In the UK, in virtue of the MDA s.23(2), civil police have powers to "stop and search" a suspect person, or any vehicle or vessel, when the officer has "reasonable" grounds for suspicion "that the person is in possession" of, or the vehicle contains, a controlled drug. Officers of HM Customs traditionally have still wider powers under their warrant to enter any premises where they have reason to believe they may find smuggled goods of any kind.

4   Usable/Measurable Amounts

A variety of UK case law addresses the *de minimis* concept. In R *v Worsell* (1969) unweighable droplets of heroin, barely seen in a tube, were held not to constitute an effective dose for use or sale; but in R *v Graham* (1970), it was accepted that the defendant had weighable scrapings of cannabis in his pockets. In *Bocking v Roberts*, as little as 20 micrograms of cannabis resin residue in a pipe, estimated colormetrically, supported conviction on Appeal; but in R *v Colyer* (1974) the same minute amount in a pipe residue was *not* considered a measurable quantity within the defendant's knowledge. On appeal, in R *v Carver* (1978), it was established that 20 micrograms residue in a smoked 'roach' and 2 mg cannabis resin scraped from a box constituted an unusable amount and would *not* support a charge of possession, although it might provide (collateral) evidence of prior possession. However, in the case of R *v Boyesen* (1980), while the Appeal Court criticised prosecution on minute amounts, the House of Lords ruled (1982) that usability was not an issue: quantity only matters in there being a sufficiently measurable amount for establishing identity of the drug and guilty knowledge of it by the defendant.

5   Prior Possession

The residual unmeasurable droplets of heroin in the Worsell case (above) would have been admissible evidence in a charge of prior possession. Similarly, while the smoke generated by a person smoking cannabis does not constitute 'cannabis' (but does contain cannabinoids controlled in a higher penalty category!), it can provide evidence of prior possession. Metabolites (unless separately 'controlled') in the accused person's urine (or blood) do not constitute possession of a controlled drug – but their presence will reveal prior consumption (cf. evidence of amphetamine taken, in R v Beet, 1977). Such metabolites may not uniquely point to one drug but may be characteristic of a family of substances, e.g. opioid derivatives arising from legitimate medicines such as codeine. Presence of metabolites in urine may also be a signal of leaching from an internal concealment, i.e. a carefully packaged drug that has been inserted in a body cavity or has been swallowed. In these circumstances, sensitive diagnostic tests will reveal a decline in metabolites if a drug (medicinal or abused) has been taken conventionally, whereas pharmacokinetic equilibrium (or, in clinical emergency, a rise in concentration) provides evidence for detention of the suspected courier for personal investigation (cf. §3.1).

2.4   Control of Premises Offences

In UK law, for charges which relate to premises on which drugs are packaged for export, or prepared, supplied, consumed or smoked, to succeed the owner must be *knowingly* in control of the premises at the material time. This concept of being 'knowingly' in control of premises did not apply prior to 1973, e.g. in the case of Miss Sweet, an Oxford landlady, when in her absence her premises were used for smoking cannabis. If a kitchen, say, is used by other persons for preparing a controlled drug, the owner of the kitchen must be proved to be aware of their intended purpose if to be convicted of control of premises used for that purpose. S.13 of the MDRegs. 1985 provides for the smoking of cannabis or cannabis resin for *the purposes of research* on premises officially approved for that purpose.

Similar considerations (of being knowingly in control) apply to allowing premises to be used for the cultivation of cannabis or otherwise for the production of a controlled drug.

2.5   Social Attitudes and Public Perceptions

1   National Surveys

Surveys of cannabis usage have been produced in many countries. These address frequency of recreational use, distinction of gender or occupation of user, and form of drug abused.

Significant debate in the United Kingdom on abuse of cannabis may be said to have begun a century ago with the exhaustive report in 1894 of the British Indian Hemp Commission. Since then, recreational use of cannabis has become by far the greatest non-medicinal usage of any controlled drug in the UK. Smoking cannabis is recognised as a leisure activity in a significant cross-section of younger people. It seems likely that one third of the population in the 16–29 age range have "tried" cannabis at some time (Ramsay and Percy, 1996), compared with 14% for amphetamine, 9% for lysergide ('LSD') and 6% for MDMA ('Ecstacy'). The same survey contrasted 43% (50% male, 35% female) of that age range who had tried any prohibited drug at some time in their life, with half that proportion (22%) in the 30–55 age group. Cannabis was involved in about 80% of the 115,000 drug seizures in 1995, and in most of the 82,000 'possession' offences (Anon, 1996). Of these cases, 52% resulted in a 'caution', in 22% a fine was imposed and 8% a custodial sentence. Using multivariate analysis and constructing regression models, it appears that cannabis use in the 20–29 age band correlates with a white or Afro-Caribbean (but not Asian) male, spending his evenings out in pubs or elsewhere, and frequently unemployed and living alone in poor housing.

In Spain, a national survey in 1980 (Rodriguez and Anglin, 1987) revealed that 20% of the population aged over 12 had tried cannabis "at some time" and 5% "regularly". In a follow-up survey in 1985 the proportions were 21% "at some time" and 12% weekly or more frequently. Between 1974–1984, university users in Barcelona doubled, from 9.6 to 20%; in Oviedo in 1986 (Lopez-Alvarez et al., 1989), 7% had used it in the previous month and 17% in the previous year. An even choice of product was noted in another university survey: in Valencia in 1975, 10% favoured herb and 12% resin.

In Mexico, a twice-yearly survey "Information Requirements System on Drugs" (Ortiz, et al., 1989), reported that of 16–19 year-olds in 1986, 64% had experimented with cannabis and 42% admitted regular use. Only solvent abuse had comparable popularity in this age range; the cannabis users were mostly male and from the lowest socio-economic class.

Gender-based selection was also noted in a study by Pela (1989). He reported that smoking cannabis was an essentially male phenomenon for Nigerian students in the 1960s and 1970s, whereas in the 1980s, perhaps reflecting a changing view of female education, use of cannabis by women was increasing, although taken usually by the oral route, e.g. the herb in soups and the oil in fizzy drinks. Studies of usage in other African countries may be found in Asuni and Pela (1986).

University experience in Central India was researched by Khan and Unnitham (1979). In a self-reporting survey of 4,300 students at 27 colleges in Jabalpur, 6.3% favoured 'bhang' [herb] in food while 2.1% smoked 'ganja' [resin] or charas; former use was admitted by 10.3 and 2.9%.

In contrast, in Malaysia in the 12 years from 1975 to 1986, heroin was by far the most available drug of abuse (80%), with much less cannabis seized, and synthetic drugs were very uncommon. This is not surprising, given the geographical proximity to South East Asian sources, but penalties for trafficking are very severe. Their National Drug Abuse Monitoring System recorded that during this period known addicts rose 11-fold, from 68 to 755 per million inhabitants.

For the same decade in Singapore, seizures of cannabis products and of opiates and opium, in 1975 were 196 and 667 respectively, representing 22% and 74% of all drugs of abuse. In 1986 the corresponding proportions were 790 (26%) and 1891 (61%). Fuller data are given in Dutt and Lee (1991).

2   Substitution of Cannabis Products

Deliberate substitution of alternative drugs of abuse can become a serious social problem. In Italy in 1974 cannabis disappeared from the market and was rapidly replaced by heroin. At about the same time, when police activity in Stockholm caused supplies of stimulants to dry up, heroin began to appear in Sweden (Hartnoll *et al.*, 1989).

In undisclosed substitution, the comminuted herbal material most frequently passed-off as cannabis is henna. Following a (temporary) decline in 1975 in illicit cannabis imports to the UK, there were some large interceptions of 'fake' resin: one of 40 kg wholly comprised compacted hanna (LGC, 1976). Herbal simulations reported at that time included one based on hops, and an ingenious presentation of stinging nettles steeped in cannabis oil, with added bird-seed to give the product verisimilitude. Coffee powder and chopped parsley have also been substituted, and even *Datura strammonium* (Corrigan, 1979), which no doubt explained Irish reports of atropine poisoning and hallucinations, and similar reports in Great Britain (Ballantyne *et al.*, 1976). Laboratory and field colour tests – see §3.3.3 – can distinguish genuine cannabis from simulations (de Faubert Maunder, 1974).

3   Global Seizures

Global patterns reflect the dominance of areas of cannabis trafficking. In Japan in 1985, seizures under the Cannabis Control Act comprised 16.1 kg resin, 104 kg herb and 10 kg oil (Tamura,1989). The corresponding numbers of persons arrested in connection with these products were 206, 919 and 148 respectively: most were "white-collar" workers and students under severe commercial or academic competitive pressure. This pattern was relatively constant over a 10 year period. The cannabis products were mostly imported from the USA but there was some from local cultivation.

Official laboratory records of analysis of local seizures of drugs of abuse provide similar evidence of global differences. There was a roughly constant proportion of 1 : 3 for cannabis to opiates in seizures examined by the Singapore Department of Scientific Services: see the summary in Table 5. This pattern is very different from that found in northern Europe. In the Republic of Ireland, reports by the State Chemist for the years 1968–1978 disclose an increase in positive identification of cannabis from just 10 items (ca. 28% of all drugs positively identified) to 525 (82%) in eleven years (Corrigan,1979). Over the same period, opiates declined proportionately from 63% (23 items, mostly morphine and synthetics) to 9% (57 items) in 1978.

For comparison, import data for England and Wales based on analyses by the Laboratory of the Government Chemist, reveal a roughly constant proportion, but rapidly increasing numbers (and weights), of cannabis products from seizures by HM Customs. Positive identifications of cannabis rose from 154 (77% of all Controlled Drug identifications) in 1969, to 2,581 (76%) in 1978, and, by 1987, 3,797 items (76%) amounting altogether to 16 tonnes of cannabis products. There is a similar rising trend in UK domestic (i.e. police) seizures of cannabis in the later period 1985–95: see Table 5.

Put in a global perspective, seizures (in tonnes) for the year 1988, as reported to 'Interpol' (Gough, 1991g) were: Europe 150 (including UK 43), North America 458, Australia and New Zealand 6.9.

Table 5

National analyses of drug seizures: Cannabis products compared with opiates and opium

Year	HMC&E [a]		UK police [b]		Eire [c]		Singapore [d]	
	Cannabis	Opiate	Cannabis	all other	Cannabis	Opiate	Cannabis	Opiate
1968	77	3			10	23		
1969	154	23			11	67		
1970	318	10			86	30		
1971	707	21			176	26		
1972	1197	26			272	113		
1973	1240	56			310	8		
1974	1166	42			376	92	167	264
1975	2059	92			389	73	196	667
1976	2268	199			301	23	184	1915
1977	2165	246			328	80	196	2212
1978	2581	309			525	57	165	1862
1979	2920	620					425	1277
1980	4041	425					634	1742
1981	3993	282					768	2025
1982	3298	434					675	2434
1983	3080	560					774	1943
1984	2772	506					860	1833
1985	3320	348	25 000	5500			774	1964
1986	3662	378	26 000	5000			790	1891
1987	3797	283	27 000	4000			874	2366
1988			32 000	5000				
1989			44 000	6000				
1990			53 000	7000				
1991			62 000	8000				
1992			60 000	12 000				
1993			75 000	15 000				
1994			89 000	19 000				
1995			93 000	22 000				

'Opiate' subsumes opium, opiates (esp. heroin) and synthetic opioids

(a) HM Customs seizures analysed by the Laboratory of the Government Chemist, London (authors's data).

(b) Number of seizures (rounded) by UK police: Home Office Statistical Bulletin 25/96 (Barber, A. *et al.*, 1996).

(c) Analyses by the Irish State Chemist, 1968–1978: (Corrigan, 1979).

(d) Analyses by Department of Scientific Services, Singapore (Dutt and Lee, 1991).

Note the minor role of cannabis products in a South East Asian administration.

2.6   Legalisation

1   Public Debate

The following discussion is primarily from the UK standpoint and should be adjusted to other more, or less, relaxed national situations. In the UK, the decriminalisation of cannabis, or even its total legalisation, has been debated for decades. The apparent widespread use of cannabis in 'pop' culture festivals in the 1960s attracted media attention. In 1967, the popular musician Paul McCartney persuaded 100 or so leading cultural luminaries to append their signatures to a full-page advertisement in the London 'Times', advocating legalisation.

An expert committee on cannabis, chaired by Lady Wooton, published its Report in 1968. They considered it still necessary, in the public interest, to maintain restriction on the availability of cannabis but, while rejecting explicit decriminalisation, they favoured reducing the penalties, e.g. by transferring cannabis products to a lower control category. For example, if cannabis products had [later] been transferred to 'class C' of the [subsequent] Misuse of Drugs Act, the maximum penalties would have been halved. Their recommendation that possession of small amounts should not be punished by imprisonment was rejected by the British government of the day.

The Advisory Council on Misuse of Drugs set up under the MDA 1971, keeping sight of the problem, produced a series of unpublished guidance papers reviewing the toxicology and psychopharmacology research on cannabis, but the Council was unable to support legalising availability. A further expert group, chaired by Sir Robert Bradlaw, presented an open Report in 1981. This noted inadequacies and inconsistencies in published research, found some evidence of deleterious effects but no incontestable report of significant harm, recommended much more epidemiological studies and suggested that certain therapeutic use might prove beneficial. To bring understanding of the issues up-to-date, in 1995 the Department of Health agreed to fund a new literature review covering the period 1983–95. The new report, although completed in late-1996, was not yet publically available in April 1997.

There has also been attention in TV media in Britain. In February 1995, the Anglia TV programme 'The healing herb' addressed the therapeutic value of cannabis to certain patients but confused this main issue by portraying recreational use and availability of supplies in Amsterdam. A year later, the (British) Channel 4 service presented a composite evening's entertainment. The case for legalisation was not helped by trivialising sequences and playlets portraying future merchandising and lengthy experiences in Dutch night-life, and the following attempt at genuine studio debate was frustrated by noisy adherents applauding favourable anecdotes while unwilling to listen to serious contributions, favourable or otherwise, from physicians, parliamentarians and retired senior policemen. It is clear from such media explorations that informed debate must clearly separate medical value/clinical use from recreational use, and be supported by written and by disciplined oral evidence within a formal structure – such as a Royal Commission.

2   Recreational Use

Arguments for and against legal availability for recreational use may be summarised as follows:

The Advantages of decriminalisation rest on claims that –

(1)   the use of cannabis does not result in physical addiction: generally, cessation of use does not precipitate the acute symptoms associated with withdrawal from opiates, although some cases of mild withdrawal symptoms from cannabis have been reported;

(2)   alcohol prohibition in the USA in the 1920s encouraged gangsterism;
(3)   cannabis is less dangerous to health, and causes less overall social harm, than does, say, alcohol abuse;
(4)   prohibition of cannabis interferes with its traditional social role in certain ethnic communities.

Disadvantages that have been adduced include:

(1)   some clinical evidence of psychic dependency on cannabis;
(2)   prolonged heavy use may lead to (temporary) psychiatric disorder (see Chapter 8);
(3)   a more substantial body of evidence of documented deleterious effects and that regular use may be de-enervating and energy sapping [cf. the fate of some jazz musicians], and in some users has led to apathetic behaviour with poor work performance;
(4)   total legalisation would result in unrestricted release of a recreational drug of unproven safety judged by current safety of medicines standards;
(5)   the risk of overdose resulting from unsupervised administration of non-standardised products, potentially leading to a psychotic condition (although there is no evidence of this condition persisting beyond one year);
(6)   the risk in pregnancy of premature birth – and all smoking increases risk factors for the embryo;
(7)   the unfeasibility of keeping supplies away from younger schoolchildren;
(8)   the likelihood of minority pressure for relaxation of control on other drugs of abuse.

Two other contrary arguments, frequently heard, are –

—that regular flouting of the present law is not of itself an argument for repealing that law;
—and that, whereas Amsterdam 'Coffee shops' may legally serve the herb or resin in food or for smoking on those premises, and domestic cultivation of high potency herb seems unrestricted,
   the Netherlands authorities still respect their international obligation to enforce controls on dealing in, and import of, cannabis.

3   Therapeutic Implications

There is a growing perception of the potential Medical significance of Cannabis. Authoritative and extensive treatments of therapeutic use and pain killing applications are supplied in Chapters 7 and 8 respectively.

A variety of medicinal preparations of cannabis products are now legally available as licensed medicinal products. 'Marinol' is presented as encapsulated $2\frac{1}{2}$, 5 or 10 mg Drotebanol (THC) in sesame oil and the USP contains a monograph for capsule presentation [cf. §4.8]. Nine synthetic analogues of cannabinoids, with a variety of therapeutic indications covering all the above four categories, are listed in Table 3; their chemistry has been discussed in §1.8.

It may reasonably be concluded that there is a good case for a modern, suitably structured, critical examination of medical claims for the clinical use of natural or synthetic cannabis products. Where treatment with them is seen unequivocally to be justified, there would need to be prompt and effective consultation on appropriate mechanisms for their legal use, without prejudice to decisions on recreational usage and international obligations on control of trafficking in cannabis products.

3   FORENSIC ANALYSIS AND PROCEDURES

3.1   Detection and Sampling

The 'front line' of detection is at national boundaries where highly trained uniformed and 'plain clothes' Customs officers, vigilantly observe the demeanour of the travelling public and freight vehicle drivers. When challenging an individual, the journey starting point, route travelled and quantity and type of baggage, accompanied and unaccompanied, may all be relevant. However, at a very busy seaport such as Dover, or a major international airport like Heathrow, the sheer pressure of the vast travelling throughput inhibits intensive interception. In practice, the major seizures of cannabis and other drugs frequently arise from good prior intelligence. This may take the form of advice from international police ('Interpol') or Customs co-operation bodies, supplying forecasts of likely arrival of clandestine shipments or suspicious passengers; or it may be a distillation from computerised national surveillance of known criminals and monitoring of frequent travellers.

A variety of detection modes is available for enforcement. Frontier checks on travellers and freight may involve soft X-ray examination or magnetic resonance imaging of packages and baggage, hard X-ray or γ-ray back scattering or vapour phase GC–MS (linked gas chromatography–mass spectrometry) probes for lorry and container traffic, experienced 'rummage crews' searching shipping small and large and – even on occasion – aircraft, and spot searches of cars and caravans – including, for instance, main and supplementary fuel tanks, spare wheels, double door skins, false compartments and camping equipment. Sampling restricted atmospheres in containers can be adapted to letter-mail searching, using air sampling cartridges and measurement of standardised drift velocity in an ion mobility spectrometer (Lawrence and Elias, 1984). Lawrence (1980) has also reviewed more generally the techniques useful in the detection of controlled drugs in mail packets.

Within the country, police enforcement is likely to be concerned with the identity of the seizure and whether quantities of drugs in the possession – or under the control – of the suspect are in sufficient amount and presentation to imply 'possession with intent to supply', i.e. 'dealing', or of a lower level consistent with 'personal use', i.e. simple possession. Armed services police investigating drugs offences within military jurisdictions will have similar concerns except that a more serious view may be taken of possession even of small quantities of cannabis and other psychoactive drugs.

Dogs, often Labrador or German Shepherd breeds, have been trained in several countries to respond to the terpenoid odour of cannabis. The dogs search for the signal scent in luggage, as on airport loading tracks, or in recesses less accessible to human searchers; and the dog-handler is concurrently trained to recognise the dog's response. The source of especial canine interest is then carefully examined and, where appropriate, tested chemically. The animals need regular and frequent rest breaks to avoid search fatigue (loss of interest) [cf. US Customs Service report, 1980] and also a periodic memory refresher with genuine target substance(s). Occasional false positive scentings of cannabis can usually be resolved by the investigator conducting a field test (see next Section). However, canine confusion of patchouli, the oil of which has a heavy, peppery perfume somewhat reminiscent of smoked cannabis, could cause embarrassment when, say, raiding a nightclub. On one notable occasion, a freighter was briefly detained in Sunderland docks after the search dogs excitedly reacted to a cargo of patchouli leaf.

Internal concealment ('stuffers and swallowers') is a particularly unpleasant, and dangerous, mode of smuggling. Usually, a series of small flexible containers, such as condoms, or rubber balloons, will have been filled with an oily concentrate of cannabis, double sealed and then either lubricated and stuffed into a bodily (anal or vaginal) cavity of the courier, or – following administration of antiemetic and antidiarrheal preparations – the packages are swallowed. If the journey (usually by air) is sufficiently brief the discomfort may not be serious; but where there are traffic delays or an increase in the rate of internal seepage of the drug, there is a serious clinical risk. If there is mechanical blockage of the G-I system, or should one or more condoms rupture releasing solvent and a massive local concentration of cannabinoids, the condition may become fatal.

When a suspicious substance or mode of concealment is detected, in circumstances under the personal control of the traveller, rapid analytical support is needed to identify the drug and to justify detention of the person and/or goods pending further investigation and testing. This may take the form of suitable sampling equipment and also simple 'field' tests which can be undertaken remote from a laboratory. Where internal concealments have been suspected, non-invasive immunoassay tests may be employed. For the suspected courier there is a degrading sequence awaiting natural release of the cannabis packets; and special hygiene protocols and subsequent laboratory sterilisation procedures are needed for the forensic confirmation of the substances concealed.

3.2   Field Tests for Cannabis Products

1   Sampling Practice

This will vary between enforcement agencies. Many police forces employ professionally trained and qualified 'scene-of-crime' officers; while some Customs administrations avail of their own investigators suitably trained by scientists in the recognition and sampling of the commoner drugs. In major seizures customs or police investigators are frequently accompanied by laboratory specialists. Bulk quantities of herbal and resin forms of cannabis have a characteristic appearance and this can be supplemented by simple colour tests to reinforce the suspicions of the enforcement agent. Trace residues of cannabis products can be collected (see below) but field examination should be limited to visual examination and non-destructive testing; it is important to avoid contaminating trace samples and not to use a significant portion of the evidence before it has been subject to formal test in an analytical laboratory. Whether bulk or trace samples, all exhibits must be correctly labelled with source, date/time, and name of agent making the seizure. Otherwise, it can be particularly galling if, after good investigation and hard laboratory work, a court case is lost through faulty continuity of an evidence chain.

2 Field Sampling Procedures

In addition to conventional scene-of-crime screw cap exhibit jars and various sizes of sealable plastics envelopes, field sampling equipment usefully includes a motor car style miniature vacuum cleaner, which can be operated from a 12 V battery or through a transformer attached to a main power supply. The suction unit may be fitted with a variety of glass adapter nozzles and Soxhlet thimble catchment devices. Such equipment was recommended (LGC, 1974) for sampling traces of herbal debris and detritus, from cannabis or other drugs lodged in clothing, vehicle concealment cavities, ship and caravan lockers, furniture, floor recesses, car boots and under household and vehicle carpets. Examination of such residues may provide important evidence in establishing that more than one consignment of cannabis has been stowed in this place, and thereby support a charge of conspiracy to import, or to supply, several batches.

3 Colour Tests

Civilian and military police and Customs investigators can all be materially aided by use of a rapid sorting test for cannabis. Where clearly negative, such a test may avoid unnecessary detention of a suspect person or goods; whereas, if positive it provides grounds for further searching, sampling and laboratory analysis. If the appearance of the specimen is not characteristic, and the nature of the drug has been contested by the suspect, a positive field test completed in their presence may sometimes elicit a plea of guilty knowledge of the drug. However, while a properly conducted field test may provide strong presumptive evidence for the presence of cannabis products, subsequent laboratory confirmation should always be commended.

The following field test (de Faubert Maunder, 1974) is recommended: within two small filter-papers, folded conewise and in contact, place ca. 1 mg of suspected cannabis product and drip light petroleum (40–60° or higher boiling fraction in tropical countries) until the outer paper is wetted by the solvent. Carefully transfer the suspected material to a correspondingly labelled exhibit envelope, discard the inner filter paper and air-dry the outer and then test it with ca. 0.1 mg of dye reagent (see discussion of dyestuff in §3.3.3) and one drop of 1% $NaHCO_3$ solution. An annular red-to-violet stain develops if cannabinoids are present. Distinction of cannabis from 240 other herbal substances was reviewed by de Faubert Maunder (1969) – see comments in §3.3.3 below.

4 Field Kits

The use by enforcement agencies of a composite drug test field kit for cannabis products and other commonly encountered drugs of abuse requires a carefully controlled logical sequence of colour tests. Several schemes were considered and reported by a United Nations (1974) consultative group. The underlying philosophy and development of such a procedure, and experience in the storage and packaging of a prototype kit, has been described (LGC, 1972). This kit was subsequently patented for commercial development (de Faubert Maunder and Phillips, 1976) and marketed by a European fine chemicals company.

Many adaptations of laboratory colour tests have been considered for field use. Modifications of historic laboratory-based colour tests for cannabis products have been reviewed by Baker and Phillips (1983) and are summarised in §3.3. Phillips (1974) described portable TLC kits for field use by laboratory staff, incorporating silica coated microscope slide support, screw cap bottles with mobile phase (e.g. xylene for cannabinoids) and fluorocarbon-propelled chromogenic sprays.

3.3   Identification of Herb and Resin

1   Screening

The first stage of laboratory identification, screening for the presence of any cannabis product in the forensic sample, may have been pre-empted by a report of a positive field test. Otherwise, note is taken of the macroscopic appearance: this is highly characteristic in bulk samples of herb and slabs of resin but may be more difficult to distinguish in highly comminuted material. More persuasive examination involves light microscopy (see §3.3.2) and colour tests (see §3.3.3), with convenient confirmation by TLC (see §3.4.2) . According to the nature and quality of the sample, and the potential seriousness of the offence, quantitation by GC or HPLC of individual cannabinoid principles may be necessary.

2   Microscopy

Compendial monographs for Cannabis – or 'Indian Hemp' – list diagnostic features convenient for the identification of the medicinal herb, e.g. in the BP 1914 and the BPC 1949 (see §4.2).

A detailed exposition of microscopic features of Cannabis has been published by Jackson and Snowdon (1968). An experienced microscopist should be able to make a definitive identification of an uncontaminated herbal specimen provided that the distinctive morphology is not missing. Fairbairn (1972) listed criteria for unequivocal confirmation but use of light microscopy for identification should be supported by chromatography (e.g. Eskes *et al.*, 1973). Reliance on it as a *single* technique can not be generally recommended (Baker and Phillips, 1983). For limited complementary use in the examination of smoking residues – see §3.5. Mitosinka *et al.* (1972) examined cannabis by scanning electron microscopy and the forensic use and presentation of SEM evidence in Court has been discussed and illustrated (LGC, 1978).

3   Laboratory Colour Tests

There is an extremely extensive literature describing colour tests for presumptive identification of cannabis. Baker and Phillips (1983) summarised some of the many variants of the Duquenois (Negm) test – and favoured the Levine modification

(Butler, 1962) despite occasional false positive results with coffee. Baker and Phillips were less enthusiastic about variants of the classical Beam and Ghamravy tests; overall, they preferred both for field and for preliminary laboratory testing the colour reaction developed by de Faubert Maunder (1969, 1974). Of the many azo dyes yielding strong colours with phenolic cannabinoids, the use of Fast Blue B has been widely reported and is the most consistent and selective. However, because of alleged carcinogenic impurities in the dyestuff, for field and laboratory use, and as a TLC chromogen, in the author's laboratory this was later replaced by Fast Blue <u>BB</u> made up in a 1% w/w admixture with sodium sulphate. The dye Fast Corinth V gives essentially similar colours and has been used in a commercial version of this drugs test kit.

From about 240 herbal substances tested by de Faubert Maunder (1969), only nutmeg and mace gave a confusable, pinkish, reaction – and TLC will speedily eliminate this uncertainty. Henna, a frequent simulation of cannabis – cf. §2.5.2 – does not give a positive reaction in the field test provided that, as recommended, two filter papers are employed. Mechoulam *et al.*, in a major review (1976), considered that "by their nature" colour tests are non-specific and should be confined to field work and laboratory screening, and then followed up by other techniques for formal identification. This has always been the author's view too, particularly where prosecution is to follow. Chromatographic procedures are discussed in the next Section.

3.4 Cannabis Oils and Cannabinoids

1 Introduction

Crude mixtures of cannabinoids, particularly oily extracts of cannabis, require individual identification – and sometimes quantitation – of the controlled ingredients. TLC methods are of first choice for speed and simplicity as qualitative tests – and have some value in assessing relative levels of congeners provided it is admitted that TLC is, at best, only semi-quantitative: 5% coefficient of variation for quantity and 2–3% reproducibility of Rf values. This can be improved at least two-fold when using microsyringe fine-drop application on so-called 'high performance TLC' (HPTLC) plates coated with 5 μm micronised silica and the spot measurement improved with use of a modern micro-densitometer. However, for assured quantitation, GC or HPLC is more appropriate.

2 Thin Layer Chromatography

Most of the earlier reports on the detection of individual cannabinoids relied on TLC, although the clarity of recognition, discrimination of isomers (e.g. *delta*^{9} *v delta*^{8} THC) and quality of reference standards sometimes leaves something to be desired. A comprehensive review by Gough and Baker (1982; see pp. 314–317) incorporates a tabulation of cannabinoids identified in 39 reports. Today TLC is still a valuable first line tool for the recognition of cannabinoids.

As analytical quality assurance, two distinct solvent systems should be used to separate and identify components of a series of extracts of cannabis herb or resin. For sample preparation, dissolve a small portion in light petroleum (40–60° or a higher

boiling fraction in a tropical climate), or a convenient dilution of a cannabis oil specimen, and micropipette a $0.5\,\mu L$ sub-sample onto a 10-cm plate commercially pre-
coated with a $250\,\mu m$ silica layer; concurrently intercalate on each plate a series of
cannabinoid reference standards. The complementary pair of mobile phases favoured
by the LGC (Fowler *et al.*, 1979) was (I) chloroform (alcohol-free): 1,1-dichloroethane
15:10 and (II) xylene:dioxane 19:1. Chromogenic recognition, and some mutual
discrimination, of the cannabinoid spots is achieved by spraying with, e.g., a solution
of Fast Blue BB. Typical retention data and colour development for 7 cannabinoids and
some acid precursors in these two solvent systems and with this chromogen (Gough,
1991a) are incorporated in Table 6.

Separation of the cannabinoids may be further improved by using two distinct solvent
phases consecutively in a two-dimensional TLC mode. Figure 1A illustrates a typical
resolution of CBD, THC and CBN from other cannabinoids in an extract from a dark
Kabul resin; this 2-D chromatogram was developed with a simple combination of
mobile phases, first toluene and then 2% diethylamine in toluene. This earlier system
also achieved a good separation of precursor acids but was superseded at LGC by the
pair of mobile phases reported by Fowler *et al.* (1979).

Whichever pair of mobile phases is used, after phase-I treatment the plate is air dried,
lightly oversprayed with diethylamine, and then redeveloped orthogonally using mobile
phase-II. Additional information may be obtained if a *second plate* is run: after the first
direction development, this second plate is heated for 5 min at $150°$ to decarboxylate the
acid precursors retained near the phase-I origin, and then the second development
resolves and semiquantitatively estimates the additional THC, CBN and related
compounds when compared with an unheated plate (O'Neil *et al.*, 1985). The 2-D mode
was more efficient in the resolution of the propyl analogues from other congeners CBG
and CBCy – contrast Figures 1A and 1B – although it leaves the cluster of precursor
acids nearer to the starting point.

Such 2D-TLC systems also tolerate higher plate loadings which affords opportunity
to make semi-quantitative estimates of the minor cannabinoids present. However, the
general disadvantage of orthogonal 2-D TLC is that the specimen must be loaded into
one corner of the plate for the consecutive development stages, whence only one
sample (and no external standards) may be run at any one time.

3   *Gas Chromatography*

As already explained, thermal treatment of cannabis extracts results in decarboxylation
of precursor acid components and therefore GLC analysis loses potentially useful
diagnostic (qualitative) information. The technique has value, however, in quantitative
assessment of total THC content. This is a useful indicator of the quality of the cannabis
sample, and may have some bearing on the penalty imposed on conviction of unlawful
practice. GC has been used to determine the quantity of total THC in illicit 'cannabirum'
concoctions from the Carribean; here, the presence of aqueous ethanol and large
amounts of sugar require extraction with a non-polar solvent to protect the GC column
(see p. 319 in Gough and Baker, 1982).

For a seized cannabis oil it is possible to estimate the approximate volume of herb or resin that had been initially extracted if the laboratory has a library of typical herb and resin samples from the source country (cf. Figure 2, later). The absence or significant presence of CBD points respectively to a herbal or resin source, and chemo-taxonomic studies (see §3.6) support an assignment of country of origin.

Many GC systems for cannabis have been reported. Gough and Baker (1982; see pp. 317–319) provided an extensive survey of 58 reports of resolution of cannabinoids. The most popular systems have used dimethylsilicone (SE-30 or, with better discrimination and stability, OV-17) mobile phase, or a cyanoethylsilicone (XE60 1% coating), supported on Gas Chrom Q as stationary phase and with flame ionisation detectors. Harvey and Paton (1975) recommended chromatographing the trimethylsilyl derivative of cannabinoid phenols and Moffat (1986) quotes retention times for such derivatives of seven cannabinoids and four metabolites. A typical gas chromatogram of cannabis resin, demonstrating the separation of THC from THV, CBD and CBN, was reproduced by Gough (1991b) and relative retention times for principal cannabinoids are included in Table 6.

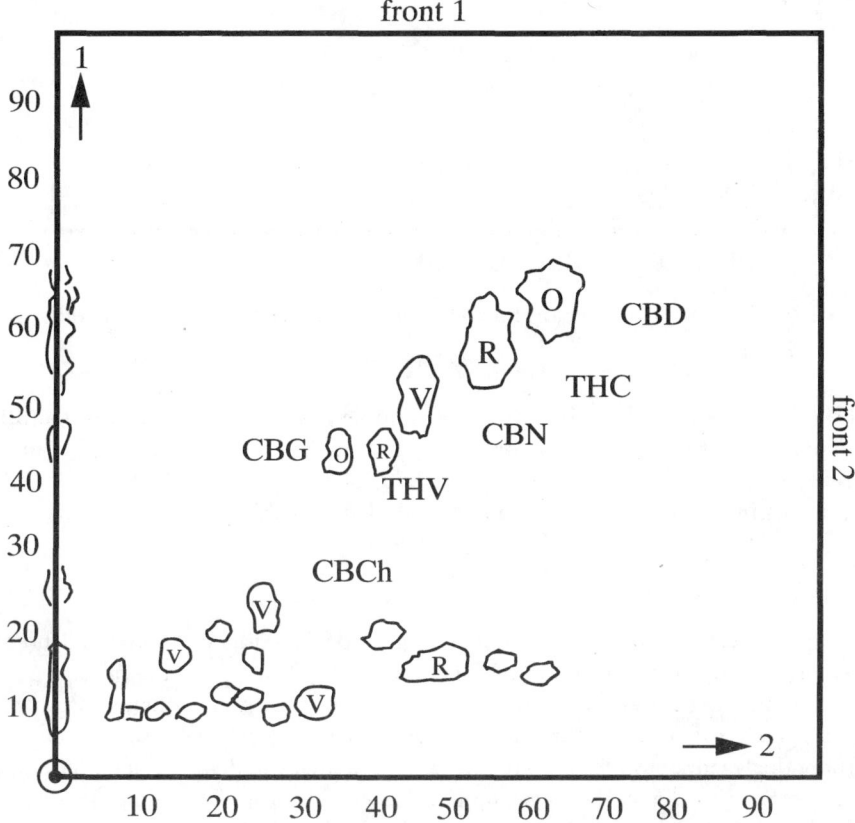

Figure 1A   Kabul resin; T1 = toluene, T2 = 2% diethylamine/toluene

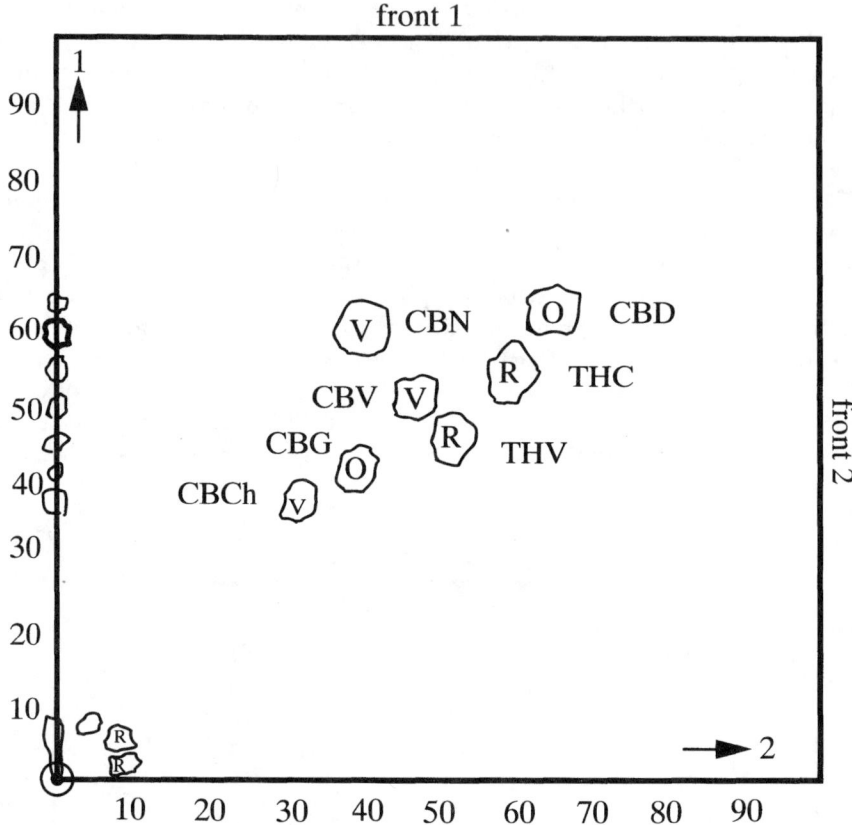

Figure 1B   LGC system, T1 = chloroform/1,1-dichloroethane 3/2, T2 = xylene/dioxane 19/1

Webb (1991) has described the use of GC-coupled–mass spectrometry (GC–MS) for qualitative examination of cannabis products, and the detection of free and conjugated metabolites in urine and other physiological fluids. An extract from his list of significant fragment ions in Electron Impact mode is included in Table 6.

4   Liquid Chromatography

HPLC is particularly useful in comparative studies of cannabis products and the earlier systems have been reviewed by Gough and Baker (1982; see pp. 319–321).

HPLC is the preferred procedure for quantitative work. Cannabis products are extracted into methanol/chloroform 4:1, conveniently accelerated by 25 min sonication. Individual cannabinoid constituents are then measured in an acidified reversed phase system, using an octadecylsilane-bonded-silica non-polar stationary phase and acetonitrile:methanol:(N/50 sulphuric acid) 9:8:7 as mobile phase (Baker *et al.*, 1980), with a UV detector set at 220 nm. Gough (1991c) illustrates a typical chromatogram

Table 6 Chromatographic retention data and mass fragments for principal cannabinoids

Cannabinoid	$R_f\backslash$T1	$R_f\backslash$T2	$R_f\backslash$T3	Colour	R_t/GC as is	R_t/GC as TMS	R_t/LC	rRF	EI mass fragments
CBD	0.63	0.65	0.05	orange-yellow	0.72	0.97	0.59	1.2	231,246,314,299,258
CBN	0.60	0.39	0.52	purple	1.25	1.03	0.89	3.1	295,238,310,119
Δ^9-THC	0.55	0.60	0.30	red	**1.00**	**1.00**	**1.00**	**1.0**	299,231,314,43,295
Δ^8-THC	—	—	—	—	—	—	1.05	0.9	221,314,248,261,193,236
CBV	0.50	0.47	—	purple	0.58	0.90	—	~3	—
THV	0.45	0.52	—	red	—	0.92	—	~1.0	—
CBG	0.41	0.39	—	orange	—	1.04	0.64	1.2	—
CBCy	—	—	—	—	—	0.97	1.10	1.8	—
CBCh	0.38	0.32	—	purple	—	—	1.40	2.7	—
CBDA	~0.2	<0.05	—	orange	—	—	0.68	1.9	—
THVA	~0.2	~0.05	—	reddish	—	—	1.09	~1.5	—
THCA	~0.2	~0.1	—	reddish	—	—	0.87	1.5	372,357,313 (Me-) 488,477,371 (TMS-)

Notes

- $R_f\backslash$T1, $R_f\backslash$T2 and $R_f\backslash$T3 are normalised R_f values in TLC systems: T1 = CHCl$_3$/MeCHCl$_2$ 15/10 and T2 = xylene/dioxane 19/1 (cf. text in §3.4.2); T3 = toluene on AgNO$_3$ treated silica plate (Moffat, 1985: p. 172).
- colour is developed after spraying TLC plate with Fast Blue BB solution.
- $R_t\backslash$GC is relative retention time, as is or as trimethylsily ethers, on OV-17 column; absolute R_t for Δ^9-THC around 9.6 min (cf. text in §3.4.3).
- $R_t\backslash$LC is relative retention time in LGC HPLC system described in §3.4.4; measured relative to Δ^9-THC; absolute R_t is around 15 min.
- rRF is relative response factor compared to UV absorbance @ 220 nm for Δ^9-THC.
- EI MS fragments from Webb (1991) with ion abundance from Moffat (1986: p. 425).
- for structures and full names of these cannabinoids, see Table 2.

revealing at least 8 peaks in an extract of Moroccan resin, the THC-acid running last. Typical retention data for principal cannabinoids are included in Table 6 but, as in all forensic chromatography, comparison should always be made with authentic specimens. Where the peaks detected do not correspond with available reference specimens of cannabinoids, they may be identified by high resolution mass spectrometry following preparative HPLC.

By combining the results of GC and HPLC quantitations it is possible to compile diagnostic 'libraries' comparing contents of THC-acid and THC with 'total' THC for an internationally representative range of cannabis specimens. Baker and co-workers have reported several such studies and Table 7 summarises their THCA/THC ratios with examples of herb from 12 countries of origin and resin from five (after Gough, 1991d).

5   NMR Spectroscopy

Dawson (1991) quotes ten papers which provide ^1H, and nine for ^{13}C, spectra of natural and synthetic cannabinoids, and which assess the ability to discriminate between such

Table 7   Cannabis herb and resin of different geographical origin: comparison of 'total' THC with THCA and THC

Country of origin	No. of specimens	Ratio THCA\THC	Cannabinoid Content % w/w		
			THCA	THC	Σ(THC)
Herb					
Colombia	1	2.9	5.6	1.9	6.8
India typeA	4	2.8	1.1	0.4	1.4
India typeB	1	2.2	8.0	3.6	10.6
Jamaica	14	2.3	2.3	1.0	3.0
Kenya	5	1.2	2.0	1.7	3.4
Nigeria	15	1.6	3.0	1.9	4.5
South Africa	10	1.8	2.2	1.2	3.1
Sri Lanka	1	0.4	0.3	0.7	1.0
Swaziland	1	0.7	1.6	2.2	3.6
Tanzania	2	1.9	2.8	1.5	4.0
Thailand	4	0.5	2.4	4.9	6.7
Zambia	2	3.6	1.7	6.2	7.7
Zimbabwe	3	2.3	2.3	1.0	3.0
Resin					
India	8	0.5	4.3	8.7	12.5
Lebanon	7	4.8	8.1	1.7	8.8
Morocco	2	6.1	6.7	1.1	7.0
Pakistan	8	0.8	2.7	3.6	6.0
Turkey	1	3.2	7.3	2.3	8.7

Notes: 'Total' THC is determined by GC, and 'actual' THC by HPLC, using systems described in §3.4. THCA as % w/w, is calculated from the difference {Σ(THC)-actual(THC)} and adjusted for THCA/THC mol.wt. ratio (i.e. × 1.14). Results are taken from reports by Baker, *et al.* (1980, 1982, 1983) and summarised by Gough (1991d).

substances. Of especial interest is the use of ^1H–^1H and ^{13}C–^{13}C homonuclear, and ^1H–^{13}C heteronuclear, correlation spectroscopy ('COSY') techniques to determine the stereochemistry of cannabinoid analogues (e.g. Offermann,1986).

6 Immunoassay Procedures

Tan and Marks (1991) discuss enzyme immunoassay (EIA) applications generally for the analysis of drugs of abuse; and they specifically refer to competitive EIA procedures for THC. Commercially available EMIT test kits rely on appropriate enzyme inhibition and comparator calibration, with results falling within a band of identification. One such kit is used in checking the presence of metabolites of cannabis products in body fluids and is helpful where enforcement agents suspect internal concealment of packages of liquid cannabis. A declining level of metabolites may indicate prior usage of cannabis; a steady state implies a potential seepage from some concealment; while increasing levels suggests the need for urgent clinical attention.

3.5 Examination of Smoking Residues

Light microscopy is not generally an appropriate technique for identifying cannabis residues which have resulted from a combustion process, or are significantly contaminated with other organic matter. However, as a non-destructive test microscopy may usefully complement an examination of incompletely carbonised plant material from a pipe bowl or from a loose rolled 'joint'. Where the amount of specimen is not too limited a preliminary indication may have been given by a Duquenois–Levine or Fast Blue BB colour test (cf. §3.3.3), but any positive result must always be confirmed by diagnostic, usually chromatographic, examination. Gough and Baker (1982, 1983) have reported detection limits for cannabinoids visualised with various chromogens.

In most instances, demonstrating the presence of THC and other cannabinoids in *two* TLC tests should furnish sufficient evidence that cannabis had been smoked, provided that reference specimens of relevant cannabinoids were applied to each plate. If the combustion of cannabis has been efficient, acid precursors should be absent but the presence of CBD in the pyrolysis residue is a valuable pointer to it having been the resin form of cannabis which had been smoked.

3.6 Chemotaxonomic Evidence of Age and Origin

1 Geographical Origin

Whether as an aid to criminal investigation of potentially related offences involving cannabis, or to provide data useful in international monitoring of the movement of cannabis products, it is helpful to assign a probable country of origin, and perhaps some estimate of the age of seized specimens. Early studies on the forensic significance of age and origin of cannabis products were described in 1972 in a Society of Economic Botany lecture at the University of Mississippi (de Faubert Maunder, 1976).

The relative amounts of cannabinoids may vary with local climate and annual variation therein, with genetic features of the seedstock, regional cultivation and harvesting traditions, and the storage history since preparation for export. Baker *et al.* (1980) summarised their accumulated data on cannabinoid content in relation to geographical origin (using TLC for comparison and HPLC for quantitation) and related these data to the physical and other chemical features of cannabis products illicitly imported into the UK from identifiable countries. They had studied many thousands of specimens from all over world and concluded that, for a sample of unknown provenance, a combination of careful visual inspection and comparative TLC analysis afforded an opinion of its probable geographical origin.

To underpin this confidence, a series of studies at LGC established that cannabis resin imports of inferred common origin exhibited a reasonably consistent pattern of cannabinoid distribution. Thus, coefficients of variation between different slabs in a given batch were in most instances no wider than intra-slab values, when measuring differences between interior and subsurface ratios of individual cannabinoids.

There may be quite striking variations in the THC content of fresh samples of herb or resin from some countries. Gough (1991e) has summarised a wealth of information reported by his co-workers, giving a range of THC content in fresh herb, resin and oils originating from 23, 9 and 7 countries respectively. Typical national ranges are illustrated in the barcharts of Figure 2.

Gough and co-workers noted some annular variation in THC content *within* particular countries of origin. Thus, compared with other parts of Africa, the THC content of Ghanaian herb seizures was appreciably lower in 1975 and 1976 but of much better quality in 1978; and Jamaican THC content was low in the 1970s but improved in the early 1980s. There is insufficient evidence to suggest whether these differences reflect changes in climate, in cultivation or in preparation of clandestine exports. The THC content of resins is usually significantly higher than that of herbal specimens but some North African resins are of lower quality than some good quality Asian herbal presentations (e.g. 'manicured Thai sticks').

King (1997) has commented on typical THC levels in hand-rolled cannabis cigarettes ('reefers') in Britain. In a survey by the UK Forensic Science Service in 1982, the mean content of resin was found to be 137 mg, or of herb 197 mg, roughly equivalent to 10 mg THC. However, a survey of drug clinics reported in 1995 suggested much higher levels: 350 mg resin or 620 mg herb, which is consistent with a THC content of 30 mg per cigarette.

2   Chemotaxonomic Studies

To validate the basis of these conclusions, a programme of chemotaxonomic studies was initiated by LGC in 1980. Cannabis plants were grown in uniform conditions in secure glasshouses in the south east of England, initially from seeds from plants of known national origin, and subsequently breeding successive generations from within their respective daughter seedstocks. Gough and co-workers (original papers Baker, *et al.*, 1982, 1983; Taylor *et al.*, 1985; later summarised by Gough, 1991f) found that the gross physical appearance (apart from being much greener than parent plants), and the chromatographic composition (determined by TLC, GC and HPLC), were for most of the UK-grown specimens, generally well correlated with their parent stock. However,

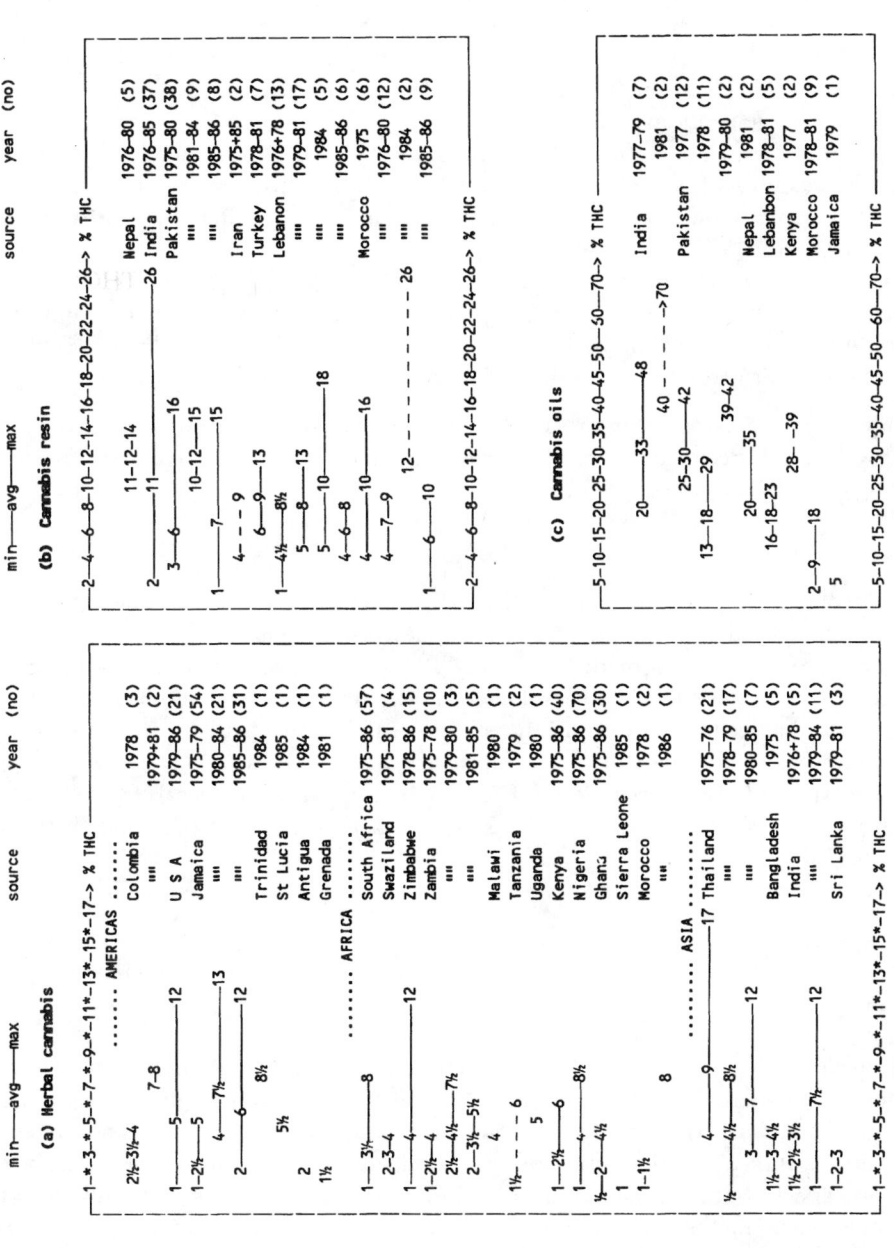

Figure 2 Global ranges of THC content (per cent.) found in fresh samples of Cannabis (a) herb, (b) resin and (c) oils. Data for similar years have been grouped and rounded to 1/2%; original references are listed in Gough (1991e)

some exceptions were reported:

1. all THC levels fluctuated annually, but in general the UK-grown plants maintained a higher ratio of THCA:THC compared with imported plants;
2. in UK plants from one of two types of Sri Lankan seedstock the CBD/THC ratio markedly declined in the first new generation but reverted to the parent ratio by 5th and 6th generations;
3. in those descendants of Sri Lankan plants in which CBD was now low or absent, CBCh was present, whereas where the CBD had been maintained at parent levels or higher, no CBCh was found: a similar reversal was noted by Fairbairn and Rowan (1975, 1977);
4. for Zambian stock, the relationship $\{THVA + THV\} > \{THCA + THC\}$ in the parent plant was reversed during six generations grown from it in England: this confirmed the observation by Boucher *et al.* (1977) of a temperate climate effect on the growth of plants from southern Africa.

The primarily chromatographic procedures that may be adopted in the direct comparison of two seized specimens where their common origin is in issue, have been reviewed by Gough and Baker (1982, 1983).

3   Age of Specimen

An estimate of the approximate age of cannabis specimens is based on the experience that, on storage, the THC and THV in cannabis products are slowly oxidised, respectively, to the more aromatic CBN and CBV. It may generally be assumed that the absence of CBN means that the sample is relatively fresh, i.e. less than 3 months have elapsed since harvesting. Unfortunately, this process is not quantitative, but for older specimens it is a crude indicator of age. Up to about three years storage can be inferred from the extent of conversion from THC to CBN, assuming a typical 'fresh' level of THC for corresponding cannabis products derived from the known or supposed region of origin.

4   QUALITY ASSURANCE

4.1   Compendial Standards

Apart from the Chinese Pharmacopoeia, there is a lack of recent pharmacopoeial monographs for herbal Cannabis and its official alcoholic presentations. From the first (1864) to the 5th (1914) editions of the British Pharmacopoeia there were monographs for 'Cannabis Indica' (cf. the definition quoted earlier, in §1.4) but the monograph for substance, and Extract and Tincture, were omitted in the next (1932) BP. Indication of the simple requirements for herbal appearance ('Characters'), content (ethanol extractable) and ash residue is given below.

The 3rd (1934) edition of the British Pharmaceutical Codex adopted a standard for 'Cannabis Indica', its Extract and its Tincture, based on that omitted from the BP 1932.

This, with small changes in the 4th (1949) edition, was the last official standard in the UK for 'Cannabis' and also formulary prescriptions for its alcoholic 'Extract' and 'Tincture'. The definition followed the BP 1914 but omitted mention of (Indian) origin and any reference to the separated resin. The monograph was dropped, without specific comment, in the next (1956) edition. The official preparation of Extract of Cannabis was by 90% ethanol percolation of powdered herb and subsequent evaporation to the ill-defined "consistence of a soft extract". Tincture of Cannabis was produced from a solution of 50 g of Extract in 1 litre of 90% ethanol.

There have been no monographs for *Cannabis* in any edition of the Pharmacopée Européenne and the International Pharmacopoeia; nor any recent monographs in the Japanese (since 1971), French or Italian Pharmacopoeias. Older versions appear in the Spanish and Portuguese Pharmacopoeias; and there was a monograph in the 2nd Indian (1966) edition but it was dropped from the 3rd (1985).

Since the first century AD, 'hemp' (Huo Ma Ren) has been included in traditional Chinese medicines. The latest English edition of the Chinese Pharmacopoeia is in 3 volumes: the third contains standards for preparations obtained from more than 500 herbs. Neither 'Cannabis' nor 'Hemp' appears in the index to the 1985 edition.

Turning to non-statutory national Herbal Pharmacopoeias, the latest British edition (B.H.P. 1996) contains no monograph for cannabis but there is a specification for 'Hemp fruit' in the Japanese Herbal Medicines Codex (JHMC).

4.2   Characters

The BP 1914 prescribed a detailed macroscopic description for the appearance of stem, lower and upper leaves, and flowers and fruit, together with presence of typical glands, hairs and cystoliths. This description was significantly extended in the BPC 1949, with references to African and American origin.

4.3   Identification – Herbal and Chemical

The BPC 1949 monograph for 'Cannabis' extended the macroscopic features given in the BP 1914 monograph for "*Cannabis Indica*". The 1949 text is still valid today –

"flattened dull green masses", more or less compacted by "adhesive resinous secretion"; compacted 'tops' of 3 to 30 cm, "comprising the upper part of the stem with ascending, longitudinally furrowed branches" and "numerous glandular trichomes";
leaves alternant palmate bracts, each bract "having two linear stipules", usually with two axilliary bracteoles, each subtending "a single pistillate flower or more or less developed fruit, occasionally containing an oily seed";
all covered with glandular trichomes and a heavy narcotic odour.

In addition, the BPC 1949 monograph provided specific details for light microscopy recognition; and also introduced an interfacial colour test, adding 15% HCl to part of a light petroleum extract of the herb (cf. §3.3). In the JHMC, 'Hemp fruit' relies on two colour tests: a red brown response with acetic anhydride/conc.sulphuric acid and a purple colour with Ninhydrin reagent.

4.4   Related Substances

Both the absolute amount, and the relative proportions, of congeneric precursor acids – notably CBDA, THCA and THVA – vary between products of different countries. The THCA/THC ratio is quite high (> 1%) in Mediterranean resins (Gough, 1991d) whereas this ratio is low (< 1%) in resins from the Indian sub-continent, which are rich in THC (cf. Table 7). As noted in §3.6, UK-grown plants had higher THCA/THC than overseas grown products. The absolute amount of THCA does not affect smoking quality given a relatively rapid decarboxylation, but is relevant to ingestion because THCA is inactive if taken orally.

Although rare as a congeneric substance in natural material, synthetically prepared delta[9]-THC may contain the delta[8] isomer, both in the thermodynamically more stable *trans* configuration. The USP monograph – see §4.8 – has a 2% limit test for the delta[8] isomer.

4.5   Adulteration – Active and Inactive

The BP 1914 set simple limit tests for minimum organic content – a minimum residue of 12.5% w/w following overnight extraction with 90% ethanol, filtered and dried at 100° – and maximum mineral residue – not more than 15% w/w ash. The BPC 1949 accepted a 10% minimum extractable organic residue but reduced the limit for ash (acid insoluble) to 5% and introduced a limit for foreign organic matter of not more than 2% and not more than 10% of fat stems (exceeding 3 mm diameter) and fruits. The JHMC requires bracts "to be absent" and sets an ash limit of 7%.

Adulteration, or frank substitution, in illicit supplies of cannabis was discussed in §2.5.2.

4.6   Assay Methods

On the assumption that delta[9]-*trans*-THC is the main active principle of cannabis, assay of natural products and preparations rests on assay of this constituent. The HPLC system favoured by LGC for forensic purposes (Baker *et al.*, 1980) was described in §3.4.4; the different USP procedure is referred to in §4.8 below.

4.7   Potency Standards, Reference Material

The BPC 1949, in its General Notices, recognised that the consistency of "soft extract" could not be defined and the inability (at that time) to standardise them; yet doses were recommended for 16 to 60 mg of Extract of Cannabis. The only oblique strength requirement for the Tincture of Cannabis was a 'weight per ml' range of 0.842–0.852 at 20°. The recommended dose range was 0.3–1.0 ml of Tincture. With so much variation in botanical sources, the only sure measure of potency is to determine the delta[9]-*trans*-THC content. A suitable Chemical Reference Material, including the USP Reference Standard (USPRS), for standardising the assay procedure needs conservation similar to that specified in §4.9 below.

4.8 Quality of Non-herbal Medicinal Preparations

The USP XXII, 6th Supplement, introduced a monograph for 'Dronabinol', containing a minimum of 95% of the (−)-delta^9-*trans* stereoisomer of THC; and a truncated monograph suitable for a Capsule preparation containing 90–110% thereof. The parent substance monograph was slightly elaborated in the following, 7th, Supplement of USP XXII.

The Identity relies on (i) comparison of spots with USPRS by TLC, using hexane/dichloromethane 1/1 mobile phase, and visualised with Fast Blue BB chromogen; and (ii) comparison with the retention time of the USPRS in the HPLC test used for assay.

In the Assay, the reversed phase HPLC system uses methanol/water/tetrahydrofuran 71/24/5 as mobile phase and UV detector set at 228 nm.

The Related Substances test limits the content of delta^8 isomer by including in the assay procedure a reference solution containing a USP RS for each isomer; this same solution can be used as a system suitability check on the ability to resolve and quantify both isomers.

4.9 Storage Desiderata

The USP requires that the Dronabinol USPRS, and the USPRS for its delta^8 isomer, should be packed in tight, light-resistant glass containers, with an inert atmosphere, and stored in 'a cool place'. This storage requirement is appropriate for other cannabinoid reference materials where there is risk of oxidation; the fully aromatic CBN and CBV can be expected to be more stable but use of light-resistant glass containers would still be preferable.

With bulk specimens, packaging and store-room ventilation conditions should be adjusted to ensure that herbal material does not become mouldy; and resin blocks should be checked for a propensity to self-heating, which accelerates the oxidative degradation of THC and THV. Particular care should be taken with storage of forensic exhibits, to minimise the loss of representative character in possible future re-examination of case material; this also helps to minimise potential loss in the recorded weight of retained exhibits of cannabis products.

In the event that uniformly rolled cigarettes, standardised for potency, should become generally available, then light-resistant packaging and air-tight jars would be advisable.

REFERENCES

Adams, R. and Baker, B.R. (1940). A method for the synthesis of a tetrahydrocannabinol which possesses marihuana activity. *J. Amer. Chem. Soc.*, **62**, 2405.

Anon (1996). Statistics of drugs seizures, and offenders dealt with, 1995. *Home Off. Stat. Bull.*, **25**, 23 Nov.

Asuni, T. and Pela, A.O. (1986). Drug abuse in Africa. *Bull. Narcot.*, **38**(1), 55–64.

Baker, P.B. and Fowler, R. (1978). Analytical aspects of the chemistry of cannabis. *Analyt. Proc.*, 347–349.

Baker, P.B., Gough, T.A. and Taylor, B.J. (1980). Illicitly imported cannabis products: some physical and chemical features indicative of their origin. *Bull. Narcot.*, **32**(2), 31–40.

Baker, P.B., Gough, T.A. and Taylor, B.J. (1982). Physical and chemical features of cannabis plants grown in England and Northern Ireland from seeds of known origin (Part I). *Bull. Narcot.*, **34**(1), 27–36.

Baker, P.B., Gough, T.A. and Taylor, B.J. (1983). *idem* (Part II): second generation studies. *Bull. Narcot.*, **35**(1), 51–62.

Baker, P.B. and Phillips, G.F. (1983). The forensic analysis of drugs of abuse. *Analyst*, **108**: pp. 779–780 (field-tests); pp. 782–785 (cannabis).

Butler, W.P. (1962). Duquenois-Levine test for Marihuana. *J. Ass. Off. Anal. Chem.*, **45**, 597.

Ballantyne, A., Lippiett, P. and Park, J. (1976). Herbal cigarettes for kicks. *Brit. Med. J*, 1539–40.

Boucher, F., Paris, M and Cosson, L. (1977). Evidence of two chemical types in South African *Cannabis sativa. Phytochem.*, 16, 1445.

Clarke, E.G.C. and Robinson, A.E. (1970). When is 'cannabis' resin? *Med. Sci. and Law*, **10**, 139.

Corrigan, D. (1979). Identification of drugs of abuse in the republic of Ireland during 1968–78. *Bull. Narcot.*, **31**(2), 57–60.

Dawson, B.A. (1991). In *The Analysis of Drugs of Abuse*, Gough, T.A., ed., Wiley, Chichester UK, pp. 297–298.

de Faubert Maunder, M.J. (1969). A simple and specific test for cannabis. *J. Ass. Publ. Analyt.*, **7**(3), 24–30.

de Faubert Maunder, M.J. (1974). An improved procedure for field testing of cannabis. *Bull. Narcot.*, **26**(4), 19–26.

dc Faubert Maunder, M.J. (1976). The forensic significance of the age and origin of cannabis. *Med. Sci. Law*, **16**(2),78–90.

de Faubert Maunder, M.J. and Phillips, G.F. (1976). Improvements in or relating to the detection of drugs. *Br. Pat.* 1 426 177.

Dutt, M.C. and Lee, T.K. (1991). In *The Analysis of Drugs of Abuse*, Gough, T.A., ed., Wiley, Chichester, UK, pp. 408–409.

Eskes, D., Verwey, A.M.A. and Witte, A.H. (1973). TLC and GC analysis of hashish samples. *Bull. Narcot.*, **25**(1), 41.

Fairbairn, J.W. (1972). The trichomes and glands of *Cannabis sativa* L. *Bull. Narcot.*, **24**(4), 29.

Fairbairn, J.W. and Rowan, M.G. (1977). Cannabinoid patterns in seedlings of *C. sativa* and their use in determination of chemical race. *J. Pharm. Pharmacol.*, **29**, 491–4.

Fowler, R., Gilhholey, R.A. and Baker, P.B. (1979). The TLC of cannabinoids. *J. Chromat.*, **171**, 509–511.

Gough, T.A. (1991). In *The Analysis of Drugs of Abuse*, Gough, T.A., ed., Wiley, Chichester UK: a.(tlc) pp. 537–8; b.(gc) pp. 543 et seq.; c.(lc) pp. 548–51; d.(THCA:THC ratios) pp. 550–1; e.(country/origin) pp. 544–7; f.(UK-grown) pp. 552–7; g.(seizures) p. 481.

Gough, T.A. and Baker, P.B. (1982). Identification of major drugs of abuse using chromatography. *J. Chrom. Sci.*, **20**, 289–329.

Gough, T.A. and Baker, P.B. (1983). Identification of major drugs of abuse using chromatography: an update. *J. Chrom. Sci.*, **21**, 145–153.

Hartnoll, R. *et al.* (1989). A multi-city study of drug misuse in Europe. *Bull. Narcot.*, **41**(1), 11.

Harvey, D.J. and Paton, W.D.M. (1975). Use of TMS and homologous derivatives for the separation and characterisation of mono and dihydroxycannabinoids by GC-MS. *J. Chromat.*, **109**, 73–80.

Jackson, B.P. and Snowdon, D.W. (1968). *Powdered Vegetable Drugs*, Churchill, London: p. 62.

Khan, M.Z. and Unnitham, N.P. (1979). Association of socio-economic factors with drug abuse among Indian students in an Indian town. *Bull. Narcot.*, **31**(2), 61–69.

King, L.A. (1997). Drug content of powders and other illicit preparations in the UK. *Forens. Sci. Int.*, **85**,135–147.

Lawrence, A.H. and Elias, L. (1984). The application of air sampling and ion mobility spectrometry to narcotics detection. *Bull. Narcot.*, **37**(1), 3–16.

Lawrence, A.H. (1980). Revenue Canada report, PCS-6156-LSS.

LGC (1972). Report of the Government Chemist for 1971, HMSO, London: pp. 18–22.

LGC (1974). Report of the Government Chemist for 1973, HMSO, London: p. 63 and Plate 3.

LGC (1976). Report of the Government Chemist for 1975, HMSO, London: pp. 62–63.

LGC (1978). Report of the Government Chemist for 1977, HMSO, London: p. 62.

Lopez-Alvarez, M.J. *et al.* (1989). Extent and patterns of drug use by students at a Spanish university. *Bull. Narcot.*, **41**(2), 117–119.

Mechoulam, R., McCallum, N.K. and Burstein, S. (1976). Recent advances in the chemistry and biochemistry of cannabis. *Chem. Rev.*, **76**, 75.

Mitosinka, G.T., Thornton, G.I.I. and Hayes, T.L. (1972). Examination of cystolithic hairs of cannabis and other plants by SEM. *J. Forens. Sci. Soc.*, **12**, 521.

Moffat, A.C. (1986). Clarke's isolation and identification of drugs, 2nd. edn. Pharm. Press, London: p. 197.

Navaratnam, V. and Foong, K. (1989). Development and application of a system for monitoring drug abuse: Malaysian experience. *Bull. Narcot.*, **41**(2), 52–65.

Offermann, W. *et al.* (1986). Stereostructure of a synthetic cannabinoid. *Tetrahedron*, **42**, 2215–2219.

O'Neil, P.J., Phillips, G.F. and Gough, T.A. (1985). The detection and characterisation of controlled drugs imported into the UK. *Bull. Narcot.*, **37**(1): see pp. 29–31.

Ortiz, A., Romano, M. and Soriano, A. (1989). Development of an information reporting system on illicit drug use in Mexico. *Bull. Narcot.*, **41**(1), 41–52.

Pela, A.O. (1989). Recent trends in drug use and abuse in Nigeria. *Bull. Narcot.*, **41**(2), 103–107.

Phillips, G.F. (1991). In *The Analysis of Drugs of Abuse*, Gough, T.A., ed., Wiley, Chichester UK, p. 491.

Phillips, G.F. (1973). The legal description of cannabis and related substances. *Med. Sci. Law*, **13**, 141.

Phillips, G.F. (1974). In Methodicum Chimicum, Korte, F. ed., Academic Press, London, 1B, 903.

Ramsay, M. and Percy, A. (1996). Drugs misuse declared: results of the 1994 British Crime Survey, *Home Office Res. Study* 151.

Rodriguez, M.E. and Anglin, M.D. (1987). The epidemiology of illicit drug use in Spain. *Bull. Narcot.*, **39**(2), 67–69.

Tamura, M. (1989). Japan: stimulant epidemics past and present. *Bull. Narcot.*, **41**(2), 83–93.

Tan, K. and Marks, V. (1991). In The analysis of drugs of abuse, Gough, T.A., ed., Wiley, Chichester UK, p. 318.

Taylor, B.J., Neal, J.D. and Gough, T.A. (1985). Physical and chemical features of cannabis plants grown from seeds of known origin (Part III): third and fourth generation studies. *Bull. Narcot.*, **37**(4), 75–81.

United Nations (1974). Report of Consultant Group, UN Div. of Narcotic Drugs, MNAR/2/74.

United Nations (1983). Multilingual dictionary of narcotic drugs and psychotropic substances under international control, UN, New York: pp. 28–31.

United States Customs Service (1980). Customs detector dog program, Dept. of Treasury, Washington, DC.

Valentine, M.D. (1995). Δ^9-THC acetate from acetylation of cannabis oil. *Sci. and Justice*, **36**(3), 195–197.

Webb, K.S. (1991). In The analysis of drugs of abuse, Gough, T.A., ed., Wiley, Chichester UK, pp. 190–194.

5. NON-MEDICINAL USES OF *CANNABIS SATIVA*

DAVID T. BROWN

School of Pharmacy, University of Portsmouth, Portsmouth, Hampshire, UK

INTRODUCTION

The plant *Cannabis sativa* has been providing man with a range of his most basic needs for centuries (Conrad, 1994a). We know that hemp – the fibrous extract of *C. sativa*, was used for clothing in ancient Egypt, at least as early as 1,200 years BC and the use of the plant as a source of rope is well documented in many cultures down the centuries (see Chapter 1, this volume, for a full account). The seeds from the plant have been subjected to various treatments to provide food and the fibre has also been used from early times as a major paper making material; indeed, early editions of the Gothenburg and King James Bibles were published on such paper and much later, the first two drafts of the American Declaration of Independence. The new president of the United States, George Washington was to be found exhorting his head gardener to: "Make the most of the Indian hemp seed...and sow it everywhere" (Washington, 1794).

These peaceful uses were not the only ones however. From the 17th century onwards, the British Royal Navy – at the time the most powerful navy in the world – relied heavily on hemp for ropes, rigging and caulking. In the mid-1800s, a typical 44-gun man of war might inventory some 60 tons of hemp rope, rigging and anchor cable, often impregnated with tar to improve the already excellent resistance to rotting encouraged by constant exposure to sea water; not to mention the hemp-derived oakum, forced between the planks to make her watertight. The sails were made of 'canvas' a derivation from an Arabic word for hemp. Soldiers' and sailors' clothing and their battle flags were likely to be made of hemp material also. The original 'Levi's' jeans were made from recycled hemp sail cloth and in World War 2, hemp was widely culti-vated in the US and Germany performing many vital functions, from fire hoses to parachute webbing.

CULTIVATION

This chapter is not intended to be an agricultural or horticultural manual for those interested in growing the plant for legitimate commercial gain. Chapter 2 provides a fairly detailed historical account of cultivation and subsequent processing of *C. sativa* and the interested reader is referred to this text and its accompanying references (see also: Conrad, 1994b; Judt, 1995). For those requiring further information, references are provided in the bibliography to organisations which may provide additional advice in this area.

The reader is referred to Chapter 2 for a general description of the conditions necessary for growth of strains of *C. sativa* which are low in psychoactive cannabinoids.

It can be seen that the plant is not fastidious; indeed, the cannabis plant requires little care and attention yet under moderately intensive conditions, provides one of the longest and most versatile cellulose fibres of any plant. It has been shown that under sustainable growth conditions, on an acre for acre basis, hemp produces four times as much fibre pulp as wood (Dewey and Merill, 1916) and the yield is 200% better than cotton – a crop which requires intensive pesticide treatment to succeed.

The plant is a rapid grower, attaining a height of 10–12 feet in 12–14 weeks. Under normal conditions, the seed yield is from 12–15 bushels per acre with an average of 16–18. Twenty percent of the plant is fibre and depending on strain, growth conditions and processing, the fibre yield can be two to three times that of flax or cotton, in a range of 400–2500 pounds per acre with a mean of approximately 1000 pounds (Dewey, 1916). One acre of hemp can produce 10 dry tonnes of animal feed, including stalk and foliage; this yield may be increased with intensive fertilisation.

It has been argued by environmentalists that hemp and other products from *C. sativa* can be produced with a favourable environmental impact: for example, hemp requires minimal herbicides and pesticides and the plant has a very long tap root which discourages soil erosion.

As far as illicit growth of the plant is concerned, then the methods used to cultivate cannabis are as ingenious as they are devious. Clandestine, domestic cultivation operations are unearthed (and summarily dismantled by the authorities) with monotonous regularity. Sophisticated systems have been discovered only after many months of undetected operation without, apparently, knowledge of close neighbours. Because high-intensity light is a requirement, ambitious growers have resorted to the theft of sources of more or less the correct specifications from places as bizarre as 100 ft up a floodlighting pylon at a soccer stadium and the external illuminations of historic buildings. Techniques for the cultivation of herbal cannabis are described elsewhere (Conrad, 1994c; Rosenthal, 1984).

This chapter does not describe the medicinal or recreational uses of Cannabis (see Chapters 1, 6–8 for this), but seeks rather to provide an overview of the plant as a contemporary source of a range of useful materials and provides some insight into the social, geo-political and economical influences which shape our attitudes to use of *C. sativa* in this way.

CURRENT USES

As early as 1938, the American periodical *Popular Mechanics* published an article entitled 'New Billion-Dollar Crop' in which it was claimed that 25,000 products could be manufactured from hemp. This may have been an imaginative over-estimate then; but in reality, the diversity of applications is stunning enough (see Table 1).

A brief description of the major uses is given below. Textiles and fine writing papers can be made from the long bast fibres. The hurds (the tougher core of the stem) can be ground down into a powder which can be used for a variety of products, including fibreboard, panelling, plywood, cavity wall insulation, packaging and even babies' nappies.

Table 1 Modern applications for *C. sativa* (hemp)

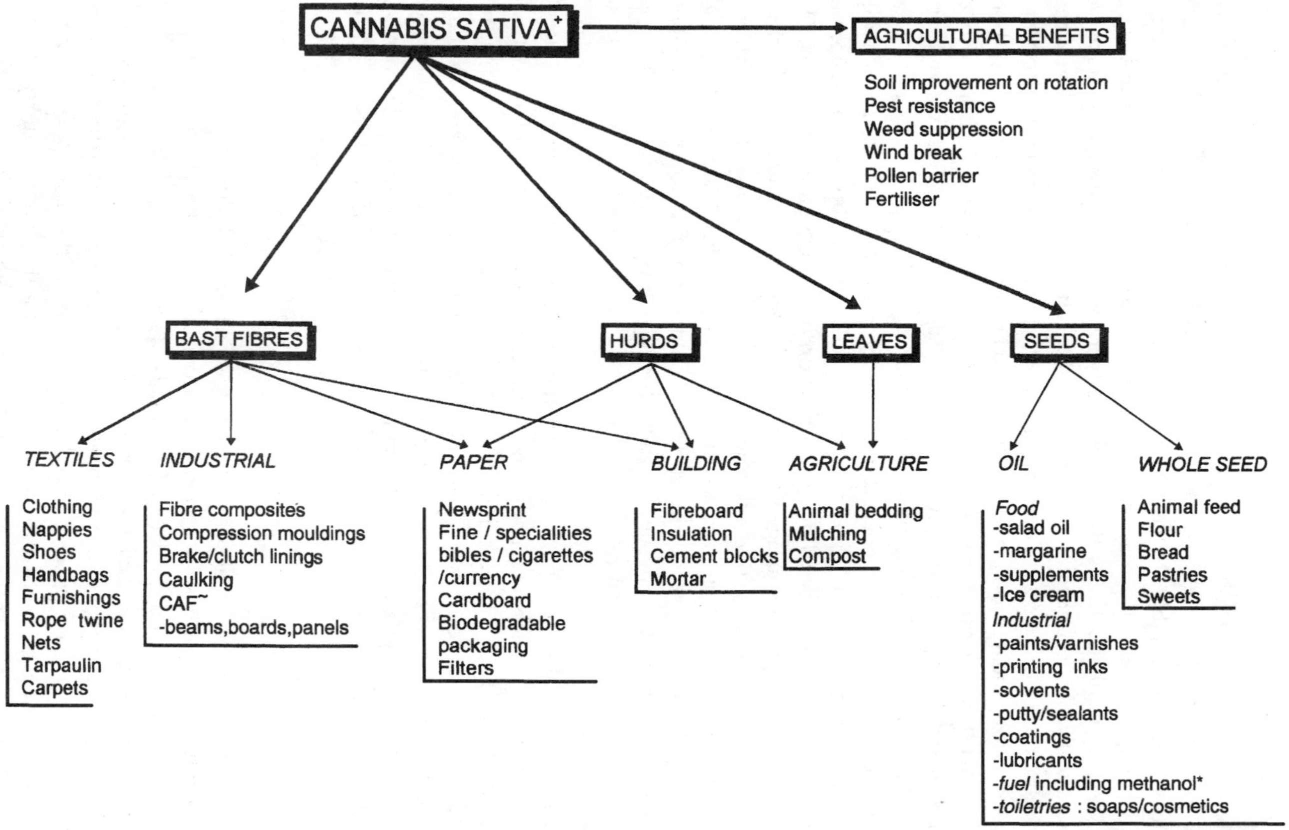

CANNABIS SATIVA⁺

AGRICULTURAL BENEFITS

Soil improvement on rotation
Pest resistance
Weed suppression
Wind break
Pollen barrier
Fertiliser

BAST FIBRES

TEXTILES

Clothing
Nappies
Shoes
Handbags
Furnishings
Rope twine
Nets
Tarpaulin
Carpets

INDUSTRIAL

Fibre composites
Compression mouldings
Brake/clutch linings
Caulking
CAF~
-beams,boards,panels

PAPER

Newsprint
Fine / specialities
bibles / cigarettes
/currency
Cardboard
Biodegradable
packaging
Filters

HURDS

BUILDING

Fibreboard
Insulation
Cement blocks
Mortar

LEAVES

AGRICULTURE

Animal bedding
Mulching
Compost

SEEDS

OIL

Food
-salad oil
margarine
-supplements
-ice cream
Industrial
-paints/varnishes
-printing inks
-solvents
-putty/sealants
-coatings
-lubricants
-*fuel* including methanol*
-*toiletries* : soaps/cosmetics

WHOLE SEED

Animal feed
Flour
Bread
Pastries
Sweets

* In addition to the above uses, hemp may also be used as a boiler fuel and to produce energy through pyrolysis (Osburn, 1989).

~ Compressed agricultural fibre.

⁺ See Chapter 7, this volume, for a summary of the medicinal applications of cannabis.

Textiles

Before the industrial revolution, hemp was a major European crop for textile manu-facture. However, the invention of machinery capable of extracting and processing the fibre from cotton (notably the cotton gin) saw a rapid expansion of cotton at the expense of hemp, where heavy manual work was still required to extract the long bast fibres. It was not until the 1930s that machines were built which could extract hemp fibre economically. An economic process for manufacturing paper was developed at about the same time (see below). These methods arrived at a time when the cotton and associated chemical and petrochemical industries were extremely powerful and some have argued that it was for this reason that legislation – ostensibly anti-drugs in nature, in the form of the Marihuana Tax Act of 1937 – was passed which effectively prohi-bited hemp farming in the US (Conrad, 1994d; Herer, 1991).

Hemp made a brief re-emergence during the Second World War, particularly in America and Germany, when imported fibre was in short supply. In 1943, some 250,000 acres were turned over to hemp production in the US alone (Hopkins, 1951); even school children were encouraged to plant their own hemp patch to help the war effort, the youngsters being proud members of a local '4-H club'. The populace of Germany was exhorted to a similar extent (Reich Nutritional Institute 1943) and in 1943, some 24,700 acres was under hemp (*hanf*) cultivation. But when the war ended, hemp farming permits were cancelled in the US and hemp production all but ceased. A total ban meant that all legal production in the US had ceased by 1957; cultivation is still illegal to this day.

Permits for hemp cultivation have been issued in a number of EC member states and the plant is grown on a much larger scale in countries such as China and Hungary where cultivation has never been banned. This is largely in recognition of the fact that hemp textiles offer a wide range of uses from everyday, sports and protective clothing to carpeting and home furnishings. It is claimed that hemp fibres are stronger, more lustrous and absorbent and are more mildew-resistant than cotton fibres. They may also be blended with cotton, to give fabrics and clothing with the advantages of both raw materials.

Paper

Hemp fibres are among the longest and strongest of natural cellulose fibres. They make excellent quality paper for books, magazines and stationery; the shorter fibres make newsprint, tissue paper and packaging materials. Hemp has a low lignin content, requiring less aggressive chemical bleaching. The paper produced is resistant to age-related yellowing which occurs with wood-derived paper and hemp paper is amenable to recycling. Production of paper derived from hemp in the European Community has been spearheaded in Germany and in France; in the latter country, Kimberly Clark (a US Fortune 500 company) operates a mill producing paper for bibles and cigarettes.

Rope

As mentioned above, hemp fibre makes a strong, rot-resistant rope. Indeed, up until 1937 and the Marihuana Tax Act, it is estimated that 70–90% of all rope used in the

US was made from hemp. In countries where cultivation is not restrained, this traditional use is still widespread; but in most Western countries, modern synthetics and other plant sources such as jute and sisal are used.

Oil as a Foodstuff

Oil has been expressed from hemp seeds and used for cooking by many cultures. More recently, analysis of commercial samples of cold-pressed hemp seed oil has revealed high levels of polyunsaturated essential fatty acids: alpha linolenic acid – omega 3 – (19–25% of total oil volume); linoleic acid – omega 6 – (51–62%) and gamma linoleic acid (1.6%). These compounds are termed polyunsaturated fatty acids and it is a widely held view that their consumption, in place of saturated fats may have wide ranging health implications in, for example, the prevention of the development of coronary heart disease associated with consumption of the latter. Imported seeds have to be sterilised by law in many countries to prevent propagation. The best oil appears to be obtained from seeds exposed to a sterilisation process which does not involve excessive heat. At least one commercial supply is available, which is described as 'green, delicious, but perishable – but which can be kept in the freezer for one year without spoiling'.

Hemp seed has been used as a foodstuff, both for animals and man, for centuries. Most commercial bird seed mixes contain hemp seeds. After oil extraction, the crushed seed is high in protein (approximately 25%), making it a potentially valuable agricultural animal feed. The seeds are also high in trace elements and vitamin A. After oil extraction, the crushed seed may be ground to flour and used to make bread, cakes, pastas and biscuits. The seeds can be mixed with other ingredients to make a wide range of foods, from soup to sweets, non-dairy cheese, butter and ice cream.

Hemp as a Fuel

Traditional uses of hemp as an energy source are described in Chapter 1. Clearly burning any unwanted material can provide heating for domestic use in some countries in the absence of, or as a substitute for wood. One modern, but not altogether unexpected twist to this was the recent observation by the State Energy and Minerals Minister for New South Wales, Australia that large quantities of confiscated cannabis could be burnt in the state's electricity generators, on the grounds that it was cheaper than coal and gave about the same yield in energy.

Hemp seeds contain approximately 40% by weight of a combustible oil which was traditionally used as a lantern fuel in a number of countries.

It has been suggested that the whole hemp plant might be commercially viable as a source of 'biomass' – a term used to describe all biologically-produced matter – from which to produce fuels such as charcoal and methanol by a process known as pyrolysis (Usborn, 1989).

Paints and Resins

Hemp seed oil is a good drying agent and until the late 1930s, linseed and hemp oils were the basis for the majority of all resins, paints, shellacs and varnishes. Cheaper petroleum-based alternatives took over at this time and it seems unlikely that hemp oil will re-emerge in this application, except perhaps in 'designer' ranges of fashionable products.

Cosmetics

Oil extracted from hemp seeds has been used as a basis for lip balm, salves, soaps and massage oil. There appear to be no particular difficulties associated with processing the oil for use in this way.

Plastics

Research has shown that hemp hurds may be processed to give cellophane packaging material in much the same way as other rich sources of cellulose. They can also be blended with recycled plastics to provide a compound for injection mouldings. The seed oil may also be converted into a plastic resin. These uses are largely experimental and are unlikely to be widespread while petrochemical derivatives remain widely available and relatively cheap. One advantage of hemp derived products is that it might be possible to develop materials which are 100% biodegradable.

HEMP CULTIVATION AROUND THE WORLD

There is a growing awareness of the economic potential for hemp products, principally textiles and clothing. The main market place is the US, with a turnover above $50 million followed by Germany (approximately DM20 million); other countries with a stake are Spain, Austria, Switzerland, Australia, Canada, France and Norway. Egypt, India, Portugal, Thailand, the Ukraine and most former Soviet Bloc countries, including Hungary, Poland, Yugoslavia and the Czech republic also produce hemp. Major hemp-growing countries today include China, France, Holland, Hungary and Russia. Although banned for commercial cultivation, Australia, Canada and Germany allow selected farms to grow hemp for research purposes whilst currently restricting general, local production. In the EC, hemp farmers are allowed to grow strains certified to contain 0.3% THC (tetrahydrocannabinol) or less. Hemp seed is licensed for export; France is a major supplier of seeds for these low-THC varieties.

As far as the European Community Agricultural Policy is concerned, hemp subsidies are available for both hemp fibre and the seed. Some individual countries are discussed below.

Australia

In 1991, Australia began growing hemp for paper; but with the exception of carefully monitored research projects, hemp cultivation is banned. There is a significant industrial lobby for legalisation, noticeably in Tasmania.

Canada

Like the US, Canada was a major hemp farming nation until the late 1930s when this was prohibited. Legislation to permit the widespread cultivation of industrial hemp has now been introduced. Processing and manufacturing plants and retail outlets for imported, hemp-derived goods already exist.

China

China has been growing hemp, unabated for the last 6000 years and has a vast internal market for hemp products. It is currently the biggest exporter of hemp paper and textiles in the world.

France

France granted its first licence for hemp production in 1960. In 1994, it produced in excess of 10,000 tons of industrial hemp. Experimental, lightweight cement ('Isochanvre') has been produced by combining hemp fibre with lime. Some 300 houses have been constructed of this material and insulated with a hemp fibre at a price which is claimed to be comparable with conventional building materials.

Germany

Hemp cultivation was banned in Germany in 1982; however experimental crops have been produced recently under licence. The fibre has been used to manufacture rope, textiles, cigarette papers and the hurds have been incorporated into composite board and insulation material. Production processes based on imported hemp are at advance stages of development and in 1994, sales for hemp products exceeded DM20 million. Interest among German farmers in the reintroduction of local hemp farming is increasing.

Holland

Local production for paper is being evaluated by the Dutch government and cultivation is increasing, in parallel with the development of processing equipment.

Hungary

Hungary was a major cultivator and supplier of hemp products to the former Soviet Union and still exports widely. Products include upholstery, heat insulation, interior decoration and packaging materials. Hemp-based textiles are widely exported to many countries, including the United States.

Poland

Poland currently grows hemp for fabric and manufactures composite boards for the construction industry.

Romania

Currently the largest source of commercial hemp in Europe, Romania has devoted greater than 40,000 acres to hemp cultivation. Much of the crop is exported to Hungary for processing (see above) prior to re-export to Western markets.

Russia

The former Soviet Union was the largest cultivator and exporter of hemp in the world. Indeed, the Vavilov Scientific research Institute in St Petersburg still holds the largest hemp seed collection in the world, including many rare species not found in other seed banks. Today, Russia consumes most of its own hemp products including rope and CAF (compressed agricultural fibre) board.

Spain

This country exports hemp pulp for paper (notably for cigarette papers and bibles) and produces rope and textiles for domestic consumption.

Ukraine

This state has large quantities of hemp growing wild and harvest of this resource is under way. Farming permits have also been issued.

United Kingdom

The early 1990s saw new agricultural initiatives in Europe, to investigate sustainable alternative crops to alleviate the food mountains being produced on the farms of Europe. As a result of a Home Office lobby by some UK farmers, the first licences for growing hemp with a low THC content were granted in 1992/1993, under the ruling that the crop was being grown for 'special purposes' or 'in the public interest'. The number of farms cultivating hemp is still small, but paper and textile markets are being developed, with government aid aimed at developing new markets for natural fibres, including hemp and flax. In June 1996, some 6,000 hectares were cultivated (compared to 1,482,000 which were designated as set-aside to preserve the status quo). This represented a small start indeed; the majority of the raw hemp processed in the UK is still imported, mainly from China and Hungary.

United States

Although the cultivation of hemp has been actively discouraged in the past, there is a growing demand for textiles and other products made from more environmentally friendly, 'biosustainable' crops than cotton or wood. With the exception of the flowers, leaves, hashish resin and fertile seed, it is legal to import raw hemp products for processing. The number of companies manufacturing hemp products from imported hemp fibre has mushroomed in the last five years so that at present, there are over 200 companies offering a wide range of hemp products in a multi-million dollar business. In this environment, legislative attitudes to local hemp cultivation may change.

SUMMARY

The annual world paper requirement has risen from 14 million tons in 1913 to 250 million tons in the 1990s. Many argue that this cannot be sustained, even with a significant increase in paper recycling, and that alternative sources of fibre must be found. In 1994, October 26th, the London Financial Times reported that "... fibre hemp ... is making a comeback in Europe and the US as an ecologically friendly raw material for clothing and paper".

It is true to say that at present, most hemp markets are in their infancy. Even with the advent of facilitating legislation, hemp is a crop which is unfamiliar to most farmers and even in developed countries, farm machinery will have to be adapted, or designed from scratch, in order to tend and harvest large-scale plantings. One could liken hemp to an ageing but accomplished and versatile actor, waiting in the wings to give a vintage performance; but at the same time, ready to take to the stage with a few, varied and perhaps, surprising new roles, some of which may be written with this particular performer in mind.

REFERENCES

Conrad C. (1994a) The many histories of hemp. In: *Hemp – Lifeline to the Future.* Creative Xpressions Publications, Novato, California, USA, pp. 5–21.

Conrad, C. (1994b) Overview of hemp farming techniques. *Ibid.,* pp. 167–175.

Conrad, C. (1994c) The agriculture of herbal cannabis. *Ibid.,* pp. 175–183.

Conrad, C. (1994d) A bright promise assassinated. *Ibid.,* pp. 37–54.

Dewey, L.H. (1916) Hemp. In: *US Department of Agriculture Year Book.* United States Agriculture Department, Washington DC, USA.

Dewey, L.H. and Merill, J.L. (1916) Hemp hurds as paper making material. Bulletin 404. United States Department of Agriculture, US Government Printing Office, Washington DC, USA.

Herer, J. (1991) *Hemp and the Marijuana Conspiracy: The Emperor Wears No Clothes.* HEMP Publishing, Van Nuys, California, USA.

Hopkins, J.F. (1951) *History of the Hemp Industry in Kentucky.* University of Kentucky Press, Lexington, Kentucky.

Judt, M. (1995) Hemp (*Cannabis sativa* L) – salvation for the earth and for the paper makers. *Agro Food Ind. Hi-tech.,* **6**(4), 35–37.

Osburn, L. (1989) Energy farming in America. Access Unlimited, Frazier Park, California, USA.

Reich Nutritional Institute (1943) Die Lustige Hanffibel. Reich Nutritional Institute, Berlin, Germany.

Rosenthal, E. (1984) *Marijuana Growers Handbook; Indoor/Greenhouse Edition.* Quick American Publishing Company, San Francisco, California, USA.

Washington, G. (1794) Note to Mount Vernon Gardener. In: *Writing of George Washington,* **33**, 270. US Library of Congress, Washington DC, USA.

BIBLIOGRAPHY

Roulac, J.W. (Ed.) Industrial Hemp; Hemptech, Oja California, 1995.

Hemp Product Producers around the World

BACH (Business Alliance for Commerce in Hemp), P.O. Box 71903, Los Angeles, California 90071-0093, USA.

Hemp Industries Association, P.O. Box 9068, Chandler heights, Arizona AZ 85227, USA.

Hemp Union, 24 Alnaby Road, Hull HU1 2PA, England.
Hemptech – Industrial Hemp Information Specialists, P.O. Box 820, Oja, California
    93024-0820, USA.
International Hemp Association, Postbus 75007, 1070 AA Amsterdam, The Netherlands.
International Kenaf Association, P.O. Box 7, Ladonia, Texas TX 75449, USA.
Isochanvre, Le Verger, F-72260, Rene, France.

The Internet

A single search engine revealed the presence of over 200 websites devoted to the
cultivation of hemp and hemp products. This illustrates that the interest in this area is
truly international and the reader is encouraged to consult this resource for further
information.

6. ADVANCES IN CANNABINOID RECEPTOR PHARMACOLOGY

ROGER G. PERTWEE

Department of Biomedical Sciences, Institute of Medical Sciences, University of Aberdeen, Foresterhill, Aberdeen AB25 2ZD, UK

1 INTRODUCTION

This review summarizes current knowledge about cannabinoid receptors and their ligands. It concentrates particularly on the distribution pattern of these receptors, their effector systems, the pharmacological and physiological effects they may mediate, the pharmacology, distribution, formation, release and fate of the endogenous cannabinoid receptor agonist, anandamide, and the state of play regarding the development of selective cannabinoid receptor agonists and antagonists and of inhibitors of anandamide synthesis and metabolism. The possible physiological significance of anandamide is also discussed as is the existence of other endogenous cannabinoid receptor agonists. The review begins with a brief account of the molecular biology of cannabinoid receptors. However, the emphasis throughout is on the pharmacology of these receptors.

2 CLONING OF CANNABINOID RECEPTORS

2.1 Cannabinoid CB_1 Receptors

The gene encoding the cannabinoid CB_1 receptor was first cloned by Matsuda *et al.* (1990) from a rat cerebral cortex cDNA library using an oligonucleotide probe based on the sequence that encodes part of the bovine substance K receptor. Rat CB_1 receptor cDNA proved to be 5.7 kilobases in length with a predicted 473 amino acid product. Subsequently, human CB_1 cDNA was isolated from human brain stem and testis cDNA libraries (Gérard *et al.*, 1990, 1991) and mouse CB_1 cDNA from a C57BL/6 mouse brain cDNA library (Chakrabarti *et al.*, 1995). These have predicted protein products of 472 and 473 amino acids respectively. The human CB_1 gene has been genetically mapped to the q14–q15 region of chromosome 6 (Caenazzo *et al.*, 1991; Hoehe *et al.*, 1991) and the mouse CB_1 gene to proximal chromosome 4, a location at which other homologues of human 6q genes occur (Stubbs *et al.*, 1996; Onaivi *et al.*, 1996a). The genomic location of the rat CB_1 receptor has yet to be determined.

There is a high level of similarity between both the nucleotide sequences and the predicted amino acid sequences of human, rat and mouse CB_1 receptors. More specifically, nucleotide sequences of human and rat are 90% identical, those of human and mouse 91% identical and those of rat and mouse 96% identical (Gérard *et al.*, 1990, 1991; Chakrabarti *et al.*, 1995). The predicted amino acid sequences of human and rat

CB_1 receptors show 97.3% homology, differing in only 13 residues (Gérard, 1990, 1991). Those of the human and mouse CB_1 receptors show 97% homology and those of the rat and mouse CB_1 receptors 99% homology (Chakrabarti et al., 1995).

The cannabinoid CB_1 receptor has a predicted architecture that is characteristic for all known G-protein coupled receptors (Onaivi et al., 1996a). Thus there are seven hydrophobic stretches of 20–25 amino acids that are believed to form transmembrane alpha helices and to be separated by alternating extracellular and intracellular peptide loops. There is also a C-terminal intracellular peptide domain that is presumably coupled to a G-protein complex and an N-terminal extracellular domain. Bramblett et al. (1995) have constructed a 3-dimensional model of the human CB_1 receptor that shows the likely orientation of its transmembrane helices. According to this model, the degree of exposure to membrane lipids is least for helix 3, slightly greater for helices 2 and 7 and considerably greater for helices 1 and 4. The N-terminal domain which is unusually long (116 amino acids) and the C-terminal domain both contain potential N-linked glycosylation sites. The human CB_1 receptor has three of these at the N-terminal and one at the C-terminal end, the rat receptor three at the N-terminal and two at the C-terminal end and the mouse receptor two at the N-terminal and two at the C-terminal end (Onaivi et al., 1996a). The predicted amino acid sequences of human, rat and mouse CB_1 receptors are markedly different from those of all other known G-protein-coupled receptors (Matsuda and Bonner, 1995).

2.2   Subtypes of Cannabinoid CB_1 Receptors

A spliced variant of CB_1 cDNA has been isolated from a human lung cDNA library (Shire et al., 1995; Rinaldi-Carmona et al., 1996a). This, the $CB_{1(a)}$ receptor, is a truncated and modified form of the CB_1 receptor that results from the excision of a 167 base pair intron within the sequence encoding the N-terminal tail of the receptor. The extracellular N-terminal region of the $CB_{1(a)}$ receptor is shorter than that of the CB_1 receptor by 61 amino acids (55 vs 116 amino acids). Moreover, the predicted first 28 amino acids in the N-terminal region of the $CB_{1(a)}$ receptor are totally different from those in the same region of the CB_1 receptor, containing a greater proportion of hydrophobic residues. As a result, the $CB_{1(a)}$ receptor lacks two of the three potential N-linked glycosylation sites present in the N-terminal region of the human CB_1 receptor.

Onaivi et al., (1996b) have detected three distinct CB_1 mRNAs in C57BL/6 mouse brain, but only one CB_1 cDNA. Brain tissues from two other mouse strains (ICR and DBA/2) were found to contain just a single CB_1 mRNA. Yamaguchi et al. (1996) have cloned two receptors with high homology to the human CB_1 receptor from the Puffer Fish (Fugu rubripes) by screening a Fugu genomic library in a bacteriophage using a[32]P labelled oligonucleotide probe under low stringency conditions. The deduced amino acid sequences of these two Puffer Fish receptors are 66.2% identical. Both Puffer Fish receptors are predicted to contain 7 lengths of 20 to 25 hydrophobic amino acids separated by hydrophilic regions, suggesting that like other cannabinoid receptors, they are coupled to G-proteins. One of the receptors has 469 amino acids and shows 72.2% homology to the human CB_1 receptor (93.2% within the transmembrane domains) and 34.9% homology to the human CB_2 receptor (Section 2.3). The other Puffer fish receptor has 471 amino acids and shows 59.0% homology to the human CB_1 receptor (81.5% within the transmembrane domains) and 31.7% homology to the human CB_2 receptor.

2.3 Cannabinoid CB_2 Receptors

The cannabinoid CB_2 receptor was first cloned by Munro *et al.* (1993) who obtained the cDNA encoding this receptor from a human promyelocytic leukaemic line, HL60, by the use of degenerate primers and polymerase chain reaction. Like the cannabinoid CB_1 receptor, the CB_2 receptor is a member of the superfamily of G-protein coupled receptors. However, it is smaller than the CB_1 receptor, having only 360 amino acids. Also, there is only a 44% homology between the predicted amino acid sequences of the human CB_1 and CB_2 receptors, this value rising to 68% if the transmembrane regions only are compared. More recently, Shire *et al.* (1996) cloned the mouse CB_2 receptor. This they did using radiolabelled human CB_2 cDNA to screen a murine spleen cDNA library. Human and mouse CB_2 receptors show far less homology than human and mouse CB_1 receptors. In particular, the deduced amino acid sequence of the mouse CB_2 receptor differs from that of the human CB_2 receptor in 60 residues (82% identity) and the mouse CB_2 receptor is 13 residues shorter than the human CB_2 receptor (at the C-terminus). Although human-mouse differences in amino acid content are to be found throughout the CB_2 receptor, most are in the extra-membrane regions especially at the N-terminus. Human and mouse CB_2 receptors have fewer potential N-linked glycosylation sites than human and mouse CB_1 receptors with just one in the N-terminal region and none at the C-terminus (Shire *et al.*, 1996; Onaivi *et al.*, 1996a). The genomic location(s) of the human and mouse CB_2 receptors have still to be reported.

3 LIGANDS FOR CANNABINOID RECEPTORS

3.1 Cannabinoid Receptor Agonists

These can be classified into four chemical groups: classical, nonclassical, eicosanoid and aminoalkylindole (Martin *et al.*, 1995; Pertwee, 1993, 1995, 1997). The structures of important members of each of these groups are shown in Figures 1–7, 9 and 11).

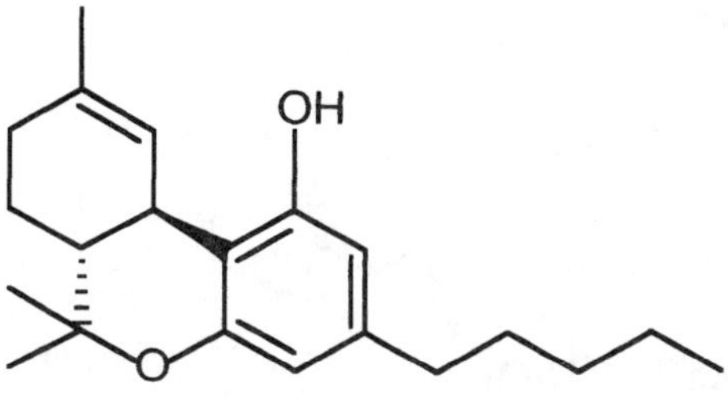

Figure 1 Structure of the classical cannabinoid receptor agonist, (−)-delta-9-tetrahydrocannabinol. This is the main psychotropic constituent of cannabis

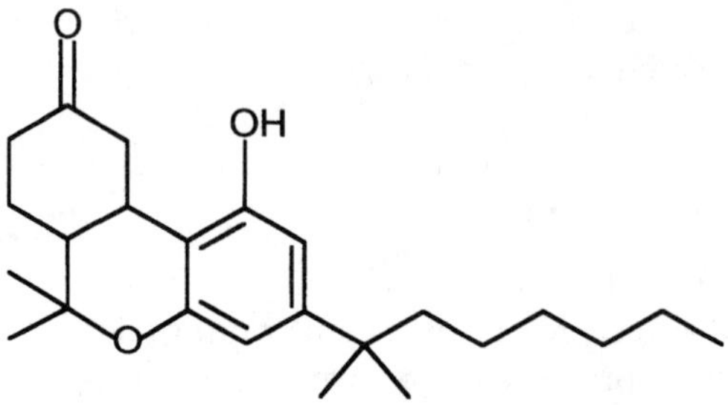

Figure 2 Structure of nabilone (Cesamet), a synthetic analogue of delta-9-tetrahydrocannabinol

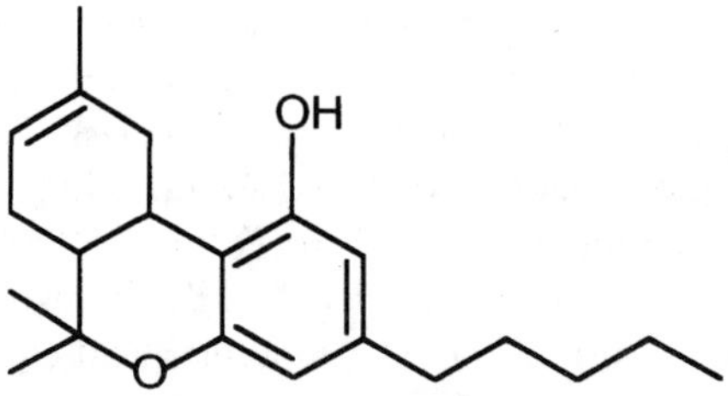

Delta-8-tetrahydrocannabinol

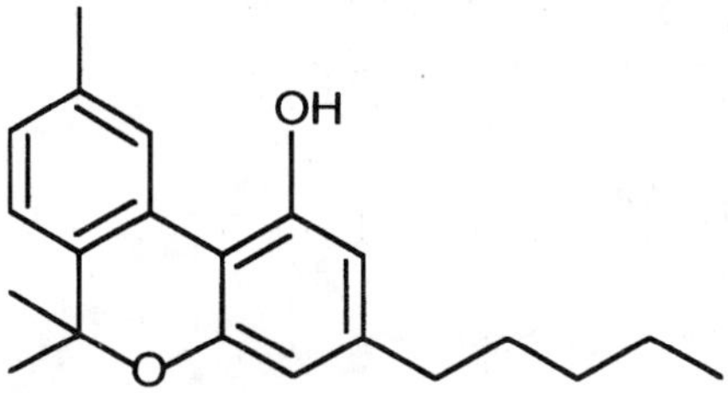

Cannabinol

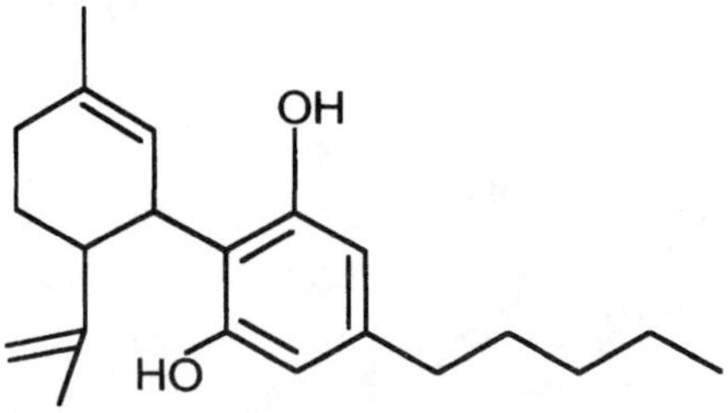

Cannabidiol

Figure 3 Structures of the cannabis constituents, delta-8-tetrahydrocannabinol, cannabinol and cannabidiol

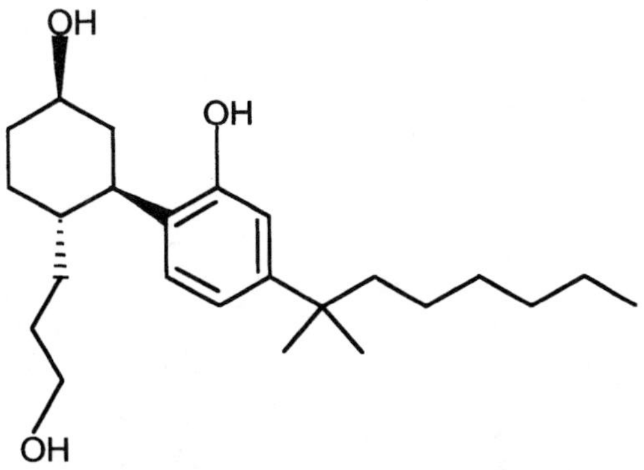

Figure 4 Structure of the nonclassical cannabinoid receptor agonist, CP 55, 940. The less active (+)-enantiomer of this compound is CP 56, 667

Many cannabinoid receptor agonists contain chiral centres and exhibit marked stereoselectivity in both binding assays and functional tests (Martin *et al.*, 1995; Pertwee, 1993, 1995, 1997). Among the classical and nonclassical cannabinoids it is the (−)-enantiomers that have the greater activity. However, for the aminoalkylindoles, the (+)-enantiomers are the more active. Certain eicosanoid cannabinoid receptor agonists also show significant stereoselectivity (Abadji *et al.*, 1994).

The classical group of cannabinoid receptor agonists are dibenzopyran derivatives. Of these, delta-9-tetrahydrocannabinol (delta-9-THC), the main psychotropic constituent of cannabis, and nabilone, a synthetic analogue of delta-9-THC, are of particular interest (Figures 1 and 2). This is because they are currently the only two cannabinoid receptor agonists that it is permissible to use as therapeutic agents. Nabilone (Cesamet) is licensed in the UK for use against nausea and vomiting provoked by anti-cancer drugs and delta-9-THC (Marinol) can be given clinically in the USA both as an anti-emetic and to combat weight loss in AIDS patients by stimulating appetite (Hollister, 1986; Pertwee, 1995; Beal *et al.*, 1995). Delta-9-THC is also widely used as a cannabinoid receptor agonist in pharmacological experiments.

CP 55,940 is one of many nonclassical cannabinoid receptor agonists to have been synthesized by Pfizer (Figure 4). These compounds are bicyclic or tricyclic analogues of delta-9-THC that lack a pyran ring. In its tritiated form, CP 55,940 is widely used as a probe for cannabinoid receptors. Indeed, it was binding assays performed with [³H]CP 55,940 that first demonstrated the presence of specific high-affinity cannabinoid binding sites in the brain (Devane *et al.*, 1988), a crucial step in the discovery of functional cannabinoid receptors. More recently, certain classical cannabinoids have also been labelled with tritium for use as cannabinoid receptor probes. These are [³H]dimethyl-heptyl analogues of 11-hydroxy-delta-9-THC and 11-hydroxy-hexahydrocannabinol (Devane *et al.*, 1992a; Thomas *et al.*, 1992).

The prototypic member of the eicosanoid group of cannabinoid receptor agonists is arachidonoylethanolamide (anandamide) (Figure 5). This is an endogenous cannabinoid receptor agonist, initially found in pig brain (Devane *et al.*, 1992b) and subsequently in

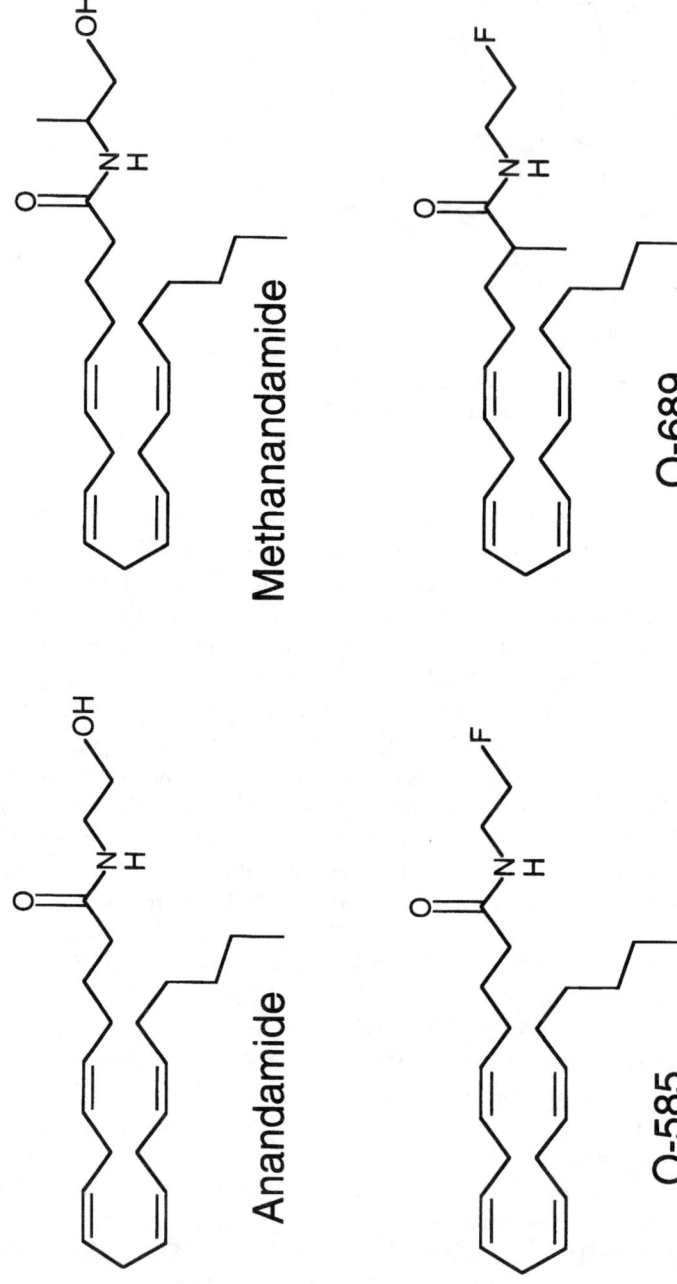

Figure 5 Structure of the endogenous cannabinoid receptor agonist, anandamide, and of three of its synthetic analogues that have greater metabolic stability and show selectivity for cannabinoid CB_1 receptors (see text and Table 1 for further details)

several other tissues (Section 7.1). Additional eicosanoid cannabinoid receptor agonists have been detected in pig or rat brain (Hanuš *et al.*, 1993; Pertwee *et al.*, 1994; Mechoulam *et al.*, 1995; Sugiura *et al.*, 1995). These are 2-arachidonoyl glycerol, which was first found in canine small intestine (Mechoulam *et al.*, 1995), homo-γ-linolenoylethanolamide and docosatetraenoylethanolamide (Figure 6). Experiments with rat brain membranes have shown 2-arachidonoyl glycerol to bind far less readily than anandamide to CB_1 receptors ($K_i = 4.8\,\mu M$ and 52 nM respectively) (Devane *et al.*, 1992b; Mechoulam *et al.*, 1995). It is also much less potent than anandamide as an inhibitor of electrically-evoked contractions of the mouse isolated vas deferens (Mechoulam *et al.*, 1995). However, these results may at least in part reflect a greater susceptibility of 2-arachidonoyl glycerol than anandamide to enzymic hydrolysis by these preparations. K_i values for 2-arachidonoyl glycerol and anandamide determined in binding assays with COS cells are much closer: 472 and 252 nM respectively in cells transfected with CB_1 receptors and 1400 and 581 nM respectively in CB_2 receptor transfected cells (Mechoulam *et al.*, 1995). Homo-γ-linolenoylethanolamide, docosatetraenoylethanolamide and anandamide have been reported by Hanuš *et al.* (1993) to have similar affinities for CB_1 receptors in rat brain membranes ($K_i = 53.4$, 34.4 and 52 nM respectively) and the potency of anandamide in the mouse vas deferens (52.7 nM) is only about twice that of the other two fatty acid amides (Pertwee *et al.*, 1994). On the basis of molecular modelling studies, Thomas *et al.* (1996) have concluded that eicosanoid and classical cannabinoids are pharmacophorically similar in that it is possible to superimpose anandamide on the delta-9-THC molecule such that the oxygen of the arachidonoyl carboxyamide lies over the pyran oxygen, the hydroxyl group of the arachidonoyl ethanol over the phenolic hydroxyl

Homo-γ-linolenoylethanolamide

7,10,13,16-Docosatetraenoylethanolamide

2-Arachidonoyl glycerol

Figure 6   Structures of three endogenous cannabinoid receptor agonists

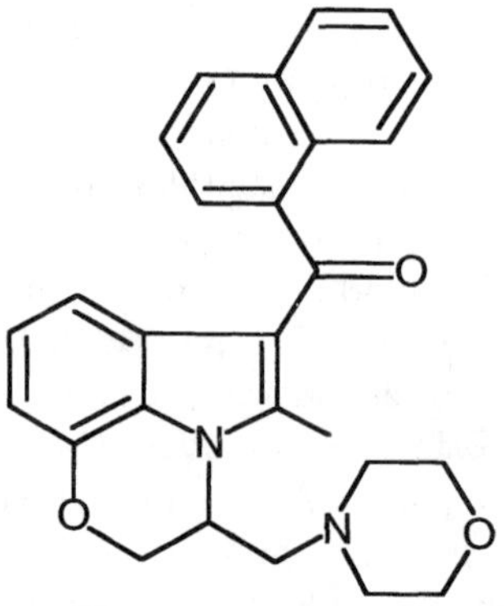

Figure 7 Structure of the aminoalkylindole cannabinoid receptor agonist, WIN 55,212. (+)-WIN 55,212 (WIN 55,212-2) is more potent than (−)-WIN 55,212 (WIN 55,212-3)

group, the five terminal arachidonoyl carbons over the hydrophobic pentyl side chain and the arachidonoyl polyolefin loop over the tricyclic ring system. Because anandamide is susceptible to enzymic hydrolysis (Deutsch and Chin, 1993; Koutek et al., 1994; Hillard et al., 1995b), in vitro assays of this agent are often carried out in the presence of an amidase inhibitor such as phenylmethylsulfonyl fluoride (Abadji et al., 1994; Childers et al., 1994; Pinto et al., 1994; Adams et al., 1995; Felder et al., 1995; Hillard et al., 1995a; Pertwee et al., 1995a; Song and Bonner, 1996; Petitet et al., 1996). The finding that anandamide is the substrate of an endogenous amidase has stimulated the development of several analogues that are less susceptible to enzymic hydrolysis. Among these are (R)-(+)-arachidonoyl-1'-hydroxy-2'-propylamide (methanandamide) and 2-methylarachid-onoyl-(2'-fluoroethyl)amide (O-689) (Abadji et al., 1994; Adams et al., 1995) (Figure 5).

Aminoalkyindoles with cannabimimetic properties were developed by Sterling Winthrop (see Martin et al., 1995). One of these, WIN 55,212-2 (Figure 7), is often used experimentally as a cannabinoid receptor agonist and, in its tritiated form, has also been used as a cannabinoid receptor probe (Jansen et al., 1992; Kuster et al., 1993). The amino-alkyindoles are quite different in structure from classical, nonclassical and eicosanoid cannabinoids and, indeed, their mode of binding to cannabinoid CB_1 receptors also seems to differ from that of other types of cannabinoid receptor ligand. Thus, Song and Bonner (1996) have shown that when lysine is replaced by alanine at position 192 of the CB_1 receptor, the ability of a classical cannabinoid (HU-210), a nonclassical cannabinoid (CP 55,940) and an eicosanoid cannabinoid (anandamide) to interact with this receptor is markedly reduced or abolished whilst that of WIN 55,212-2 remains unaffected. Further evidence that aminoalkylindoles differ from other cannabinoids in the way in which they interact with cannabinoid receptors has been obtained by Petitet et al. (1996) for CB_1 receptors and by Shire et al. (1996) for CB_2 receptors. Although WIN 55,212-2 may differ from other types of cannabinoids in its mode of attachment to the recognition sites of cannabinoid receptors, there seems to be considerable overlap in the space occupied at these sites by all known types of ligand for these receptors. Thus, [³H]WIN 55,212-2 is readily displaced from CB_1 and CB_2 receptors by classical, nonclassical and eicosanoid cannabinoids and [^{3}H]CP 55,940 is readily displaced from such receptors by WIN 55,212-2 (see Pertwee, 1997).

3.2 Cannabinoid Receptor Antagonists

3.2.1 Cannabinoid CB$_1$ Receptor Antagonists

Two compounds have been reported to behave as competitive, surmountable CB$_1$ receptor antagonists (Figure 8). One of these is LY320135 which has 16.5 times greater affinity for CB$_1$ than CB$_2$ receptors (Table 1). The pharmacology of this compound has yet to be reported in detail. The other compound is SR141716A (Rinaldi-Carmona et al., 1994). This potently displaces [^{3}H]CP 55,940 from specific binding sites, binds at least 57 times more readily to CB$_1$ than CB$_2$ receptors and lacks significant affinity for a wide range of noncannabinoid receptors (Rinaldi-Carmona et al., 1994; Showalter et al., 1996; Table 1). It is effective as an antagonist both in vivo and in vitro (Section 6) and is widely used as an experimental tool. K_d values of SR141716A for antagonism of WIN 55,212-2, CP 55,940 and delta-9-THC in the mouse isolated vas deferens are 2.4, 0.64 and 2.66 nM respectively (Pertwee et al., 1995c).

There are several reports that SR141716A produces effects that are opposite in direction to those produced by cannabinoid receptor agonists (cf. Sections 5 and 6). More specifically, when administered alone, SR141716A has been found to produce hyperkinesia in mice (Compton et al., 1996), provoke signs of increased arousal in rats (Santucci et al., 1996), improve social short-term memory in rats and mice (Terranova et al., 1996), augment cyclic AMP production in cells transfected with cannabinoid receptors (Felder et al., 1995), increase the amplitude of electrically-evoked contractions of isolated tissue preparations (Coutts et al., 1995; Coutts and Pertwee, 1996; Pertwee and Fernando, 1996; Pertwee et al., 1996b) and enhance electrically-evoked neurotransmitter release in rat hippocampal slices (acetylcholine), the myenteric plexus of guinea-pig small intestine (acetylcholine) and guinea-pig retinal discs (noradrenaline and dopamine) (Coutts and Pertwee, 1996, 1997; Gifford and Ashby, 1996; Schlicker et al., 1996). These effects may be an indication that an endogenous cannabinoid receptor agonist is being released to produce cannabimimetic tone that is susceptible to reversal by SR141716A. Alternatively, cannabinoid receptors may exist in two interchangeable states, the one precoupled to and the other uncoupled from the effector system. It could then be that SR141716A shows activity by itself because it is an inverse agonist rather than a pure antagonist, binding preferentially to the receptors in the uncoupled state and so shifting the equilibrium away from the receptors in the precoupled state.

3.2.2 Other Cannabinoid Receptor Antagonists

Other compounds that have been reported to produce a surmountable attenuation of certain cannabinoid-induced effects are WIN 56,098, 6-bromopravadoline (WIN 54,461), 6-iodopravadoline (AM630) and 6'-cyanohex-2'-yne-delta-8-THC (O-823) (Figure 8). Of these, the least potent antagonist is WIN 56,098 with a K_d of 1.85 µM for antagonism of delta-9-THC-induced inhibition of electrically evoked contractions of the mouse isolated vas deferens (Pacheco et al., 1991). WIN 54,461 shows greater potency

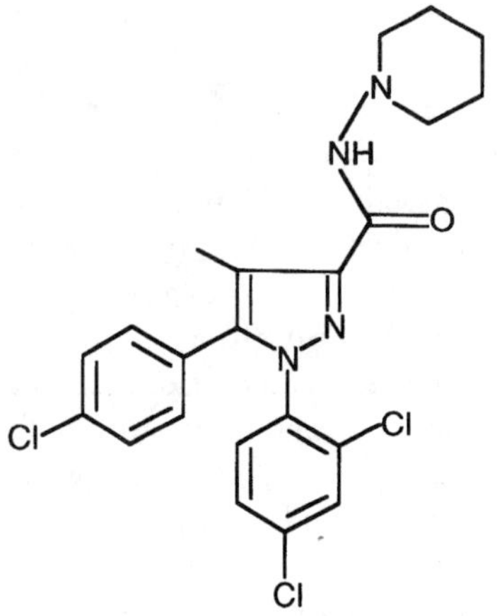

LY320135

SR141716A

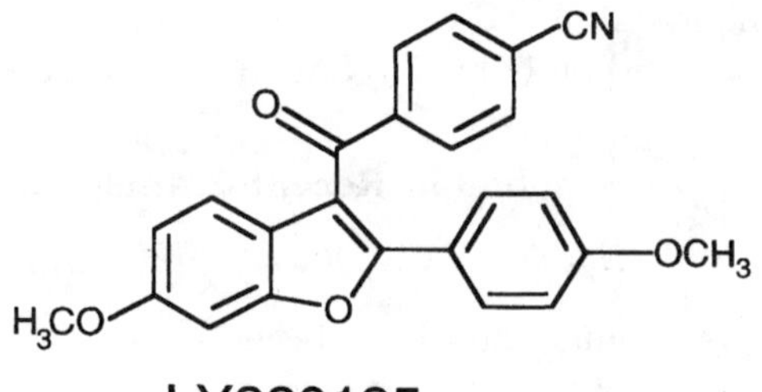

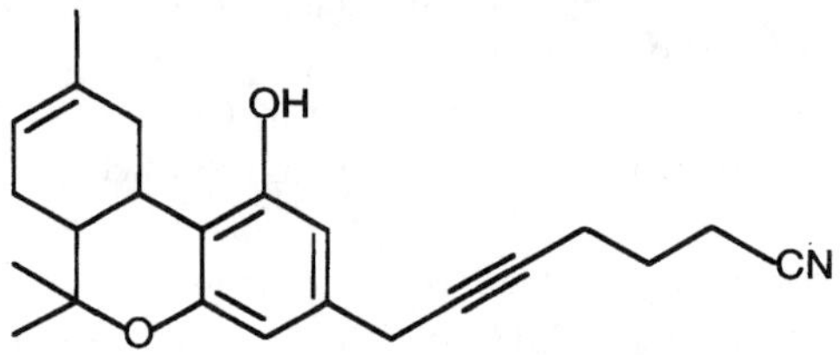

O-823

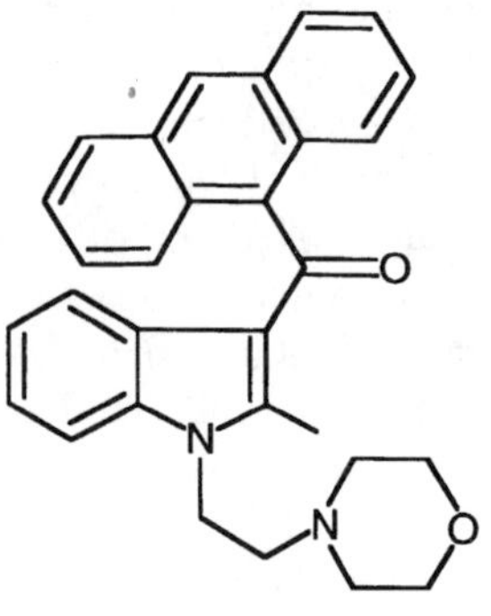

WIN 56,098 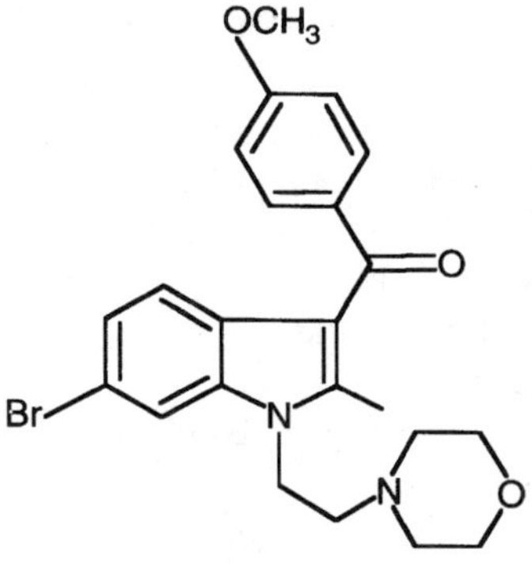   WIN 54,461 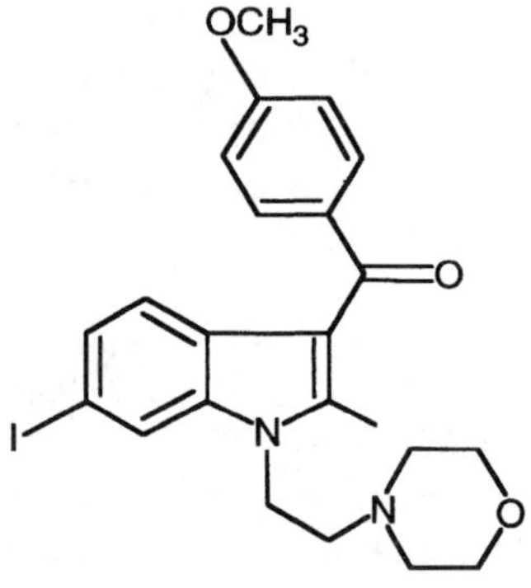   AM630

Figure 8 Structures of compounds which behave as cannabinoid receptor antagonists or partial agonists (see text and Table 1 for further details)

as an antagonist in this assay system, with K_d values against WIN 55,212-2 and delta-9-THC of 50 and 316 nM respectively (Casiano *et al.*, 1990; Eissenstat *et al.*, 1995). AM630 is also more potent than WIN 56,098 as an antagonist, mouse vas deferens experiments with WIN 55,212-2, CP 55,940 and delta-9-THC yielding K_d values of 36.5, 17.3 and 14 nM, respectively (Pertwee *et al.*, 1995b). However, in the myenteric plexus-longitudinal muscle preparation of guinea-pig small intestine, AM630 has no detectable

Table 1 The abilities of certain ligands to bind to CB_1 and CB_2 receptors

Compound	CB_1K_i values (nM)	CB_2K_i values (nM)	Affinity ratio (CB_1/CB_2)	Affinity ratio (CB_2/CB_1)	Reference
Delta-9-THC	41	36	0.88	1.14	Showalter et al. (1996)
	53	75.3	1.42	0.7	Felder et al. (1995)
HU-210	0.73	0.22	0.3	3.3	Showalter et al. (1996)
	0.06	0.52	8.67	0.12	Felder et al. (1995)
CP 55, 940	0.58	0.69	1.2	0.84	Showalter et al. (1996)
	3.7	2.55	0.69	1.45	Felder et al. (1995)
WIN 55, 212-2	1.89	0.28	0.15	6.75	Showalter et al. (1996)
	62.3	3.3	0.05	18.9	Felder et al. (1995)
Anandamide	89	371	4.2	0.24	Showalter et al. (1996)
	543	1940	3.57	0.28	Felder et al. (1995)
LY320135*	203	3340	16.5	0.06	Fahey et al. (1995)
O-689	5.7	132	23.2	0.04	Showalter et al. (1996)
O-585	8.6	324	37.7	0.03	Showalter et al. (1996)
Methanandamide	20	815	40.75	0.03	Khanolkar et al. (1996)
SR 141716A*	12.3	702	57	0.02	Showalter et al. (1996)
	11.8	973	82.5	0.01	Felder et al. (1995)
Cannabinol	308	96.3	0.31	3.2	Showalter et al. (1996)
	1130	301	0.27	3.75	Felder et al. (1995)
1-deoxy-delta-8-THC-DMH	23	2.9	0.13	7.9	Huffman et al. (1996)
JWH-O15	383	13.8	0.04	27.8	Showalter et al. (1996)
JWH-O51	1.2	0.032	0.03	37.5	Huffman et al. (1996)
L-759,633	15,850	20	0.001	793	Gareau et al. (1996)
L-759,656	>20,000	19.4	<0.001	>1000	Gareau et al. (1996)
Compound 9	2043	14	0.007	146	Gallant et al. (1996)

*LY320135 and SR141716A are antagonists (see text). Some of the other listed compounds have been reported to behave as full or partial agonists at CB_1 (Pertwee, 1993, 1995, 1997; Martin et al., 1995) and/or CB_2 receptors (Pertwee, 1997; Felder et al., 1995; Bayewitch et al., 1995, 1996; Bouaboula et al., 1996; Slipetz et al., 1995). Values of K_i (dissociation constant) were determined in competitive binding assays with [^{3}H]CP 55,940. See Figures 1, 3–5 and 7–11 for the chemical structures of the compounds listed.

antagonist action, behaving instead as a weak CB_1 receptor agonist (Pertwee *et al.*, 1996b). O-823 too has mixed agonist-antagonist properties (Pertwee *et al.*, 1996a), results from experiments with the mouse isolated vas deferens and myenteric plexus-longitudinal muscle preparation of guinea-pig small intestine suggesting that it behaves as a potent partial agonist when cannabinoid receptor reserve is high but as a potent antagonist when receptor reserve is low ($K_d = 0.3$ nM for antagonism of CP55,940). The *in vivo* pharmacology of O-823 and AM630 remains to be explored. However, there is already evidence that WIN 54,461 does not show antagonist properties *in vivo* (Eissenstat *et al.*, 1995). Important advances announced during the proof stage of this book have been the development of a selective and potent CB_2 receptor antagonist, SR144528 (Barth *et al.*, 1977; Rinaldi-Carmona *et al.*, 1988), and the discovery that methyl arachidonyl fluorophosphonate (Section 7.5) is an insurmountable cannabinoid receptor antagonist (Fernando and Pertwee, 1997).

3.3   Cannabinoid Receptor Agonists with Selectivity for CB_1 or CB_2 Receptors

Many established cannabinoids exhibit little difference in their affinities for CB_1 and CB_2 receptors (Table 1). These include delta-9-THC, CP 55,940 and anandamide. However, there are several recently developed compounds that do show significant selectivity for CB_1 or CB_2 receptors (Table 1). Apart from the antagonists, SR141716A and LY320135 (Section 3.2.1), compounds with greater affinity for CB_1 than CB_2 receptors include three synthetic analogues of anandamide: methanandamide, O-585 and O-689 (Figure 5). All these compounds are agonists. Compounds with significantly greater affinity for CB_2 than CB_1 receptors include JWH-015, JWH-051 and the Merck Frosst compounds shown in Figure 9 (L-759,633 and L-759,656) and Figure 10. WIN 55,212-2 also exhibits modest selectivity for cannabinoid CB_2 receptors. Although there are reports that JWH-015 and JWH-051 behave as CB_1 receptor agonists *in vivo* or *in vitro* (Huffman *et al.*, 1996; Griffin *et al.*, 1997), their activity in an established bioassay for CB_2 receptor agonists has still to be reported. Also still to be announced are the pharmacological properties of the Merck Frosst compounds at both CB_1 and CB_2 receptors. Whilst CP 55,940 and WIN 55,212-2 are undoubtedly CB_1, $CB_{1(a)}$ and CB_2 receptor agonists (Rinaldi-Carmona *et al.*, 1996a; Pertwee, 1997), there is uncertainty as to whether delta-9-THC and anandamide can activate CB_2 receptors (Sections 5.1 and 6.2.2) although none that these ligands can serve as agonists for CB_1 or $CB_{1(a)}$ receptors (Sections 5 and 6 and Rinaldi-Carmona *et al.*, 1996a).

Even though potent, selective CB_1 and CB_2 receptor ligands have been developed, most binding data come from experiments that have been performed with radiolabelled probes having similar affinities for CB_1 and CB_2 receptors ([^{3}H]CP 55,940, [^{3}H]WIN 55,212-2 and the [^{3}H]dimethylheptyl analogue of 11-hydroxy-hexahydrocannabinol) (Table 1; Devane *et al.*, 1992a; Bayewitch *et al.*, 1995). Some binding experiments have also been performed with the [^{3}H]dimethylheptyl analogue of 11-hydroxy-delta-9-THC (see Pertwee, 1997). The relative affinity of this probe for CB_1 and CB_2 receptors has yet to be reported. It is worth noting, therefore, that its delta-8-THC analogue, HU-210, binds more or less equally well to CB_1 and CB_2 receptors (Table 1). The CB_1-selective ligand, [^{3}H]SR 141716A, is now available, but relatively few binding experiments have

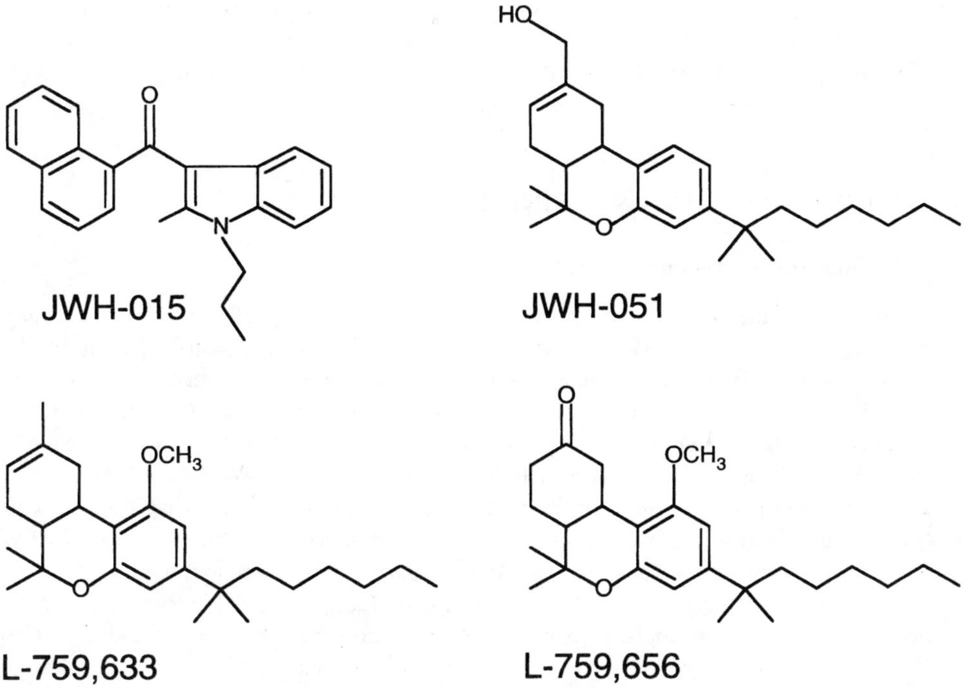
HO
O
N
JWH-015
JWH-051
OCH3
O
OCH3
L-759,633
L-759,656

Figure 9 Structures of cannabinoid receptor ligands showing selectivity for cannabinoid CB_2 receptors (see text and Table 1 for further details)

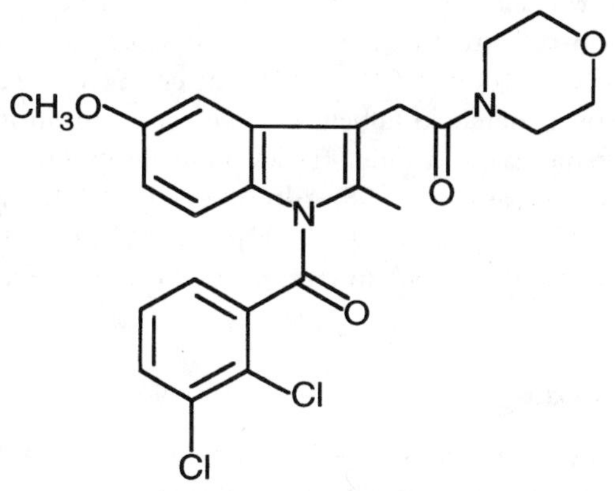
CH3O
N
O
O
N
O
Cl
Cl

Figure 10 One of a series of indoles with high affinity and selectivity for CB_2 receptors (see text, Table 1 and Gallant *et al.* (1996) for further details). It is listed in Table 1 as Compound 9

been performed with this compound. No radiolabelled CB_2-selective ligands have yet been produced. As to the question of whether ligands can be developed with significantly different affinities for CB_1 and $CB_{1(a)}$ receptors, existing binding data indicate that the CB_1 to $CB_{1(a)}$ receptor affinity ratio is 10.1 for SR141716A, 3–4 for delta-9-THC,

CP 55,940 and WIN 55,212-2 and 0.83 for anandamide and that the rank order of affinity is CP 55,940 > SR141716A > delta-9-THC > WIN 55,212-2 > anandamide for CB_1 receptors and CP 55,940 > delta-9-THC > SR141716A > anandamide > WIN 55,212-2 for $CB_{1(a)}$ receptors (Rinaldi-Carmona et al., 1996a).

4   DISTRIBUTION OF CANNABINOID RECEPTORS

4.1   Cannabinoid Receptor mRNA

Although by far the highest concentrations of CB_1 and $CB_{1(a)}$ mRNA are to be found in the CNS (Galiègue et al., 1995; Shire et al., 1995), it has been possible, largely by the application of reverse transcription coupled to the polymerase chain reaction, to demonstrate the presence of both these mRNAs in many peripheral tissues. Outside the CNS, the highest levels of human CB_1 mRNA are in pituitary gland and immune cells, particularly B-cells and natural killer cells (Galiègue et al., 1995). As detailed by Pertwee (1997), other peripheral tissues of human, dog, rat and/or mouse that contain CB_1 mRNA include immune tissues (tonsils, spleen, thymus, bone marrow), reproductive tissues (ovary, uterus, testis, vas deferens, prostate gland), gastrointestinal tissues (stomach, colon, bile duct), superior cervical ganglion, heart, lung, urinary bladder and adrenal gland. $CB_{1(a)}$ mRNA is thought to exist as a minor transcript, the ratio of $CB_{1(a)}$ to CB_1 mRNA in humans never exceeding 0.2 and, in kidney, bile duct and certain areas of infant or 2-year old brain, diminishing to 0.02 or less (Shire et al., 1995, and Section 4.3).

  CB_2 mRNA occurs mainly in immune tissues, for example human, rat and mouse spleen, human leukocytes (B cells > T cells), human tonsils and rat peritoneal mast cells (Das et al., 1995; Facci et al., 1995; Galiègue et al., 1995). Levels of human CB_2 mRNA are particularly high in B-cells, natural killer cells and spleen as well as in tonsils, where they are similar to those of human CB_1 mRNA in cerebellum (Galiègue et al., 1995). CB_2 mRNA has also been detected, albeit at lower concentrations, in human thymus gland, bone marrow, pancreas and lung. Its levels in peripheral tissues greatly exceed those of CB_1 mRNA (Galiègue et al., 1995). Although CB_2 mRNA has not been detected in human or rat brain (Munro et al., 1993; Galiègue et al., 1995), there is one report of its presence together with CB_1 mRNA in cultures of mouse cerebellar granule neurones (Skaper et al., 1996).

4.2   Cannabinoid Binding Sites

The presence of specific cannabinoid binding sites within the CNS has been demonstrated both by autoradiography and by binding assays performed with membrane preparations obtained from tissue homogenates (Bidaut-Russell et al., 1990; Herkenham et al., 1990, 1991b; Kuster et al., 1993; Rinaldi-Carmona et al., 1996b; Section 4.3). These must be CB_1(and $CB_{1(a)}$) binding sites as CB_2 receptors are not expressed within the CNS (Section 4.1). Cannabinoid binding sites have also been detected in certain tissues outside the CNS. Using an autoradiographic technique, Lynn and Herkenham (1994) detected specific [^{3}H]CP 55,940 binding sites in rat immune tissues (spleen, lymph nodes, Peyer's patches and leukocytes) and in rat anterior pituitary gland although not in a wide range of other rat tissues including thymus, reproductive tissues (ovary, uterus, testis, vas deferens, prostate gland), gastrointestinal tract, heart, lung, urinary bladder and adrenal gland. Results from binding assays with [^{3}H]CP 55,940 using membrane preparations obtained from tissue homogenates have confirmed the presence of specific cannabinoid binding sites in spleen cell membranes (Kaminski *et al.*, 1992; Rinaldi-Carmona *et al.*, 1994) and also demonstrated the presence of such binding sites in guinea-pig small intestine and pregnant mouse uterus (Paterson and Pertwee, 1993; Das *et al.*, 1995). Most published binding data provide no information about the types of cannabinoid receptors present in peripheral tissues as they have been obtained with receptor probes that bind equally well to CB_1 and CB_2 receptors or whose relative affinities for CB_1 and CB_2 binding sites are unknown. An exception is an investigation carried out by Rinaldi-Carmona *et al.* (1994). They found that, unlike delta-9-THC, CP 55,940 and WIN 55,212-2 that have approximately the same affinities for CB_1 and CB_2 binding sites (Table 1), the CB_1-selective ligand SR 141716A did not readily displace [^{3}H]CP 55,940 from rat splenic binding sites, indicating these sites to be predominantly 'non-CB_1' and hence presumably CB_2. The finding that the tissue concentration of CB_1 receptors is much less outside than within the CNS does not necessarily imply that peripheral CB_1 receptors are unimportant as, in some peripheral tissues at least, these receptors may be confined to discrete regions such as nerve terminals (Sections 4.3. and 6) that form only a small part of the total tissue mass.

4.3   Distribution of Cannabinoid Receptors within the CNS

Results obtained in autoradiographic studies with rat brain and spinal cord indicate that the distribution pattern of specific binding sites for cannabinoids within the CNS is heterogeneous, unlike that for any other known receptor type and consistent with the known ability of cannabinoid receptor agonists to impair cognition and memory, to alter motor function and movement and to relieve pain (Herkenham *et al.*, 1990, 1991b). The highest concentrations of cannabinoid binding sites in rat brain (4–6.4 pmol/mg protein) are in the substantia nigra pars reticulata, the entopeduncular nucleus, the globus pallidus, the lateral caudate-putamen, the ependymal and subependymal zones at the centre of the olfactory bulb and the molecular layer of the cerebellum. Other areas of rat brain quite rich in cannabinoid binding sites (2–4 pmol/mg protein) include the hippocampus, cerebral cortex, intrabulbar anterior commissure, nucleus accumbens and septum. Among the areas of rat brain less densely populated with cannabinoid binding sites are (a) the central gray substance, the area postrema and the caudal nucleus of the solitary tract (1–2.4 pmol/mg protein), (b) the amygdala, thalamus, habenula, preoptic area and hypothalamus (<2 pmol/mg protein) and (c) much of the brain stem (<1 pmol/mg protein). Regions of rat spinal cord that are richest in cannabinoid binding sites are lamina X and the substantia gelatinosa (ca 1 pmol/mg protein). Similar conclusions about the central distribution of cannabinoid receptors can be drawn from binding data that derive (a) from other autoradiographic studies with rat tissue (Jansen *et al.*, 1992; Mailleux and Vanderhaeghen, 1992a; Thomas *et al.*, 1992; Rinaldi-Carmona *et al.*, 1996b), (b) from autoradiographic studies with human tissue (Herkenham *et al.*, 1990; Mailleux *et al.*, 1992; Mailleux and Vanderhaeghen, 1992b; Glass *et al.*, 1993, 1997;

Westlake *et al.*, 1994) or with rhesus monkey, dog and guinea-pig tissue (Herkenham *et al.*, 1990) and (c) from experiments performed with membrane preparations obtained from homogenates of tissue taken from discrete areas of rat brain (Bidaut-Russell *et al.*, 1990; Rinaldi-Carmona *et al.*, 1996b) or spinal cord (Welch *et al.*, 1995).

There are many similarities between the central distribution of cannabinoid binding sites and CB_1 mRNA (Mailleux and Vanderhaeghen, 1992a; Rubino *et al.*, 1994; Westlake *et al.*, 1994). Where differences do occur, these can be attributed to a spatial separation between nerve terminals bearing cannabinoid receptors and the cell bodies that contain the mRNA responsible for expressing these receptors. Brain regions in which cannabinoid receptors seem to be present on fibres and nerve terminals that project from another part of the brain include the substantia nigra pars reticulata, entopeduncular nucleus and globus pallidus (Herkenham *et al.*, 1991a).

The only types of cannabinoid receptor that have so far been detected in the brain are CB_1 and $CB_{1(a)}$. The ratio of human $CB_{1(a)}$ to CB_1 mRNA in brain tissue varies with both brain area and age. Values of this ratio are 0.14–0.2 in adult occipital cortex, striatum, substantia nigra and 2-year-old brain stem, 0.08 in infant cortex + cerebellum, 0.02 or less in adult frontal cortex and < 0.005 in infant brain stem (Shire *et al.*, 1995). The importance of cannabinoid receptors should not be judged solely by their distribution pattern as there is evidence from binding experiments with GTP[γ-^{35}S] that there are regional differences in the efficiency with which these receptors are coupled to their effector mechanisms in the brain (Sim *et al.*, 1995).

5   EFFECTOR SYSTEMS

5.1   Adenylate Cyclase and Mitogen-Activated Protein Kinases

It is now well-established that cannabinoid receptor agonists can inhibit the production of cyclic AMP and the more recent observation, that these agonists can activate mitogen-activated protein (MAP) kinases, is also generally accepted (see Childers and Deadwyler, 1996; Pertwee, 1988; 1997). The evidence that these effects are mediated by cannabinoid receptors and, indeed, that CB_1, $CB_{1(a)}$ and CB_2 receptors are all coupled through G-proteins to adenylate cyclase and MAP kinases is summarized in Tables 2–4. This evidence derives from experiments with tissues expressing cannabinoid receptors either after transfection or naturally. Of particular importance are the findings that

(a)   cannabinoids show concentration-dependence and high potency as modulators of adenylate cyclase and MAP kinase activity in tissues containing CB_1, $CB_{1(a)}$ or CB_2 receptors (Tables 2–4),

(b)   cannabinoid receptor agonists inhibit adenylate cyclase (Matsuda *et al.*, 1990; Gérard *et al.*, 1991; Felder *et al.*, 1993; Vogel *et al.*, 1993; Slipetz *et al.*, 1995; Song and Bonner, 1996) and activate MAP kinases (Bouaboula *et al.*, 1995a,b, 1996) in cultured cells transfected with genetic material encoding cannabinoid CB_1 or CB_2 receptors but not in cannabinoid receptor-free cells of the same lines and

(c) both effects are attenuated by submicromolar concentrations of SR141716A in tissues containing CB_1 or $CB_{1(a)}$ receptors (Tables 2–4).

Since CB_1, $CB_{1(a)}$ and CB_2 receptors are all members of the superfamily of G-protein coupled receptors, another important observation is that the modulatory effects of cannabinoids on the activities of adenylate cyclase and MAP kinases can be attenuated by pertussis toxin, an agent which is known to block $G_{i/o}$-protein mediated processes by inducing ADP-ribosylation. Such attenuation has been observed for modulation of adenylate cyclase in tissues endowed with CB_1, $CB_{1(a)}$ or CB_2 receptors (Tables 2 and 3) and of MAP kinase activity in cell lines expressing CB_1 or CB_2 receptors (Table 4). Experiments with tissues containing CB_1 or CB_2 receptors have also shown that cannabinoids exhibit appropriate stereoselectivity as inhibitors of adenylate cyclase (Tables 2 and 3). Whether such stereoselectivity can be detected for the activation of MAP kinases remains to be demonstrated. It is noteworthy that cannabinoid-induced activation of MAP kinases through CB_1 or CB_2 receptors seems not to depend on the ability of cannabinoids to inhibit cyclic AMP production and, also, that this activation has been shown to lead to an enhancement of *Krox 24* expression (Bouaboula *et al.*, 1995a,b, 1996).

There is evidence to suggest that delta-9-THC and anandamide, which bind more or less equally well to CB_1 and CB_2 receptors (Felder *et al.*, 1995; Mechoulam *et al.*, 1995; Bayewitch *et al.*, 1996; Showalter *et al.*, 1996), are significantly less effective in activating CB_2 receptors than CB_1 receptors. More particularly, there are some reports that both compounds are effective inhibitors of cyclic AMP production in cell lines expressing CB_1 receptors (Felder *et al.*, 1993, 1995; Vogel *et al.*, 1993; Barg *et al.*, 1995; Bayewitch *et al.*, 1996) but produce negligible inhibition in cell lines containing only CB_2 receptors (Bayewitch *et al.*, 1995, 1996; Slipetz *et al.*, 1995). Indeed, in one set of experiments, delta-9-THC was found to antagonize CB_2 receptor-mediated inhibition of adenylate cyclase (Bayewitch *et al.*, 1996). In other investigations with CB_2 receptor-containing cell lines, however, delta-9-THC and anandamide have been found to behave as agonists, the measure again being inhibition of cyclic AMP production (Felder *et al.*, 1995; Shire *et al.*, 1996). The reason for this discrepancy remains to be established. However, one possible explanation that merits exploration is that both these agents have relatively low CB_2 efficacies and so elicit detectable CB_2 receptor-mediated responses only in biological systems that are particularly well-populated with this receptor type.

5.2 Arachidonic Acid

Shivachar *et al.*, (1996) have obtained evidence that activation of cannabinoid $CB_1/CB_{1(a)}$ receptors can lead to the mobilization of arachidonic acid. They found delta-9-THC and anandamide (0.5–5 µM) to produce increases in the level of free [^{3}H]arachidonic acid in prelabelled rat cortical astrocytes and that these increases could be attenuated by pertussis toxin and by SR141716A, albeit only at the rather high concentrations of 1 and 10 µM. Results from earlier investigations indicate that cannabinoids can also mobilize arachidonic acid without acting through cannabinoid receptors (Reichman *et al.*, 1991; Felder *et al.*, 1992, 1993, 1995; see also Howlett, 1995).

Table 2 Inhibition of cyclic AMP formation by cells transfected with CB_1, $CB_{1(a)}$ or CB_2 receptors or by certain untransfected cell lines

	Human or rat CB transfected cultured cells				Human, rat or mouse CB transfected cultured cells		Untransfected cultured cells				
	CB_1			$CB_{1(a)}$	CB_2						
	CHO	COS	293	CHO	CHO	COS	N18TG2	U373 MG	EL4.IL-2	Cer G	Hippo
Added cannabinoid agonists show:											
pharmacological activity	Yes	Yes	Yes	Yes	Yes	Yes	Yes[a]	Yes	Yes	Yes	Yes
pharmacological activity that is dose related	Yes	Yes	Yes	Yes	Yes	ND	Yes	Yes	Yes	Yes	ND
high potency[b]	Yes	Yes	Yes	Yes	Yes	Yes	Yes	Yes	No	Yes	ND
appropriate structure-activity relationships	Yes[c]	ND	Yes[c]	Yes[d]	Yes[c,e]	ND[f]	Yes[c]	Yes[c]	ND	Yes[c]	ND
appropriate stereoselectivity[c]	Yes	ND	Yes	ND	Yes	ND	Yes	ND	ND	Yes	Yes
susceptibility to antagonism by SR141716A[g]	Yes	ND	ND	Yes	No	ND	ND	Yes	ND	ND	ND
susceptibility to attenuation by pertussis toxin	Yes	ND	ND	Yes	Yes	ND	Yes	Yes	ND	Yes	Yes
resistance to antagonism by certain non-cannabinoid receptor antagonists	ND	ND	ND	ND	ND	ND	Yes[h]	ND	ND	ND	ND
The presence has been demonstrated of:											
specific cannabinoid binding sites	Yes	Yes	Yes	Yes	Yes	Yes	Yes	Yes	ND	ND	ND
genetic material capable of expressing cannabinoid CB_1 and/or CB_2 receptors	CB_1	CB_1	CB_1	$CB_{1(a)}$	CB_2	CB_2	CB_1	CB_1[i]	CB_2[j]	ND	ND
Endogenous cannabinoids behave like cannabinoid receptor agonists	Yes	ND	Yes	Yes	k	ND	Yes	ND	ND	ND	ND
References	1	2	3	4	5	2	6	7	8	9	10

[a] No pharmacological activity observed in the absence of GTP (Howlett, 1985). [b] At least some agonists effective at $< 1 \mu M$. [c] Potency correlates with affinity for cannabinoid CB_1 receptors. [d] For CP 55,940, delta-9-tetrahydrocannabinol and anandamide but not WIN 55,212-2, potency correlates well with affinity for cannabinoid CB_1 receptors. [e] Potency correlates with affinity for cannabinoid $CB_{1(a)}$ receptors. [f] Delta-9-tetrahydrocannabinol showed activity in some experiments (Felder et al., 1995; Shire et al., 1996) but little or no activity in other experiments

(Slipetz et al., 1995; Bayewitch et al., 1995; 1996). [f] Delta-9-tetrahydrocannabinol behaved as a weak partial agonist. [g] SR141716A produced antagonism at a concentration < 1 μM. [h] Naloxone, atropine, yohimbine. [i] CB$_2$ mRNA not detected. [j] CB$_1$ mRNA not detected. [k] Anandamide was found to be active in some experiments (Felder et al., 1995; Shire et al., 1996) but not in others (Bayewitch et al., 1995).

Cer G = rat cerebellar granule cells. Hippo = hippocampal cells. ND = not determined.

References: 1 – Matsuda et al. (1990); Felder et al. (1991); Gérard et al. (1991); Felder et al. (1992, 1993, 1995); Vogel et al. (1993); Rinaldi-Carmona et al. (1994, 1996a); Barg et al. (1995); Bouaboula et al. (1995b). 2 – Bayewitch et al. (1995). 3 – Song and Bonner (1996). 4 – Rinaldi-Carmona et al. (1996a). 5 – Rinaldi-Carmona et al. (1994); Slipetz et al. (1995); Felder et al. (1995); Bayewitch et al. (1995, 1996); Bouaboula et al. (1996); Shire et al. (1996). 6 – Howlett and Fleming (1984); Howlett (1984, 1985, 1987); Howlett et al. (1986, 1988, 1990); Vogel et al. (1993); Pinto et al. (1994); Fride et al. (1995); Daaka et al. (1996); Bayewitch et al. (1996). 7 – Bouaboula et al. (1995a). 8 – Condie et al. (1996). 9 – Pacheco et al. (1993). 10 – Deadwyler et al. (1995).

See also Mackie et al. (1995) for cyclic AMP data from CB$_1$ transfected AtT-20 cells and Howlett et al. (1986), Rowley and Rowley (1990), Hirst and Lambert (1995), Ho and Zhao (1996) for cyclic AMP data from untransfected NG108-15 rat/mouse hybrid cells (these express both mouse and rat CB$_1$ cDNA), rat pituitary tumour GH4C1 cells, human leukaemia ML2 cells and SH-SY5Y cells.

Table 3 Inhibition of cyclic AMP formation by various *in vitro* preparations

	Rat striatal or cerebellar membranes	Rat striatal slices[c]	Mouse spleen cells	Membrane preparations of mouse		Mouse vas deferens	Human lymphocytes
				uterus	embryo		
Added cannabinoid receptor agonists show:							
pharmacological activity	Yes[d]	Yes	Yes	Yes	Yes	Yes	Yes
pharmacological activity that is dose related	Yes	ND	Yes	Yes	ND	Yes	Yes
high potency[a]	Yes	No	No	Yes	ND	Yes	No
appropriate structure-activity relationships[b]	Yes	ND	ND	Yes	Yes	ND	ND
appropriate stereoselectivity[b]	Yes	ND	ND	ND	ND	Yes	ND
susceptibility to antagonism by SR141716A	ND	ND	ND	ND	ND	Yes[e]	ND
susceptibility to attenuation by pertussis toxin	ND	Yes	Yes	Yes	Yes	ND	Yes
resistance to antagonism by certain non-cannabinoid receptor antagonists	ND	Yes[f]	ND	ND	ND	ND	ND
The presence has been demonstrated of:							
specific cannabinoid binding sites	Yes	Yes	Yes	Yes	Yes	ND	ND
genetic material capable of expressing cannabinoid CB_1 and/or CB_2 receptors	ND	ND	CB_1 & CB_2	CB_1[g]	CB_1 & CB_2	CB_1 & CB_2[h]	ND[i]
Endogenous cannabinoids behave like cannabinoid receptor agonists	Yes	ND	Yes	Yes	ND	ND	ND
References	1	2	3	4	5	6	7

[a] At least some agonists effective at $<1\,\mu M$. [b] Potency correlates well with affinity for cannabinoid CB_1 receptors. [c] Some experiments also carried out with rat cerebrocortical, hippocampal and cerebellar slices (Bidaut-Russell *et al.*, 1990). [d] No pharmacological activity was observed in the absence of GTP. [e] SR141716A produced antagonism at a concentration $<1\,\mu M$. [f] Naloxone and spiperone. [g] CB_2 mRNA not detected. [h] CB_2-like mRNA detected. [i] Presence of both CB_1 and CB_2 receptors is likely (see Pertwee, 1997).

References: 1 – Pacheco *et al.* (1991, 1993, 1994); Childers *et al.* (1994). 2 – Bidaut-Russell *et al.* (1990); Bidaut-Russell and Howlett (1991). 3 – Schatz *et al.* (1992); Kaminski *et al.* (1992, 1994); Das *et al.* (1995); Lee *et al.* (1995); Mechoulam *et al.* (1995). 4 – Das *et al.* (1995). 5 – Paria *et al.* (1995). 6 – Pertwee *et al.* (1996b); Griffin *et al.*, 1997. 7 – Diaz *et al.* (1993).

Table 4   Activation of MAP kinase

	Human CB transfected cultured cells			Human untransfected cultured cells	
	CB_1	$CB_{1(a)}$	CB_2		
	CHO	CHO	CHO	WI-38	U373 MG
Added cannabinoid receptor agonists show:					
pharmacological activity	Yes[a]	Yes	Yes[a]	Yes	Yes
pharmacological activity that is dose related	Yes	Yes	Yes	Yes	Yes
high potency[b]	Yes	Yes	Yes	No[c]	Yes
appropriate structure-activity relationships[d]	Yes	ND	ND	ND	ND
appropriate stereoselectivity[d]	ND	ND	ND	ND	ND
susceptibility to antagonism by SR141716A[e]	Yes	Yes	ND	ND	ND
susceptibility to attenuation by pertussis toxin	Yes	ND	Yes	Yes	ND
The presence has been demonstrated of:					
specific cannabinoid binding sites	Yes	Yes	Yes	ND	Yes
genetic material capable of expressing cannabinoid CB_1 and/or CB_2 receptors	CB_1	$CB_{1(a)}$	CB_2	ND	CB_1^f
Endogenous cannabinoids behave like cannabinoid receptor agonists	ND	ND	ND	Yes	ND
References	1	2	3	4	5

[a] Not susceptible to attenuation by dibutyryl- or 8-bromo-cyclic AMP in combination with 3-isobutyl-1-methylxanthine. [b] At least some agonists effective at $< 1\,\mu M$. [c] Only anandamide was used (in the absence of any inhibitor of enzymic hydrolysis). [d] Potency correlates well with affinity for cannabinoid CB_1 receptors. [e] SR141716A produced antagonism at a concentration $< 1\,\mu M$. [f] CB_2 mRNA not detected.

MAP kinase = mitogen-activated protein kinase. ND = not determined. Cannabinoid-induced activation of MAP kinase and *Krox 24* expression has also been observed in HL60 cells but not in Daudi cells; both these cell lines naturally express cannabinoid CB_2 mRNA (Bouaboula *et al.*, 1996).

References: 1 – Bouaboula *et al.* (1995b); Rinaldi-Carmona *et al.* (1996a). 2 – Rinaldi-Carmona *et al.* (1996a). 3 – Bouaboula *et al.* (1996). 4 – Wartmann *et al.* (1995). 5 – Bouaboula *et al.* (1995a, 1995b).

5.3   Ion Channels

Experiments with various cultured cell lines have shown cannabinoid receptor agonists to inhibit N- and P/Q-type calcium channels and activate A-type and inwardly rectifying potassium channels (Table 5). The available data indicate these effects are mediated by CB_1 receptors coupled to inhibitory G-proteins but not by CB_2 receptors. In particular,

(a)   cannabinoids show concentration-dependence and appropriate stereoselectivity and potency (effective at $< 1\,\mu M$) as modulators of inward calcium currents,

Table 5 Effects on ion channel conductance

	Rat CB$_1$ transfected cultured cells				Untransfected cultured cells					Rat nodose ganglion cells
	Rat SCG	AfT-20	AfT-20	Xenopus oocytes	NG108-15	N18	rat hippocampus	rat hippocampus	NIE-115	
Ion channel[a]	Ca^{2+}	Ca^{2+}	K$_{ir}$	K$_{ir}$	Ca^{2+}	Ca^{2+}	Ca^{2+}	K$^+$(A-type)	Na$^+$	5-HT$_3$ receptor
Increased/decreased conductance	−	−	+	+	−	−	−	+	−	−
Added cannabinoid receptor agonists show:										
pharmacological activity	Yes[b]	Yes	Yes	Yes	Yes[c]	Yes	Yes	Yes[d]	Yes	Yes[e]
pharmacological activity that is dose related	Yes	Yes	Yes	Yes	Yes	Yes	Yes	Yes	Yes	Yes
high potency[f]	Yes	Yes	Yes	Yes	Yes	Yes	Yes	Yes	Yes	Yes
appropriate structure-activity relationships[g]	ND	ND	Yes	ND	ND	ND	Yes	Yes	ND	ND
appropriate stereoselectivity[g]	ND	ND	Yes	ND	Yes	ND	Yes	Yes	ND	Yes
susceptibility to antagonism by SR141716A[h]	Yes	Yes	Yes	No	ND	ND	Yes	ND	ND	ND
susceptibility to attenuation by pertussis toxin	Yes	Yes	Yes	ND	Yes	Yes	ND	Yes	ND	ND
resistance to antagonism by certain non-cannabinoid receptor antagonists	ND	ND	ND	ND	ND	ND	ND	ND	ND	Yes[i]
The presence has been demonstrated of:										
specific cannabinoid binding sites	j	Yes	Yes	Yes	ND	ND	ND	ND	ND	ND
genetic material capable of expressing cannabinoid CB$_1$ and/or CB$_2$ receptors	CB$_1$	CB$_1$[k]	CB$_1$[k]	CB$_1$	CB$_1$	ND	ND	ND	ND	ND
Endogenous cannabinoids behave like cannabinoid receptor agonists	ND	Yes	Yes	ND	ND	Yes	ND	ND	ND	Yes
References	1	2	2	3	4	5	6	7	8	9

[a] N- and P/Q-types of calcium channel are most susceptible to cannabinoid-induced inhibition (Caulfield and Brown, 1992; Mackie and Hille, 1992; Mackie et al, 1993, 1995; Pan et al, 1996; Twitchell and Mackie, 1996). [b] GDP-β-S abolished the ability of WIN 55,212-2 to inhibit Ca^{2+} currents. [c] Not susceptible to attenuation by dibutyryl- or 8-chlorophenylthio-cyclic AMP in combination with 3-isobutyl-1-methylxanthine. [d] Forskolin, 3-isobutyl-1-methylxanthine and 8-bromo-cyclic AMP produced shifts in potassium current in the same direction to those produced by cannabinoids (Deadwyler et al, 1995), whereas GTP-γ-S produced shifts in potassium current opposite in direction to those produced by cannabinoids (Deadwyler et al, 1993). [e] Pharmacological activity unaffected by Sp-cyclic AMP or GDP-β-S. [f] At least some agonists effective

at $< 1 \mu M$. [g]Potency correlates well with affinity for cannabinoid CB_1 receptors. [h]The ability of cannabinoids to modulate calcium and potassium currents is attenuated by the cannabinoid receptor antagonist, LY320135 (Fahey et al., 1995). [i]Naltrexone. [j]Experiments with anandamide yielded inconsistent results. [k]Cannabinoids did not modulate Ca^{2+} or K_{ir} currents in AtT-20 cells transfected with human CB_2 receptors (Felder et al., 1995).

SCG = superior cervical ganglion neurones; Ca^{2+} = high-voltage activated (inward) Ca^{2+} currents; K_{ir} = inwardly rectifying K^+ currents; K^+ (A-type) = outward A-type currents; Na^+ = inward Na^+ currents; 5-HT$_3$ receptor = 5-HT$_3$ receptor inward currents; + = increase; − = decrease; ND = not determined or unclear.

References: 1 – Pan et al. (1996). 2 – Felder et al. (1995); Mackie et al. (1995). 3 – Henry and Chavkin (1995). 4 – Matsuda et al. (1990); Caufield and Brown (1992); Mackie and Hille (1992). 5 – Mackie et al. (1993). 6 – Twitchell and Mackie (1996). 7 – Deadwyler et al. (1993, 1995). 8 – Turkanis et al. (1991a,b). 9 – Fan (1995).

outward A-type potassium currents and inwardly rectifying potassium currents in cell lines containing native or transfected CB_1 receptors (Table 5),

(b) the ability of cannabinoids to produce these effects can be attenuated by pertussis toxin (Table 5),

(c) cannabinoid receptor agonists modulate calcium currents or inwardly rectifying potassium currents in cultured cells transfected with CB_1 receptors but not in cannabinoid receptor-free cells of the same lines (Henry and Chavkin, 1995; Mackie et al., 1995; Pan et al., 1996) or in cells transfected with CB_2 receptors (Felder et al., 1995) and

(d) the ability of cannabinoids to modulate calcium currents or inwardly rectifying potassium currents is readily attenuated by the CB_1-selective cannabinoid receptor antagonists, SR141716A (200 nM) and LY320135 (Table 5).

There are indications that cannabinoid receptors may also play a part in cannabinoid-induced inhibition of 5-HT_3 receptor-mediated inward currents in rat nodose ganglion neurones but none as yet that cannabinoid receptors mediate cannabinoid-induced inhibition of inward sodium currents in neuroblastoma cells (Table 5). The available evidence suggests that the modulatory effect of cannabinoids on A-type potassium channels stems from their ability to inhibit adenylate cyclase and so suppress intracellular levels of cyclic AMP (Deadwyler et al., 1995; Hampson et al., 1995; Childers and Deadwyler, 1996) but that cannabinoids do not act through this enzyme to modulate calcium currents (Mackie and Hille, 1992). Finally, Poling et al. (1996) have obtained evidence that in murine B82 fibroblasts transfected with CB_1 cDNA, *Shaker*-related voltage-gated potassium channels can be inhibited by certain cannabinoid receptor ligands, including delta-9-THC and anandamide ($EC_{50} = 2.4$ and 2.7 µM respectively). However, as the inhibitory effect of anandamide was attenuated by neither SR141716A nor pertussis toxin it is unlikely that *Shaker*-related voltage-gated potassium channels are coupled to any known type of cannabinoid receptor.

6   EFFECTS MEDIATED BY CANNABINOID RECEPTORS

6.1   *In Vivo* Effects

The role of cannabinoid receptors has so far been investigated in only a few of the many effects that cannabinoid receptor agonists are known to elicit (Paton and Pertwee, 1973a,b; Dewey, 1986; Martin, 1986; Pertwee, 1988). The strategy has been to look for susceptibility to antagonism by SR141716A and has, therefore, focussed on effects mediated by CB_1 (and $CB_{1(a)}$) receptors. In particular, SR141716A has been reported to block the hypokinetic, cataleptic, antinociceptive and hypothermic effects of delta-9-THC or WIN 55,212-2 in mice (Rinaldi-Carmona et al., 1994; Dutta et al., 1995; Compton et al., 1996; Reche et al., 1996) and the discriminative stimulus effects of one or other of these cannabinoids in rats, rhesus monkeys and pigeons (Wiley et al., 1995; Mansbach et al., 1996; Pério et al., 1996). These are all effects that have long been exploited for the bioassay of cannabinoid receptor agonists (Martin et al., 1995; Pertwee,

1997). Another *in vivo* effect attenuated by SR141716A is anandamide-induced hypotension in anaesthetized rats (Varga *et al.*, 1995). Investigations into the part played by CB_2 receptors in the *in vivo* pharmacology of cannabinoids have still to be carried out.

6.2 *In Vitro* Effects

As already discussed (Section 5), there is evidence that activation of cannabinoid CB_1, $CB_{1(a)}$ and/or CB_2 receptors can lead to inhibition of adenylate cyclase, stimulation of MAP kinases, mobilization of arachidonic acid and modulation of calcium and potassium currents. To these effects may be added inhibition of neurotransmitter release in certain tissues, inhibition of the uptake of γ-aminobutyric acid (GABA) into central neurones, inhibition of 5-hydroxytryptamine (5-HT) release from mast cells, inhibition of hippocampal long term potentiation and inhibition of electrically-evoked contractions of certain isolated smooth muscle preparations. Of these *in vitro* effects, the most widely exploited for the purpose of quantitative bioassay is the ability to inhibit (a) cyclic AMP production in tissues expressing cannabinoid receptors either naturally or after transfection (CB_1, $CB_{1(a)}$ and CB_2), and (b) electrically-evoked contractions of the myenteric plexus-longitudinal muscle preparation (MPLM) of guinea-pig small intestine (CB_1) or of the mouse isolated vas deferens (CB_1 and possibly also CB_2). Cannabinoid binding assays are also frequently used (CB_1, $CB_{1(a)}$ and CB_2).

In the sections which follow, no reference is made to $CB_{1(a)}$ receptors. It should be noted, however, that whilst the attenuation of a cannabinoid-induced effect by submicromolar concentrations of SR141716A is a good indication that this effect is not CB_2 receptor-mediated, such data do not exclude possible mediation of the effect by $CB_{1(a)}$ receptors.

6.2.1 *Inhibition of Neurotransmitter Release*

Cannabinoid receptor agonists readily inhibit electrically-evoked release of (a) acetylcholine from rat hippocampal slices (Gifford and Ashby, 1996), (b) acetylcholine from guinea-pig MPLM (Coutts and Pertwee, 1996, 1997; Pertwee *et al.*, 1996b; see also Pertwee, 1988), (c) noradrenaline from rat isolated atria and rat vasa deferentia (Ishac *et al.*, 1996) and (d) dopamine and noradrenaline from guinea-pig retinal discs (Schlicker *et al.*, 1996). The evidence that these effects are mediated by cannabinoid CB_1 receptors is summarized below.

(a) In hippocampal slices, MPLM, retinal discs and atria, cannabinoids inhibit transmitter release in a concentration-related fashion and exhibit high potency (effective at $<1\,\mu M$) (Coutts and Pertwee, 1996, 1997; Gifford and Ashby, 1996; Ishac *et al.*, 1996; Pertwee *et al.*, 1996b; Schlicker *et al.*, 1996).

(b) It has also been demonstrated that, in MPLM and retinal discs, WIN 55,212-2 is markedly more potent as an inhibitor of transmitter release than its (−)-enantiomer, WIN 55,212-3 (Coutts and Pertwee, 1997; Schlicker *et al.*, 1996). This rank order of potency is in accord with the relative affinity of this enantiomeric pair for specific CB_1 (and CB_2) binding sites (Felder *et al.*, 1992; Jansen *et al.*, 1992; Slipetz *et al.*, 1995).

(c)   The effects of cannabinoids on transmitter release can be attenuated by SR141716A, at a concentration of 1 µM or less (hippocampal slices, MPLM, retinal discs and atria), or at a concentration of 10 µM (vasa deferentia) (Coutts and Pertwee, 1996, 1997; Gifford and Ashby, 1996; Ishac et al., 1996; Pertwee et al., 1996b; Schlicker et al., 1996).

(d) CB_1 mRNA is present in rat hippocampus and in rat (and mouse) vas deferens (Mailleux and Vanderhaeghen, 1992a; Ishac et al., 1996; Griffin et al., 1997). In addition, it has been detected in rat superior cervical ganglia (Ishac et al., 1996) which contain the cell bodies of sympathetic neurones projecting to the heart.

(e)   Rat hippocampus and guinea-pig MPLM contain specific high-affinity cannabinoid binding sites (Herkenham et al., 1990, 1991b; Paterson and Pertwee, 1993).

There is also indirect evidence from electrophysiological experiments with rat cultured neurones that activation of presynaptic CB_1 receptors inhibits glutamate release in the hippocampus (Shen et al., 1996).

6.2.2   Inhibition of GABA Uptake into Central Neurones and of 5-HT Release from Mast Cells

There is some evidence that delta-9-THC can act through CB_1 receptors to inhibit the uptake of GABA into slices of rat globus pallidus (Maneuf et al., 1996) and that WIN 55,212-2, nabilone and delta-8-THC can act through CB_2 receptors to inhibit dinitrophenylated human serum albumin-induced release of preloaded [³H]5-HT from rat RBL-2H3 cultured mast cells (Facci et al., 1995). In particular,

(a)   the inhibitory effects of these cannabinoids on GABA uptake and 5-HT release are concentration-related (Facci et al., 1995; Maneuf et al., 1996),

(b)   the effect of delta-9-THC on GABA uptake can be attenuated by SR141716A, albeit at the rather high concentrations of 30 and 100 µM (Maneuf et al., 1996),

(c)   specific cannabinoid binding sites are present in rat globus pallidus (Section 4.3) and RBL-2H3 cells (Facci et al., 1995),

(d)   RBL-2H3 cells express CB_2 but not CB_1 receptors (Facci et al., 1995) and

(e)   cannabidiol (Figure 3), which is not a cannabinoid receptor ligand, has been found not to inhibit 5-HT release from RBL-2H3 cells at concentrations of up to 60 µM (Facci et al., 1995).

It noteworthy that the concentrations of cannabinoids required to inhibit GABA uptake into slices of globus pallidus (50 µM) or 5-HT release from mast cells (EC_{50} > 2 µM) are higher than is to be expected for effects that are receptor-mediated. Similarly, the concentrations of SR141716A used to antagonize delta-9-THC-induced inhibition of GABA uptake (see above) are well above those usually required to attenuate in vitro responses to cannabinoids (< 1 µM) (see Pertwee 1997). It is also noteworthy that Facci et al. (1995) found that 5-HT release from RBL-2H3 cells was not inhibited by anandamide, which instead, attenuated the inhibitory effects of nabilone and WIN 55,212-2. (Reports that anandamide has affinity for cannabinoid CB_2 receptors but little CB_2 efficacy were discussed in Section 5.1). Facci et al. (1995) also found that 5-HT release

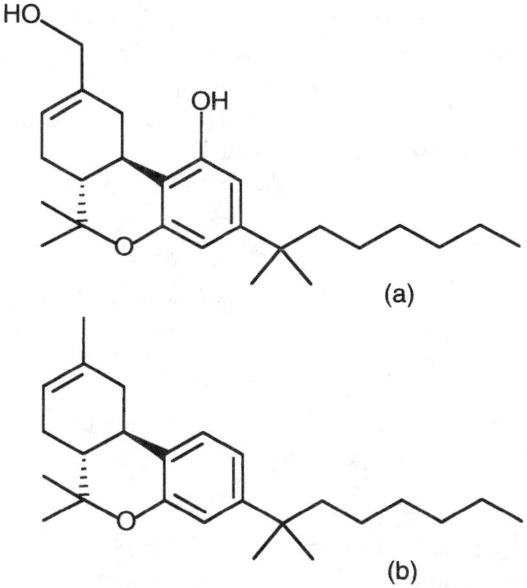

Figure 11 Structures of two synthetic analogues of delta-8-tetrahydrocannabinol (a) HU-210 and (b) (−)-1-deoxy-delta-8-tetrahydrocannabinol-dimethylheptyl. The less active (+)-enantiomer of HU-210 is HU-211

from RBL-2H3 cells could be inhibited by palmitoylethanolamide and that low concentrations of this N-acylethanolamine readily displaced [^{3}H]WIN 55,212-2 from specific binding sites on RBL-2H3 cell membranes. In contrast to the second of these findings, Showalter *et al.* (1996) have found that palmitoylethanolamide has little affinity for CB_2 receptors.

6.2.3 Inhibition of Long Term Potentiation

There have been several reports that cannabinoid receptor agonists can inhibit long term potentiation in rat hippocampal slices (Nowicky *et al.*, 1987; Collins *et al.*, 1994, 1995; Collin *et al.*, 1995; Terranova *et al.*, 1995). That this effect may be mediated by cannabinoid CB_1 receptors is indicated by the observations listed below.

(a) Cannabinoids can inhibit hippocampal long term potentiation in a concentration-related manner and exhibit high potency (effective at $\leqslant 1\,\mu M$) (Nowicky *et al.*, 1987; Collins *et al.*, 1994, 1995; Terranova *et al.*, 1995).

(b) The classical cannabinoid receptor agonist, HU-210 (Figure 11), is more potent as an inhibitor of hippocampal long term potentiation than its (+)-enantiomer, HU-211 (Collins *et al.*, 1994). This rank order of potency corresponds to the relative affinity shown by this enantiomeric pair for specific CB_1 (and CB_2) binding sites (Howlett *et al.*, 1990; Bayewitch *et al.*, 1995; Slipetz *et al.*, 1995; Showalter *et al.*, 1996; see also Pertwee, 1997).

(c)   The inhibitory effect of cannabinoids on hippocampal long term potentiation can be attenuated by submicromolar concentrations of SR141716A (Collins *et al.*, 1995; Terranova *et al.*, 1995).

(d)   Cannabinoid CB$_1$ mRNA and specific high-affinity cannabinoid binding sites are present in rat hippocampus (Herkenham *et al.*, 1990, 1991b; Mailleux and Vanderhaeghen, 1992a).

6.2.4   Inhibition of Electrically-Evoked Contractions

Cannabinoid receptor agonists inhibit electrically-evoked contractions of a range of isolated smooth muscle preparations. These are guinea-pig MPLM (Pertwee *et al.*, 1992, 1996b; see also Pertwee, 1988), mouse vas deferens (Pacheco *et al.*, 1991; Pertwee *et al.*, 1992, 1995c; Kuster *et al.*, 1993), mouse bladder (Pertwee and Fernando, 1996), mouse MPLM (Pertwee *et al.*, 1993a) and guinea-pig and rat vas deferens (Pertwee *et al.*, 1993a). The evidence that these effects are mediated by cannabinoid CB$_1$ receptors can be summarized as follows.

(a)   In all these preparations, cannabinoids inhibit electrically-evoked contractions in a concentration-related fashion, are highly potent (effective at $<1\,\mu M$) and exhibit stereoselectivity of the sort expected for cannabinoid receptor agonists (Pacheco *et al.*, 1991; Pertwee *et al.*, 1992, 1993a, 1995c, 1996b; Kuster *et al.*, 1993; Pertwee and Fernando, 1996).

(b)   Experiments with guinea-pig MPLM and with mouse vas deferens and bladder have shown that the inhibitory effects of cannabinoids on evoked contractions of these tissues can be readily attenuated by submicromolar concentrations of SR141716A (Rinaldi-Carmona *et al.*, 1994; Pertwee and Fernando, 1996; Pertwee *et al.*, 1995c, 1996b). The ability of SR141716A to antagonize cannabinoids in mouse MPLM and the vas deferens of guinea-pig and rat has still to be investigated.

(c)   Cannabinoid CB$_1$ mRNA has been detected in rat and mouse vasa deferentia (Ishac *et al.*, 1996; Griffin *et al.*, 1997) and specific high-affinity cannabinoid binding sites in guinea-pig MPLM (Paterson and Pertwee, 1993).

(d)   Vasa deferentia obtained from delta-9-THC-pretreated mice show tolerance to the inhibitory effects of delta-9-THC and other cannabinoids on electrically-evoked contractions but not to certain noncannabinoid twitch inhibitors that also act through G protein coupled receptors (clonidine and μ, δ and κ opioid receptor agonists) (Pertwee *et al.*, 1993b; Pertwee and Griffin, 1995). Nor is the onset of cannabinoid tolerance in the mouse vas deferens accompanied by any detectable change in the sensitivity of this tissue to the contractile transmitters that are thought to mediate the twitch response (Pertwee and Griffin, 1995).

There is evidence, at least for guinea-pig MPLM, mouse vas deferens and mouse bladder, that the cannabinoid receptors through which cannabinoids inhibit electrically-evoked contractions are located on prejunctional nerve terminals (see Pertwee, 1997). Here they most probably modulate the release of contractile neurotransmitters (see Section 6.2.1). These are acetylcholine for guinea-pig MPLM (Cowie *et al.* 1978), noradrenaline and ATP for mouse vas deferens (Stjärne and Åstrand 1985; Von

Kügelgen *et al.*, 1989) and noradrenaline and acetylcholine for rodent bladder (Burnstock *et al.*, 1972; Brown *et al.*, 1979; Boland *et al.*, 1993). It has recently been found that the mouse vas deferens contains both CB_1 and CB_2-like mRNA, that the CB_2-selective ligands JWH-015 and JWH-051 potently inhibit electrically-evoked contractions of this preparation and that this effect is not attenuated by SR141716A except at a concentration ($10\,\mu M$) expected to block CB_2 as well as CB_1 receptors (Griffin *et al.*, 1996, 1997). These observations raise the possibility that CB_2 receptors may share the ability of CB_1 receptors to mediate inhibition of electrically-evoked contractions of the mouse vas deferens.

7 DISTRIBUTION, FORMATION, RELEASE AND FATE OF ANANDAMIDE

7.1 Distribution

Tissues in which anandamide has been found are human, cow, sheep, pig and rat brain, human and rat spleen, human heart and rat skin and testis (Devane *et al.*, 1992b; Schmid *et al.*, 1995; Felder *et al.*, 1996; Sugiura *et al.*, 1996c). The capacity to synthesize and/or hydrolyse anandamide has also been observed in a range of tissues (Sections 7.2 and 7.4). Within the brain, anandamide has so far been detected in hippocampus, striatum, cerebellum and thalamus, its concentrations between areas (measured before the onset of any significant postmortem changes) varying more widely in human brain (25 pmol/g tissue in the cerebellum to 148 pmol/g tissue in the hippocampus) than in rat brain (20–29 pmol/g tissue) (Felder *et al.*, 1996). Basal levels of anandamide in fresh tissue tend to be markedly less than those of other endogenous *N*-acylethanolamines. Thus Schmid *et al.* (1995) have found anandamide to make up only 0.9% of all *N*-acylethanolamines in fresh pig brain (173 pmol anandamide/g wet weight), 1.1% in fresh cow brain (up to 115 pmol anandamide/g wet weight) and to be undetectable in fresh sheep brain. Similarly, Sugiura *et al.* (1996b) have reported there to be 4.3 pmol of anandamide/g wet weight in fresh rat brain. This accounts for 0.7% of total brain *N*-acylethanolamines and contrasts markedly with the corresponding percentage values for palmitoylethanolamide and stearoylethanolamide which are 50.6% and 19.4% respectively.

7.2 Formation

Two enzyme catalysed pathways have been proposed for the biosynthesis of anandamide. In one, the immediate precursor is *N*-arachidonoylphosphatidylethanolamine, the hydrolysis of which (to anandamide) may be catalysed by phospholipase D (Di Marzo *et al.*, 1994, 1996a,b; Schmid *et al.*, 1995; Cadas *et al.*, 1996; Sugiura *et al.*, 1996b,c). Evidence for this mode of synthesis has been observed in rat brain and testis microsomes (Sugiura *et al.*, 1996b,c), in N18TG2 mouse neuroblastoma cells and J774 mouse macrophages (Di Marzo *et al.*, 1996a,b) and in primary cultures of rat brain striatal and cortical neurones and of mouse brain cortical neurones (Di Marzo *et al.*, 1994; Cadas *et al.*, 1996; Hansen *et al.*, 1997). Further support for this biosynthetic pathway comes from the observation that production of anandamide (and other *N*-acylethanolamines) by N18TG2 mouse neuroblastoma cells, J774 mouse macrophages and cultures of rat brain neurones prelabelled with [^{3}H] or [^{14}C]ethanolamine can be increased by exogenous phospholipase D (Di Marzo et al., 1994, 1996a). Evidence for phospholipase D-induced production of substance(s) with anandamide-like activity has also been obtained in electrophysiological experiments with rat brain slices (Poling et al., 1996). In the second proposed pathway, the immediate precursors are arachidonic acid and ethanolamine. Formation of anandamide from these compounds has been detected in rat brain homogenates (Deutsch and Chin, 1993), bovine brain P$_2$ membranes (Devane and Axelrod, 1994), rat brain microsomes (Sugiura et al., 1996b), cytosolic and microsomal fractions of rabbit brain (Kruszka and Gross, 1994), rabbit kidney and liver microsomes (Kruszka and Gross, 1994) and pregnant mouse uterine microsomes (Paria et al., 1996). This condensation reaction is also catalysed by a partially purified enzyme obtained from pig brain microsomes (Ueda et al., 1995a). Within bovine brain, the reaction has been reported to occur more rapidly in hippocampus than in thalamus, striatum, frontal cortex, pons, cerebellum or medulla (Devane and Axelrod, 1994).

There is evidence from experiments with rat brain microsomes that the enzyme catalysing the condensation of arachidonic acid and ethanolamine lacks selectivity, this preparation readily catalysing the incorporation of ethanolamine into a range of different fatty acids (Sugiura et al., 1996b). However, greater selectivity for arachidonic acid has been observed in cytosolic and microsomal fractions of rabbit brain (Kruszka and Gross, 1994) and bovine brain hippocampal membranes (Devane and Axelrod, 1994). In brain preparations, the condensation of arachidonic acid and ethanolamine seems to require rather high concentrations of both reactants, particularly ethanolamine for which reported apparent K_m values are 50 mM (pig brain) and 135 mM (rat brain) (Schmid et al., 1995; Ueda et al., 1995a, Sugiura et al., 1996b). Consequently, it could well be that unless there are special metabolic pools for arachidonic acid and ethanolamine, this reaction normally proceeds in the opposite direction (Sugiura et al., 1996b). This may not be true for all tissues as Paria et al. (1996) have obtained results from experiments with a microsomal preparation of pregnant mouse uterus that led them to conclude that, in mouse uterus at least, anandamide can be formed from arachidonic acid ($K_m = 3.8 \mu M$) and ethanolamine ($K_m = 1.2$ mM) under physiological conditions. It is possible, therefore, that there are multiple pathways for anandamide synthesis that vary with both species and cell type (Paria et al., 1996).

7.3 Release

In line with the hypothesis that anandamide may not be stored but rather synthesized on demand, experiments with whole cell preparations have yielded results that point to the presence of a mechanism for activating anandamide synthesis and release in neurones. Thus there is evidence from experiments with primary cultures of rat brain striatal or cortical neurones prelabelled with [^{3}H]ethanolamine that de novo synthesis of the putative anandamide precursor, [^{3}H]N-arachidonoylphosphatidylethanolamine, and formation and release of [^{3}H]anandamide can be triggered by a Ca^{2+} ionophore (ionomycin), by high K$^+$, by a glutamate receptor agonist (kainic acid) and by potassium channel blockers (4-aminopyridine and 3,4-diaminopyridine) (Di Marzo et al., 1994;

Fontana *et al.* 1995; Cadas *et al.*, 1996). There is also evidence that Ca^{2+}-dependent formation of N-acylphosphatidylethanolamine in cortical neurones is enhanced by cyclic AMP (Cadas *et al.*, 1996). Since cannabinoid receptors are negatively coupled to adenylate cyclase, this may indicate the existence of a negative feedback mechanism through which released anandamide inhibits its own further formation by the activation of presynaptic cannabinoid receptors. It is noteworthy that ionomycin triggers the release of several N-acylethanolamines and that (as in unstimulated tissue) anandamide makes up only a very small proportion of these ethanolamides (Di Marzo *et al.*, 1994; Fontana *et al.*, 1995; Cadas *et al.*, 1996). For example, Fontana *et al.* (1995) have found the percentage composition of N-acylethanolamines recovered from cortical neurones prelabelled with [^{3}H]ethanolamine and stimulated with ionomycin to be 31.3% oleoylethanolamide, 25.7% palmitoylethanolamide, 21.7% stearoylethanolamide, 10.6% linolenoylethanolamide, 4.9% arachidonoylethanolamide (anandamide) and 1.1% linoleoylethanolamide. There are several reports that significant increases in the levels of anandamide and other N-acylethanolamines occur postmortem in unfrozen brain tissue (Hansen *et al.*, 1995; Schmid *et al.*, 1995; Felder *et al.*, 1996; Sugiura *et al.*, 1996b). Whether these postmortem changes are triggered by the increases in intracellular free calcium that are usually associated with brain ischaemia has yet to be investigated as has the question of their physiological/pathophysiological significance, if any.

7.4 Fate

Results from experiments with broken cell preparations have identified several tissues containing an enzyme that catalyses the hydrolysis of anandamide to arachidonic acid and ethanolamine. As neither of these metabolites is a cannabinoid receptor ligand, this is an inactivation process. Preparations in which anandamide hydrolysis has been observed include intact N18TG2 mouse neuroblastoma cells, N18TG2 and C6 rat glioma cell membranes, cow or rat brain homogenates, rat forebrain membranes and microsomal preparations of pig, rat or mouse brain or of N18TG2 cells or pregnant mouse uterus (Deutsch and Chin, 1993; Desarnaud *et al.*, 1995; Hillard *et al.*, 1995a,b; Maurelli *et al.*, 1995; Omeir *et al.*, 1995; Ueda *et al.*, 1995a; Paria *et al.*, 1996; Watanabe *et al.*, 1996). Homogenates of rat heart, rat skeletal muscle, HeLa cells, human larynx epidermoid carcinoma (Hep2) cells and human hepatocellular carcinoma (HepG2) cells have been reported not to hydrolyse anandamide (Deutsch and Chin, 1993). Experiments with subcellular fractions indicate that the capacity to hydrolyse anandamide resides mainly in membranes, particularly microsomal membranes, and is essentially absent from the cytosol (Deutsch and Chin, 1993; Desarnaud *et al.*, 1995; Hillard *et al.*, 1995a,b; Maurelli *et al.*, 1995; Ueda *et al.*, 1995a). Microsomal anandamide hydrolysing activity has been reported to be much higher in rat liver and brain than in rat heart, kidney, intestine, stomach, lung, spleen or skeletal muscle (Desarnaud *et al.*, 1995).

Factors that have been reported to determine the rate of enzymic hydrolysis of anandamide include temperature, pH and substrate and tissue/protein concentration (Desarnaud *et al.*, 1995; Hillard *et al.*, 1995a,b; Maurelli *et al.*, 1995; Ueda *et al.*, 1995a; Paria *et al.*, 1996; Watanabe *et al.*, 1996). Reported apparent K_m values for the enzyme vary with the broken cell preparation used, ranging from 3.4 to 67 µM (Desarnaud *et al.*, 1995; Hillard *et al.*, 1995b; Maurelli *et al.*, 1995; Ueda *et al.*, 1995a; Paria *et al.*, 1996; Watanabe *et al.*, 1996). It is unlikely from its known properties and cellular location that the enzyme that hydrolyses anandamide is cathepsin G, chymotrypsin, trypsin, deamidase (lysosomal protective protein), ceramidase, plasmin, aminopeptidase, elastase or lipoyl-X hydrolase (Hillard *et al.*, 1995b; Ueda *et al.*, 1995a).

An enzyme that catalyses the conversion of anandamide to arachidonic acid and ethanolamine was recently cloned from a rat liver cDNA library and found to be expressed in rat liver and brain and to a lesser extent, in rat spleen, lung, kidney and testis (Cravatt *et al.*, 1996). It was not detectable in rat heart or skeletal muscle. The enzyme has been named fatty acid amide hydrolase as it also hydrolyses certain other fatty acid amides when transfected into COS-7 cells (Cravatt *et al.*, 1996). The lack of absolute specificity of the cloned enzyme and a finding that the rate of the fatty acid amide hydrolysis it catalyses depends on the chain length and degree of unsaturation of the substrate (anandamide > oleoylamide ≫ myristic amide > palmitic amide > stearic amide), is consistent with data obtained using other preparations (Cravatt *et al.*, 1996; Desarnaud *et al.*, 1995; Maurelli *et al.*, 1995; Ueda *et al.*, 1995a).

Anandamide seems also to serve as a substrate for enzymes other than fatty acid amide hydrolase. Thus there are indications from experiments with mouse hepatic microsomes that cytochromes P450, particularly P450 3A, catalyse the monohydroxylation, dihydroxylation and epoxidation of anandamide (Bornheim *et al.*, 1993, 1995). Hydroxylation/epoxidation of anandamide has also been detected in experiments with mouse brain microsomes (Bornheim *et al.*, 1995), although not with rat brain microsomes (Desarnaud *et al.*, 1995). In addition, there is evidence that anandamide is converted by lipoxygenases to hydroperoxy- and hydroxy-metabolites, some of which (e.g. 15-hydroxy-5,8,11,13-eicosatetraenoylethanolamide) retain cannabimimetic activity (Hampson *et al.*, 1995; Ueda *et al.*, 1995b).

Experiments with whole cell preparations have yielded results indicating that, after its presumed neuronal release, anandamide may be rapidly removed from the extracellular space by tissue uptake processes. Thus Di Marzo *et al.* (1994) have obtained evidence from experiments with primary cultures of rat brain striatal and cortical neurones that anandamide is taken up into neurones and glia by a carrier-mediated, saturable, temperature-dependent process ($t_{1/2} = 2.5$ min). Their data also suggest that following its uptake, anandamide is intracellularly converted to arachidonic acid and ethanolamine which are then incorporated into phospholipids. It is noteworthy that, in mouse cultured cortical neurones, Hansen *et al.* (1995) have detected only a rather slow hydrolysis of added anandamide ($t_{1/2} = 2.6$ h). In line with the idea that the metabolic inactivation of anandamide takes place intracellularly, is the observation that hydrolase activity is significantly less in synaptic plasma membranes than in brain microsomes (Hillard *et al.*, 1995b).

7.5   Inhibitors of Anandamide Biosynthesis and Hydrolysis

Several agents have been identified that inhibit the enzymic hydrolysis of anandamide to arachidonic acid and ethanolamine when administered at concentrations lying in the nM or μM range (Table 6). These are phenylmethylsulfonyl fluoride (PMSF), diisopropylfluorophosphate (DFP) and *p*-bromophenylacyl bromide which are general protease inhibitors, thimerosal, p-chloromercuribenzoic acid (PCMB) and p-hydroxy-mercuribenzoate which are sulphydryl reactive agents, arachidonyl trifluoromethyl ketone (AACOCF$_3$) and methyl arachidonyl fluorophosphonate (MAFP) which are phospholipase A$_2$ inhibitors, palmitylsulfonyl fluoride, γ-linolenyl, stearyl, palmityl and myristyl trifluoromethyl ketone, arachidonic acid, ethyl 2-oxostearate, ethyl 2-oxopalmitate, oleoylethanolamide and the putative sleep-inducing factor, oleoylamide. Several of the agents listed in Table 6 bind to specific cannabinoid CB$_1$ receptors at concentrations at which they inhibit the hydrolysis of anandamide (see footnote to Table 6). Consideration of the nature of some of the compounds that inhibit anandamide hydrolysis has led Hillard *et al.* (1995b) to suggest that the enzyme which catalyses this reaction depends on disulphide bonding for its activity, has serine as part of its active site and is not cathepsin G, chymotrypsin, trypsin or lipoyl-X hydrolase. In line with this suggestion is the finding by Deutsch *et al.* (1997) that MAFP inhibits anandamide hydrolysis much more readily than it does trypsin or chymotrypsin. Deutsch *et al.* (1997) also found anandamide metabolism to be markedly more susceptible than acetylcholinesterase to inhibition by MAFP. Interestingly, the water soluble serine esterase inhibitor, 4-(2-aminoethyl)benzenesulfonyl fluoride (> 1 mM), has been reported by Hillard *et al.* (1995b) not to inhibit anandamide hydrolysis, suggesting that the active site of the hydrolase may be located in a hydrophobic region of the enzyme protein. Concentrations of PMSF, AACOCF$_3$ and certain other agents that inhibit anandamide hydrolysis have been reported also to inhibit the enzymic hydrolysis of oleoylamide to oleic acid (Table 6 and Cravatt *et al.*, 1996).

Results from experiments with pig or rat brain preparations indicate PMSF, AACOCF$_3$, DFP, PCMB and MAFP to be no less potent in inhibiting the formation of anandamide from arachidonic acid and ethanolamine than in inhibiting anandamide hydrolysis, supporting the notion that both processes are catalysed by the same enzyme (Table 6). AACOCF$_3$ has also been reported to inhibit both processes in mouse uterine microsomes, although inhibition of synthase activity by this agent was less complete than that of hydrolase activity and was not dose-related (Paria *et al.*, 1996). However, concentrations of PMSF that inhibited anandamide hydrolase activity in uterine microsomes were found not to inhibit the condensation of arachidonic acid and ethanolamine (Paria *et al.*, 1996). Instead they appeared to increase the rate of this reaction. There is also a report that the enzymic condensation of arachidonic acid and ethanolamine by rat brain homogenate is not inhibited by 1.5 mM PMSF (Deutsch and Chin, 1993).

8 PHYSIOLOGICAL SIGNIFICANCE OF ANANDAMIDE

Evidence that anandamide is a chemical mediator with the physiological role of activating cannabinoid CB$_1$ receptors when released from neurones is summarized below.

(a) Anandamide is a selective cannabinoid receptor ligand, binding readily to cannabinoid CB$_1$, CB$_{1(a)}$ and CB$_2$ receptors but not to a wide range of noncannabinoid receptors (Mechoulam and Fride, 1995; Pertwee, 1997; Section 3.3).

Table 6   Inhibitors of anandamide synthesis and metabolism

Enzyme inhibitor	Tissue preparation (Synthase/hydrolase)	Inhibitory concentration vs synthase (μM)	Inhibitory concentration vs hydrolase (μM)	Reference
PMSF	Rat brain homogenate	1500 = inactive	ND	Deutsch and Chin (1993)
	Rat brain homogenate	ND	200	Koutek et al. (1994)
	Rat brain homogenate	ND	0.9[b]	Deutsch et al. (1997)
	Rat forebrain membranes	ND	150	Hillard et al. (1995a)
	Rat forebrain membranes[a]	ND	12.9[b]	Hillard et al. (1995b)
	Rat brain microsomes	ND	10.5[b]	Lang et al. (1996)
	Partially purified pig brain enzyme	10–3000[c]	3–3000[c]	Ueda et al. (1995a)
	Neuroblastoma and glioma cell membranes	ND	1500	Deutsch and Chin (1993)
	Intact neuroblastoma cells	ND	7.8	Koutek et al. (1994)
	Neuroblastoma cell membranes[d,e]	ND	100	Maurelli et al. (1995)
	Mouse uterine microsomes	8000 = inactive	1–2000[c]	Paria et al. (1996)
PSF	Rat brain microsomes	ND	0.05[b]	Lang et al. (1996)
AACOCF$_3$	Rat brain homogenate[f]	ND	7.5	Koutek et al. (1994)
	Partially purified pig brain enzyme	0.3–30[c]	0.3–30[c]	Ueda et al. (1995a)
	Intact neuroblastoma cells[g]	ND	3.9	Koutek et al. (1994)
	Neutroblastoma cell membranes[d,e]	ND	10	Maurelli et al. (1995)
	Mouse uterine microsomes	0.001–10[h]	0.001–10[c]	Paria et al. (1996)
MAFP[i]	Rat brain homogenate	0.0015[b]	0.0025[b]	Deutsch et al. (1997)
	Intact neuroblastoma cells	ND	0.02[b]	Deutsch et al. (1997)
DFP	Rat forebrain membranes	ND	150	Hillard et al. (1995a)
	Rat forebrain membranes	ND	6.9[b]	Hillard et al. (1995b)
	Partially purified pig brain enzyme	3–300[c]	3–300[c]	Ueda et al. (1995a)
Thimerosal	Rat forebrain membranes	ND	150	Hillard et al. (1995a)
	Rat forebrain membranes	ND	17[b]	Hillard et al. (1995b)

Enzyme inhibitor	Tissue preparation (Synthase/hydrolase)	Inhibitory concentration vs synthase (μM)	Inhibitory concentration vs hydrolase (μM)	Reference
PCMB	Partially purified pig brain enzyme	0.03–1 [c]	0.03–1 [c]	Ueda et al. (1995a)
THC	Mouse brain microsomes	ND	96	Watanabe et al. (1996)
CBN	Mouse brain microsomes	ND	32	Watanabe et al. (1996)
CBD	Mouse brain microsomes	ND	32	Watanabe et al. (1996)

Abbreviations: PMSF = phenylmethylsulfonyl fluoride; PSF = palmitylsulfonyl fluoride; AACOCF$_3$ = arachidonyl trifluoromethyl ketone; MAFP = methyl arachidonyl fluorophosphonate; DFP = diisopropyl fluorophosphate; PCMB = p-chloromercuribenzoic acid; THC = delta-9-tetrahydrocannabinol; CBN = cannabinol; CBD = cannabidiol.

10 μM AACOCF$_3$, 150 μM DFP, 150 μM thimerosal, THC (<1 μM) and CBN (<1 μM) undergo significant binding to specific cannabinoid CB$_1$ receptors, whereas concentrations of PMSF, CBD and arachidonic acid that inhibit hydrolase activity do not (Koutek et al., 1994; Hillard et al., 1995a; see also Pertwee, 1997). MAFP binds irreversibly to specific cannabinoid CB$_1$ receptors (IC$_{50}$ = 0.02 μM) (Deutsch et al., 1997).

[a] Hydrolase also markedly inhibited by 100 μM arachidonic acid. [b] IC$_{50}$. [c] Dose-related effect. [d] Oleoylamide hydrolysis is also inhibited by PMSF and AACOCF$_3$. [e] [^{14}C]anandamide and [^{14}C]oleoylamide hydrolysis inhibited by 100 μM p-hydroxymecuribenzoate, p-bromophenylacyl bromide, oleoylethanolamide, anandamide (unlabelled) and oleoylamide (unlabelled) but not by 100 μM o-phenanthroline, 100 μM benzamidine, 100 μM palmitoylethanolamide, 1 mM dithiothreitol or 5 mM EDTA. [f] Hydrolase also inhibited to the same extent by 7.5 μM ethyl 2-oxostearate, ethyl 2-oxopalmitate, and by γ-linolenyl, stearyl, palmityl and myristyl trifluromethyl ketone, none of which (at 10 μM) significantly affected cannabinoid receptor binding. [g] Hydrolase also inhibited to a much smaller extent by 7.8 μM ethyl 2-oxostearate and γ-linolenyl trifluromethyl ketone. [h] Inhibition is modest and not dose-related. [i] Irreversible inhibitor of anandamide hydrolysis.

(b) The presence of anandamide has been demonstrated in brain, testis, heart and spleen (Section 7.1), all of which contain CB_1 receptors (Sections 4.1, 4.2 and 6.2.1).

(c) The capacity to synthesize anandamide (Section 7.2) has been detected in several tissues thought to contain cannabinoid receptors (Sections 4.1, 4.2 and Table 2). These are rat brain, rat testis and N18TG2 mouse neuroblastoma cells (from N-arachidonoylphosphatidylethanolamine) and bovine, pig, rabbit and rat brain and pregnant mouse uterus (from arachidonic acid and ethanolamine).

(d) Mechanisms for triggering the formation and release of anandamide have been detected in cultured striatal and cortical neurones (Sections 7.2 and 7.3). Stimuli that trigger anandamide formation in and release from these cells are ones known to induce the neuronal release of established transmitters by increasing intracellular calcium or depolarizing neuronal membranes. The neuronal release of anandamide under more physiological conditions has yet to be demonstrated.

(e) Experiments with cultured striatal and cortical neurones have also provided evidence for the existence of mechanisms for the removal of released anandamide from the vicinity of its receptors (Section 7.4). The results obtained suggest that anandamide is first removed from the extracellular space by an uptake process and then hydrolysed intracellularly. Although the presence of uptake mechanisms for anandamide in other tissue preparations remains to be investigated, it is already known that several tissues thought to contain cannabinoid receptors (Sections 4.1, 4.2 and Table 2) hydrolyse anandamide or express fatty acid amide hydrolase (Section 7.4). These are brain, spleen, testis, lung, uterus and neuroblastoma cells. Within rat brain, the distribution of anandamide hydrolysing activity is heterogeneous and broadly parallels that of cannabinoid receptors (Desarnaud et al., 1995; Hillard et al., 1995b). Since anandamide is not the only substrate for the enzyme that catalyses its hydrolysis to arachidonic acid and ethanolamine (Section 7.4), it could well be that this catabolic reaction will also prove to be detectable in tissues that lack cannabinoid receptors.

Although there is no doubt that anandamide is a CB_1 receptor agonist (Section 3.3), there are several reports that it is not a CB_2 receptor agonist, at least when the measured response is inhibition of cyclic AMP production or attenuation of 5-HT release from mast cells (Sections 5.1 and 6.2.2). Consequently, the possibility exists that the natural agonists for CB_1 and CB_2 receptors are not the same. A report that anandamide is absent from human serum, plasma or cerebrospinal fluid suggests that it does not require transport through the circulation to reach its sites of action (Felder et al., 1996). Storage sites for anandamide have not yet been demonstrated and concentrations of anandamide in various areas of human brain (Felder et al., 1996) are considerably less than those of established transmitters such as noradrenaline (in human hypothalamus) and of dopamine and 5-hydroxytryptamine (in human caudate nucleus) (Mackay et al., 1978), leaving open the possibility that anandamide is synthesized on demand rather than stored. Felder et al. (1995) have speculated that any storage of anandamide is more likely to occur within cell membranes than cytosolic vesicles. Another speculation, that after its release, anandamide may interact with cannabinoid receptors without first leaving the membrane also deserves consideration (see Herkenham, 1995).

The significance of the findings that basal levels of anandamide in the brain are markedly less than those of several other endogenous N-acylethanolamines and that agents inducing the neuronal formation and release of anandamide seem to stimulate a far greater production of other fatty acid ethanolamides (Sections 7.1 and 7.3) remains to be established. Possibly, these findings are indications of the efficiency and selectivity of the mechanisms that remove anandamide from its sites of action following its release. Also still to be established is the physiological significance of the presence in mammalian tissues of cannabinoid receptor agonists other than anandamide (Section 3.1). One of these, 2-arachidonoyl glycerol, has been detected in the brain in notably larger amounts than anandamide (Sugiura et $al.$, 1995, 1996b). However, it has less affinity both for CB_1 binding sites, even in the presence of the esterase inhibitor, diisopropylfluorophosphate, and for CB_2 binding sites (Sugiura et $al.$, 1995; Mechoulam et $al.$, 1995). Although 2-arachidonoyl glycerol is known to be a CB_1 receptor agonist, its ability to activate CB_2 receptors remains to be established (Mechoulam et $al.$, 1995; Sugiura et $al.$, 1996a). The biosynthesis of 2-arachidonoyl glycerol probably proceeds by the enzymic hydrolysis of 1-acyl-2-arachidonoyl glycerols, lyso-2-arachidonoylphosphatidylcholine and/or lyso-2-arachidonoylphosphatidylinositol (Sugiura et $al.$, 1995; Di Marzo et $al.$, 1996b).

9 CONCLUDING DISCUSSION

The discovery of cannabinoid CB_1 and CB_2 receptors and of endogenous cannabinoid receptor agonists has been followed by the emergence of evidence that one of the roles of central and peripheral CB_1 receptors is modulation of neurotransmitter release and by the development of selective CB_1 and CB_2 receptor agonists and antigonists. The availability of these agents should greatly facilitate a more complete identification of the physiological roles of cannabinoid receptors, both CB_1 and CB_2.

One important question urgently requiring resolution is that of whether anandamide and delta-9-THC can or cannot activate CB_2 receptors. If they cannot, there are important implications. One of these, as mentioned earlier, is that CB_1 and CB_2 receptors may have different natural agonists. Another is that since it is CB_2 receptors that predominate in immune tissues, the effects of anandamide, delta-9-THC (and cannabis) on immune function may be much less than those of drugs that do activate CB_2 receptors (yet to be determined). It will also be important to extend existing knowledge about the effector systems of CB_1, $CB_{1(a)}$ and CB_2 receptors, to establish whether mammalian tissues contain other types/subtypes of cannabinoid receptors and to determine the extent to which non-receptor-mediated processes contribute to the pharmacology of individual cannabinoids.

Other important goals for future research must be to identify with greater certainty those endogenous ligands that interact with cannabinoid receptors under physiological and/or pathophysiological conditions, to extend knowledge about the physiological and biochemical processes responsible for the formation, release and fate of these ligands and to establish whether such ligands are stored or synthesized on demand. Advances in these areas should facilitate the development of drugs that selectively modulate extracellular levels of endogenous cannabinoid receptor ligands.

It will also be important to elucidate more fully the part played by cannabinoid receptors and their endogenous agonists in disease states. Given the known pharmacological properties of cannabis and cannabinoid receptor agonists and the distribution pattern of cannabinoid receptors, initial experiments should perhaps focus on the role of the endogenous cannabinoid system in disorders of cognition, memory, affect, motor control and immune function. The therapeutic potential of cannabinoid receptor ligands as anti-inflammatory agents and analgesics and in the control of glaucoma, bronchial asthma, epilepsy and/or gastrointestinal motility disorders also merits attention (Hollister, 1986). Another important goal is to establish more fully the part played by cannabinoid receptors in tolerance. A summary of current information about the pathophysiology of cannabinoid receptors and about their role in the production of cannabinoid tolerance can be found elsewhere (Pertwee, 1997).

After the completion of this chapter, important additional information appeared in the literature both about the biochemistry and pharmacology of 2-arachidonoyl glycerol and about agents that can inhibit the metabolism or tissue uptake of endogeneous cannabinoids. For a brief account of these recent advances, see Pertwee (1998) Pharmacological, physiological and clinical implications of the discovery of cannabinoid receptors, *Biochem. Soc. Transactions*. In press

REFERENCES

Abadji, V., Lin, S., Taha, G., Griffin, G., Stevenson, L.A., Pertwee, R.G. and Makriyannis, A. (1994) (R)-methanandamide: a chiral novel anandamide possessing higher potency and metabolic stability. *J. Med. Chem.* **37**, 1889–1893.

Adams, I.B., Ryan, W., Singer, M., Thomas, B.F., Compton, D.R., Razdan, R.K. and Martin, B.R. (1995) Evaluation of cannabinoid receptor binding and *in vivo* activities for anandamide analogs. *J. Pharmacol. Exp. Ther.* **273**, 1172–1181.

Barg, J., Fride, E., Hanuš, L., Levy, R., Matus-Leibovitch, N., Heldman, E., Bayewitch, M., Mechoulam, R. and Vogel, Z. (1995) Cannabinomimetic behavioral effects of and adenylate cyclase inhibition by two new endogenous anandamides. *Eur. J. Pharmacol.* **287**, 145–152.

Barth, F., Rinaldi-Carmona, M., Millan, J., Derocq, J.-M., Bouaboula, M., Casellas, P., Congy, C., Oustric, D., Sarran, M., Calandra, B., Portier, M., Shire, D., Brelière, J.-C. and Le Fur, G. (1997) SR144528, a potent and selective antagonist of the CB_2 receptor. *Proc. Int. Cannabinoid Res. Soc.*, p. 11.

Bayewitch, M., Avidor-Reiss, T., Levy, R., Barg, J., Mechoulam, R. and Vogel, Z. (1995) The peripheral cannabinoid receptor: adenylate cyclase inhibition and G protein coupling. *FEBS Lett.* **375**, 143–147.

Bayewitch, M., Rhee, M.-H., Avidor-Reiss, T., Breuer, A., Mechoulam, R. and Vogel, Z. (1996) (-)-Δ^9-Tetrahydrocannabinol antagonizes the peripheral cannabinoid receptor-mediated inhibition of adenylyl cyclase. *J. Biol. Chem.* **271**, 9902–9905.

Beal, J.E., Olson, R., Laubenstein, L., Morales, J.O., Bellman, P., Yangco, B., Lefkowitz, L., Plasse, T.F. and Shepard, K.V. (1995) Dronabinol as a treatment for anorexia associated with weight loss in patients with AIDS. *J. Pain Sympt. Manag.* **10**, 89–97.

Bidaut-Russell, M., Devane, W.A. and Howlett, A.C. (1990) Cannabinoid receptors and modulation of cyclic AMP accumulation in the rat brain. *J. Neurochem.* **55**, 21–26.

Bidaut-Russell, M. and Howlett, A.C. (1991) Cannabinoid receptor-regulated cyclic AMP accumulation in the rat striatum. *J. Neurochem.* **57**, 1769–1773.

Boland, B., Himpens, B., Paques, C., Casteels, R. and Gillis, J.M. (1993) ATP induced-relaxation in the mouse bladder smooth muscle. *Br. J. Pharmacol.* **108**, 749–753.

Bornheim, L.M., Kim, K.Y., Chen, B. and Correia, M.A. (1993) The effect of cannabidiol on mouse hepatic microsomal cytochrome P450-dependent anandamide metabolism. *Biochem. Biophys. Res. Commun.* **197**, 740–746.

Bornheim, L.M., Kim, K.Y., Chen, B. and Correia, M.A. (1995) Microsomal cytochrome P450-mediated liver and brain anandamide metabolism. *Biochem. Pharmacol.* **50**, 677–686.

Bouaboula, M., Bourrié, B., Rinaldi-Carmona, M., Shire, D., Le Fur, G. and Casellas, P. (1995a) Stimulation of cannabinoid receptor CB1 induces *krox-24* expression in human astrocytoma cells. *J. Biol. Chem.* **270**, 13973–13980.

Bouaboula, M., Poinot-Chazel, C., Bourrié, B., Canat, X., Calandra, B., Rinaldi-Carmona, M., Le Fur, G. and Casellas, P. (1995b) Activation of mitogen-activated protein kinases by stimulation of the central cannabinoid receptor CB1. *Biochem. J.* **312**, 637–641.

Bouaboula, M., Poinot-Chazel, C., Marchand, J., Canat, X., Bourrié, B., Rinaldi-Carmona, M., Calandra, B., Le Fur, G. and Casellas, P. (1996) Signaling pathway associated with stimulation of CB2 peripheral cannabinoid receptor. Involvement of both mitogen-activated protein kinase and induction of Krox-24 expression. *Eur. J. Biochem.* **237**, 704–711.

Bouaboula, M., Rinaldi, M., Carayon, P., Carillon, C., Delpech, B., Shire, D., Le Fur, G. and Casellas, P. (1993) Cannabinoid-receptor expression in human leukocytes. *Eur. J. Biochem.* **214**, 173–180.

Bramblett, R.D., Panu, A.M., Ballesteros, J.A. and Reggio, P.H. (1995) Construction of a 3D model of the cannabinoid CB1 receptor: determination of helix ends and helix orientation. *Life Sci.* **56**, 1971–1982.

Brown, C., Burnstock, G. and Cocks, T. (1979) Effects of adenosine 5'-triphosphate (ATP) and β-γ-methylene ATP on the rat urinary bladder. *Br. J. Pharmacol.* **65**, 97–102.

Burnstock, G., Dumsday, B. and Smythe, A. (1972) Atropine resistant excitation of the urinary bladder: the possibility of transmission via nerves releasing a purine nucleotide. *Br. J. Pharmacol.* **44**, 451–461.

Cadas, H., Gaillet, S., Beltramo, M., Venance, L. and Piomelli, D. (1996) Biosynthesis of an endogenous cannabinoid precursor in neurons and its control by calcium and cAMP. *J. Neurosci.* **16**, 3934–3942.

Caenazzo, L., Hoehe, M.R., Hsieh, W.-T., Berrettini, W.H., Bonner, T.I. and Gershon, E.S. (1991) HindIII identifies a two allele DNA polymorphism of the human cannabinoid receptor gene (CNR). *Nucl. Acids Res.* **19**, 4798.

Casiano, F.M., Arnold, R., Haycock, D., Kuster, J. and Ward, S.J. (1990) Putative aminoalkylindoles (AAI) antagonists. *NIDA Research Monograph Series* **105**, 295–296.

Caulfield, M.P. and Brown, D.A. (1992) Cannabinoid receptor agonists inhibit Ca current in NG108-15 neuroblastoma cells via a Pertussis toxin-sensitive mechanism. *Br. J. Pharmacol.* **106**, 231–232.

Chakrabarti, A., Onaivi, E.S. and Chaudhuri, G. (1995) Cloning and sequencing of a cDNA encoding the mouse brain-type cannabinoid receptor protein. *DNA Sequence – J. Sequencing and Mapping* **5**, 385–388.

Childers, S.R. and Deadwyler, S.A. (1996) Role of cyclic AMP in the actions of cannabinoid receptors. *Biochem. Pharmacol.* **52**, 819–827.

Childers, S.R., Sexton, T. and Roy, M.B. (1994) Effects of anandamide on cannabinoid receptors in rat brain membranes. *Biochem. Pharmacol.* **47**, 711–715.

Collin, C., Devane, W.A., Dahl, D., Lee, C.-J., Axelrod, J. and Alkon, D.L. (1995) Long-term synaptic transformation of hippocampal CA1 γ-aminobutyric acid synapses and the effect of anandamide. *Proc. Natl. Acad. Sci. (USA)* **92**, 10167–10171.

Collins, D.R., Pertwee, R.G. and Davies, S.N. (1994) The action of synthetic cannabinoids on the induction of long-term potentiation in the rat hippocampal slice. *Eur. J. Pharmacol.* **259**, R7–R8.

Collins, D.R., Pertwee, R.G. and Davies, S.N. (1995) Prevention by the cannabinoid antagonist, SR141716A, of cannabinoid-mediated blockade of long-term potention in the rat hippocampal slice. *Br. J. Pharmacol.* **115**, 869–870.

Compton, D.R., Aceto, M.D., Lowe, J. and Martin, B.R. (1996) *In vivo* characterization of a specific cannabinoid receptor antagonist (SR141716A): inhibition of Δ^9-tetrahydrocannabinol-induced responses and apparent agonist activity. *J. Pharmacol. Exp. Ther.* **277**, 586–594.

Condie, R., Herring, A., Koh, W.S., Lee, M. and Kaminski, N.E. (1996) Cannabinoid inhibition of adenylate cyclase-mediated signal transduction and interleukin 2 (IL-2) expression in the murine T-cell line, EL4.IL-2. *J. Biol. Chem.* **271**, 13175–13183.

Coutts, A.A., Fernando, S.R., Griffin, G., Nash, J.E. and Pertwee, R.G. (1995) Evidence that guinea-pig small intestine contains CB1 cannabinoid receptors and possibly also a cannabinoid receptor ligand. *Br. J. Pharmacol. (Proc. Suppl.)* **116**, 46P.

Coutts, A.A. and Pertwee, R.G. (1996) Effect of cannabinoid agonists and SR141716A on acetylcholine release from the myenteric plexus of the guinea-pig small intestine. *Br. J. Pharmacol. (Proc. Suppl.)* **119**, 312P.

Coutts, A.A. and Pertwee, R.G. (1997) Inhibition by cannabinoid receptor agonists of acetylcholine release from the guinea-pig myenteric plexus. *Br. J. Pharmacol.* **121**, 1557–1566.

Cowie, A.L., Kosterlitz, H.W. and Waterfield, A.A. (1978) Factors influencing the release of acetylcholine from the myenteric plexus of the ileum of the guinea-pig and rabbit. *Br. J. Pharmacol.* **64**, 565–580.

Cravatt, B.F., Giang, D.K., Mayfield, S.P., Boger, D.L., Lerner, R.A. and Gilula, N.B. (1996) Molecular characterization of an enzyme that degrades neuromodulatory fatty-acid amides. *Nature* **384**, 83–87.

Daaka, Y., Friedman, H. and Klein, T.W. (1996) Cannabinoid receptor proteins are increased in Jurkat, human T-cell line after mitogen activation. *J. Pharmacol. Exp. Ther.* **276**, 776–783.

Das, S.K., Paria, B.C., Chakraborty, I. and Dey, S.K. (1995) Cannabinoid ligand-receptor signaling in the mouse uterus. *Proc. Natl. Acad. Sci. USA* **92**, 4332–4336.

Deadwyler, S.A., Hampson, R.E., Bennett, B.A., Edwards, T.A., Mu, J., Pacheco, M.A., Ward, S.J. and Childers, S.R. (1993) Cannabinoids modulate potassium current in cultured hippocampal neurons. *Recept. Channels* **1**, 121–134.

Deadwyler, S.A., Hampson, R.E., Mu, J., Whyte, A. and Childers, S. (1995) Cannabinoids modulate voltage sensitive potassium A-current in hippocampal neurons *via* a cAMP-dependent process. *J. Pharmacol. Exp. Ther.* **273**, 734–743.

Desarnaud, F., Cadas, H. and Piomelli, D. (1995) Anandamide amidohydrolase activity in rat brain microsomes. Identification and partial characterization. *J. Biol. Chem.* **270**, 6030–6035.

Deutsch, D.G. and Chin, S.A. (1993) Enzymatic synthesis and degradation of anandamide, a cannabinoid receptor agonist. *Biochem. Pharmacol.* **46**, 791–796.

Deutsch, D.G., Omeir, R., Arreaza, G., Salehani, D., Prestwich, G.D., Huang, Z. and Howlett, A. (1997) Methyl arachidonyl fluorophosphonate: a potent irreversible inhibitor of anandamide amidase. *Biochem. Pharmacol.* **53**, 255–260.

Devane, W.A. and Axelrod, J. (1994) Enzymatic synthesis of anandamide, an endogenous ligand for the cannabinoid receptor, by brain membranes. *Proc. Natl. Acad. Sci. USA* **91**, 6698–6701.

Devane, W.A., Breuer, A., Sheskin, T., Järbe, T.U.C., Eisen, M.S. and Mechoulam, R. (1992a) A novel probe for the cannabinoid receptor. *J. Med. Chem.* **35**, 2065–2069.

Devane, W.A., Dysarz, F.A., Johnson, M.R., Melvin, L.S. and Howlett, A.C. (1988) Determination and characterization of a cannabinoid receptor in rat brain. *Mol. Pharmacol.* **34**, 605–613.

Devane, W.A., Hanuš, L., Breuer, A., Pertwee, R.G., Stevenson, L.A., Griffin, G., Gibson, D., Mandelbaum, A., Etinger, A. and Mechoulam, R. (1992b) Isolation and structure of a brain constituent that binds to the cannabinoid receptor. *Science* **258**, 1946–1949.

Dewey, W.L. (1986) Cannabinoid Pharmacology. *Pharmacol. Rev.* **38**, 151–178.

Di Marzo, V., De Petrocellis, L., Sepe, N. and Buono, A. (1996a) Biosynthesis of anandamide and related acylethanolamides in mouse J774 macrophages and N_{18} neuroblastoma cells. *Biochem. J.* **316**, 977–984.

Di Marzo, V., De Petrocellis, L., Sugiura, T. and Waku, K. (1996b) Potential biosynthetic connections between the two cannabimimetic eicosanoids, anandamide and 2-arachidonoyl-glycerol, in mouse neuroblastoma cells. *Biochem. Biophys. Res. Commun.* **227**, 281–288.

Di Marzo, V., Fontana, A., Cadas, H., Schinelli, S., Cimino, G., Schwartz, J.-C. and Piomelli, D. (1994) Formation and inactivation of endogenous cannabinoid anandamide in central neurons. *Nature* **372**, 686–691.

Diaz, S., Specter, S. and Coffey, R.G. (1993) Suppression of lymphocyte adenosine 3′: 5′-cyclic monophosphate (cAMP) by delta-9-tetrahydrocannabinol. *Int. J. Immunopharmacol.* **15**, 523–532.

Dutta, A.K., Sard, H., Ryan, W., Razdan, R.K., Compton, D.R. and Martin, B.R. (1995) The synthesis and pharmacological evaluation of the cannabinoid antagonist SR141716A. *Med. Chem. Res.* **5**, 54–62.

Eissenstat, M.A., Bell, M.R., D'Ambra, T.E., Alexander, E.J., Daum, S.J., Ackerman, J.H., Gruett, M.D., Kumar, V., Estep, K.G., Olefirowicz, E.M., Wetzel, J.R., Alexander, M.D., Weaver, J.D., Haycock, D.A., Luttinger, D.A., Casiano, F.M., Chippari, S.M., Kuster, J.E., Stevenson, J.I. and Ward, S.J. (1995) Aminoalkylindoles: structure-activity relationships of novel cannabinoid mimetics. *J. Med. Chem.* **38**, 3094–3105.

Facci, L., Dal Toso, R., Romanello, S., Buriani, A., Skaper, S.D. and Leon, A. (1995) Mast cells express a peripheral cannabinoid receptor with differential sensitivity to anandamide and palmitoylethanolamide. *Proc. Natl. Acad. Sci. USA* **92**, 3376–3380.

Fahey, K., Chaney, M., Cullinan, G., Hunden, D., Johnson, D., Koppel, G., Felder, C.C., Joyce, K.E., Briley, E.M., Mackie, K., Crawley, J.N., Martin, B.R. and Brownstein, M.J. (1995) The synthesis and biological activity of a novel class of antagonists for the brain CB1 cannabinoid receptor. Abstract 449.9. *Soc. Neurosci. Abstracts* **21**, 1144.

Fan, P. (1995) Cannabinoid agonists inhibit the activation of 5-HT_3 receptors in rat nodose ganglion neurons. *J. Neurophysiol.* **73**, 907–910.

Felder, C.C., Briley, E.M., Axelrod, J., Simpson, J.T., Mackie, K. and Devane, W.A. (1993) Anandamide, an endogenous cannabimimetic eicosanoid, binds to the cloned human cannabinoid receptor and stimulates receptor-mediated signal transduction. *Proc. Natl. Acad. Sci. USA* **90**, 7656–7660.

Felder, C.C., Joyce, K.E., Briley, E.M., Mansouri, J., Mackie, K., Blond, O., Lai, Y., Ma, A.L. and Mitchell, R.L. (1995) Comparison of the pharmacology and signal transduction of the human cannabinoid CB_1 and CB_2 receptors. *Mol. Pharmacol.* **48**, 443–450.

Felder, C.C., Nielsen, A., Briley, E.M., Palkovits, M., Priller, J., Axelrod, J., Nguyen, D.N., Richardson, J.M., Riggin, R.M., Koppel, G.A., Paul, S.M. and Becker, G.W. (1996) Isolation and measurement of the endogenous cannabinoid receptor agonist, anandamide, in brain and peripheral tissues of human and rat. *FEBS Lett.* **393**, 231–235.

Felder, C.C., Veluz, J.S., Williams, H.L., Briley, E.M. and Matsuda, L.A. (1992) Cannabinoid agonists stimulate both receptor- and non-receptor-mediated signal transduction pathways in cells transfected with and expressing cannabinoid receptor clones. *Mol. Pharmacol.* **42**, 838–845.

Fernando, S.R. and Pertwee, R.G. (1997) Evidence that methyl arachidonyl fluorophosphonate is an irreversible cannabinoid receptor antagonist. *Br. J. Pharmacol.* **121**, 1176–1720.

Fontana, A., Di Marzo, V., Cadas, H. and Piomelli, D. (1995) Analysis of anandamide, an endogenous cannabinoid substance, and of other natural *N*-acylethanolamines. *Prostaglandins Leukotrienes and Essential Fatty Acids* **53**, 301–308.

Fride, E., Barg, J., Levy, R., Saya, D., Heldman, E., Mechoulam, R. and Vogel, Z. (1995) Low doses of anandamides inhibit pharmacological effects of Δ^9-tetrahydrocannabinol. *J. Pharmacol. Exp. Ther.* **272**, 699–707.

Galiègue, S., Mary, S., Marchand, J., Dussossoy, D., Carrière, D., Carayon, P., Bouaboula, M., Shire, D., Le Fur, G. and Casellas, P. (1995) Expression of central and peripheral cannabinoid receptors in human immune tissues and leukocyte subpopulations. *Eur. J. Biochem.* **232**, 54–61.

Gallant, M., Dufresne, C., Gareau, Y., Guay, D., Leblanc, Y., Prasit, P., Rochette, C., Sawyer, N., Slipetz, D.M., Tremblay, N., Metters, K.M. and Labelle, M. (1996) New class of potent ligands for the human peripheral cannabinoid receptor. *Bioorg. Med. Chem. Lett.* **6**, 2263–2268.

Gareau, Y., Dufresne, C., Gallant, M., Rochette, C., Sawyer, N., Slipetz, D.M., Tremblay, N., Weech, P.K., Metters, K.M. and Labelle, M. (1996) Structure activity relationships of tetra-hydrocannabinol analogues on human cannabinoid receptors. *Bioorg. Med. Chem. Lett.* **6**, 189–194.

Gérard, C., Mollereau, C., Vassart, G. and Parmentier, M. (1990) Nucleotide sequence of a human cannabinoid receptor cDNA. *Nucl. Acids Res.* **18**, 7142.

Gérard, C.M., Mollereau, C., Vassart, G. and Parmentier, M. (1991) Molecular cloning of a human cannabinoid receptor which is also expressed in testis. *Biochem. J.* **279**, 129–134.

Gifford, A.N. and Ashby, C.R. (1996) Electrically evoked acetylcholine release from hippocampal slices is inhibited by the cannabinoid receptor agonist, WIN 55212-2, and is potentiated by the cannabinoid antagonist, SR 141716A. *J. Pharmacol. Exp. Ther.* **277**, 1431–1436.

Glass, M., Faull, R.L.M. and Dragunow, M. (1993) Loss of cannabinoid receptors in the substantia nigra in Huntington's disease. *Neuroscience* **56**, 523–527.

Glass, M., Dragunow, M. and Faull, R.L.M. (1997) Cannabinoid receptors in the human brain: a detailed anatomical and quantitative autoradiographic study in the fetal, neonatal and adult human brain. *Neuroscience* **77**, 299–318.

Griffin, G., Lainton, J.A.H., Huffman, J.W. and Pertwee, R.G. (1996) Pharmacological characterization of four novel cannabinoid receptor agonists in the mouse vas deferens. *Proc. Int. Cannabinoid Res. Soc.*, p. 43.

Griffin, G., Fernando, S.R., Ross, R.A., MacKay, N.G., Ashford, M.L.J., Shire, D., Huffman, J.W., Yu, S., Lainton, J.A.H. and Pertwee, R.G. (1997) Evidence for the presence of the CB_2-like cannabinoid receptors on peripheral nerve terminals. *Eur. J. Pharmacol.* **339**, 53–61.

Hampson, A.J., Hill, W.A.G., Zan-Phillips, M., Makriyannis, A., Leung, E., Eglen, R.M. and Bornheim, L.M. (1995) Anandamide hydroxylation by brain lipoxygenase: metabolite structures and potencies at the cannabinoid receptor. *Biochim. Biophys. Acta* **1259**, 173–179.

Hampson, R.E., Evans, G.J.O., Mu, J., Zhuang, S.-Y., King, V.C., Childers, S.R. and Deadwyler, S.A. (1995) Role of cyclic AMP dependent protein kinase in cannabinoid receptor modulation of potassium "A-current" in cultured rat hippocampal neurons. *Life Sci.* **56**, 2081–2088.

Hansen, H.S., Lauritzen, L., Strand, A.M., Moesgaard, B. and Frandsen, A. (1995) Glutamate stimulates the formation of *N*-acylphosphatidylethanolamine and *N*-acylethanolamine in cortical neurons in culture. *Biochim. Biophys. Acta* **1258**, 303–308.

Hansen, H.S., Lauritzen, L., Strand, A.M., Vinggaard, A.M., Frandsen, A. and Schousboe, A. (1997) Characterization of glutamate-induced formation of *N*-acylphosphatidylethanolamine and *N*-acylethanolamine in cultured neocortical neurons. *J. Neurochem.* **69**, 753–761.

Hanuš, L., Gopher, A., Almog, S. and Mechoulam, R. (1993) Two new unsaturated fatty acid ethanolamides in brain that bind to the cannabinoid receptor. *J. Med. Chem.* **36**, 3032–3034.

Henry, D.J. and Chavkin, C. (1995) Activation of inwardly rectifying potassium channels (GIRK1) by co-expressed rat brain cannabinoid receptors in *Xenopus* oocytes. *Neurosci. Lett.* **186**, 91–94.

Herkenham, M. (1995) Localization of cannabinoid receptors in brain and periphery. In *Cannabinoid Receptors* (ed. Pertwee, R.G.), pp. 145–166. London, Academic Press.

Herkenham, M., Lynn, A.B., de Costa, B.R. and Richfield, E.K. (1991a) Neuronal localization of cannabinoid receptors in the basal ganglia of the rat. *Brain Res.* **547**, 267–274.

Herkenham, M., Lynn, A.B., Johnson, M.R., Melvin, L.S., de Costa, B.R. and Rice, K.C. (1991b) Characterization and localization of cannabinoid receptors in rat brain: a quantitative *in vitro* autoradiographic study. *J. Neurosci.* **11**, 563–583.

Herkenham, M., Lynn, A.B., Little, M.D., Johnson, M.R., Melvin, L.S., de Costa, B.R. and Rice, K.C. (1990) Cannabinoid receptor localization in brain. *Proc. Natl. Acad. Sci. USA* **87**, 1932–1936.

Hillard, C.J., Edgemond, W.S. and Campbell, W.B. (1995a) Characterization of ligand binding to the cannabinoid receptor of rat brain membranes using a novel method: application to anandamide. *J. Neurochem.* **64**, 677–683.

Hillard, C.J., Wilkison, D.M., Edgemond, W.S. and Campbell, W.B. (1995b) Characterization of the kinetics and distribution of *N*-arachidonylethanolamine (anandamide) hydrolysis by rat brain. *Biochim. Biophys. Acta* **1257**, 249–256.

Hirst, R.A. and Lambert, D.G. (1995) Do SH-SY5Y human neuroblastoma cells express cannabinoid receptors? *Biochem. Soc. Trans.* **23**, 418S.

Ho, B.Y. and Zhao, J. (1996) Determination of the cannabinoid receptors in mouse x rat hybridoma NG108-15 cells and rat GH4C1 cells. *Neurosci. Lett.* **212**, 123–126.

Hoehe, M.R., Caenazzo, L., Martinez, M.M., Hsieh, W.-T., Modi, W.S., Gershon, E.S. and Bonner, T.I. (1991) Genetic and physical mapping of the human cannabinoid receptor gene to chromosome-6q14–q15. *New Biol.* **3**, 880–885.

Hollister, L.E. (1986) Health aspects of cannabis. *Pharmacol. Rev.* **38**, 1–20.

Howlett, A.C. (1984) Inhibition of neuroblastoma adenylate cyclase by cannabinoid and nantradol compounds. *Life Sci.* **35**, 1803–1810.

Howlett, A.C. (1985) Cannabinoid inhibition of adenylate cyclase. Biochemistry of the response in neuroblastoma cell membranes. *Mol. Pharmacol.* **27**, 429–436.

Howlett, A.C. (1987) Cannabinoid inhibition of adenylate cyclase: relative activity of constituents and metabolites of marihuana. *Neuropharmacology* **26**, 507–512.

Howlett, A.C. (1995) Cannabinoid compounds and signal transduction mechanisms. In *Cannabinoid Receptors* (ed. Pertwee, R.G.), pp. 167–204. London, Academic Press.

Howlett, A.C., Champion, T.M., Wilken, G.H. and Mechoulam, R. (1990) Stereochemical effects of 11-OH-Δ^8-tetrahydrocannabinol-dimethylheptyl to inhibit adenylate cyclase and bind to the cannabinoid receptor. *Neuropharmacology* **29**, 161–165.

Howlett, A.C. and Fleming, R.M. (1984) Cannabinoid inhibition of adenylate cyclase. Pharmacology of the response in neuroblastoma cell membranes. *Mol. Pharmacol.* **26**, 532–538.

Howlett, A.C., Johnson, M.R., Melvin, L.S. and Milne, G.M. (1988) Nonclassical cannabinoid analgetics inhibit adenylate cyclase: development of a cannabinoid receptor model. *Mol. Pharmacol.* **33**, 297–302.

Howlett, A.C., Qualy, J.M. and Khachatrian, L.L. (1986) Involvement of G_i in the inhibition of adenylate cyclase by cannabimimetic drugs. *Mol. Pharmacol.* **29**, 307–313.

Huffman, J.W., Yu, S., Showalter, V., Abood, M.E., Wiley, J.L., Compton, D.R., Martin, B.R., Bramblett, R.D. and Reggio, P.H. (1996) Synthesis and pharmacology of a very potent cannabinoid lacking a phenolic hydroxyl with high affinity for the CB2 receptor. *J. Med. Chem.* **39**, 3875–3877.

Ishac, E.J.N., Jiang, L., Lake, K.D., Varga, K., Abood, M.E. and Kunos, G. (1996) Inhibition of exocytotic noradrenaline release by presynaptic cannabinoid CB_1 receptors on peripheral sympathetic nerves. *Br. J. Pharmacol.* **118**, 2023–2028.

Jansen, E.M., Haycock, D.A., Ward, S.J. and Seybold, V.S. (1992) Distribution of cannabinoid receptors in rat brain determined with aminoalkylindoles. *Brain Res.* **575**, 93–102.

Kaminski, N.E., Abood, M.E., Kessler, F.K., Martin, B.R. and Schatz, A.R. (1992) Identification of a functionally relevant cannabinoid receptor on mouse spleen cells that is involved in cannabinoid-mediated immune modulation. *Mol. Pharmacol.* **42**, 736–742.

Kaminski, N.E., Koh, W.S., Yang, K.H., Lee, M. and Kessler, F.K. (1994) Suppression of the humoral immune response by cannabinoids is partially mediated through inhibition of adenylate cyclase by a pertussis toxin-sensitive G-protein coupled mechanism. *Biochem. Pharmacol.* **48**, 1899–1908.

Koutek, B., Prestwich, G.D., Howlett, A.C., Chin, S.A., Salehani, D., Akhavan, N. and Deutsch, D.G. (1994) Inhibitors of arachidonoyl ethanolamide hydrolysis. *J. Biol. Chem.* **269**, 22937–22940.

Khanolkar, A.D., Abadji, V., Lin, S.Y., Hill, W., Taha, G., Abouzid, K., Meng, Z.X., Fau, P.S. and Makriyannis, A. (1996) Head group analogs of arachidonylethanolamide, the endogenous cannabinoid ligand. *J. Med. Chem.* **39**, 4515–4519.

Kruszka, K.K. and Gross, R.W. (1994) The ATP- and CoA-independent synthesis of arachidonoylethanolamide – a novel mechanism underlying the synthesis of the endogenous ligand of the cannabinoid receptor. *J. Biol. Chem.* **269**, 14345–14348.

Kuster, J.E., Stevenson, J.I., Ward, S.J., D'Ambra, T.E. and Haycock, D.A. (1993) Aminoalkylindole binding in rat cerebellum: selective displacement by natural and synthetic cannabinoids. *J. Pharmacol. Exp. Ther.* **264**, 1352–1363.

Lang, W., Qin, C., Hill, W.A.G., Lin, S., Khanolkar, A.D. and Makriyannis, A. (1996) High-performance liquid chromatographic determination of anandamide amidase activity in rat brain microsomes. *Analyt. Biochem.* **238**, 40–45.

Lee, M., Yang, K.H. and Kaminski, N.E. (1995) Effects of putative cannabinoid receptor ligands, anandamide and 2-arachidonyl-glycerol, on immune function in B6C3F1 mouse splenocytes. *J. Pharmacol. Exp. Ther.* **275**, 529–536.

Lynn, A.B. and Herkenham, M. (1994) Localization of cannabinoid receptors and nonsaturable high-density cannabinoid binding sites in peripheral tissues of the rat: implications for receptor-mediated immune modulation by cannabinoids. *J. Pharmacol. Exp. Ther.* **268**, 1612–1623.

Mackay, A.V.P., Yates, C.M., Wright, A., Hamilton, P. and Davies, P. (1978) Regional distribution of monoamines and their metabolites in the human brain. *J. Neurosci.* **30**, 841–848.

Mackie, K., Devane, W.A., Hille, B. (1993) Anandamide, an endogenous cannabinoid, inhibits calcium currents as a partial agonist in N18 neuroblastoma cells. *Mol. Pharmacol.* **44**, 498–503.

Mackie, K. and Hille, B. (1992) Cannabinoids inhibit N-type calcium channels in neuroblastoma-glioma cells. *Proc. Natl. Acad. Sci. USA* **89**, 3825–3829.

Mackie, K., Lai, Y., Westenbroek, R. and Mitchell, R. (1995) Cannabinoids activate an inwardly rectifying potassium conductance and inhibit Q-type calcium currents in AtT20 cells transfected with rat brain cannabinoid receptor. *J. Neurosci.* **15**, 6552–6561.

Mailleux, P. and Vanderhaeghen, J.-J. (1992a) Distribution of neuronal cannabinoid receptor in the adult rat brain: a comparative receptor binding radioautography and *in situ* hybridization histochemistry. *Neuroscience* **48**, 655–668.

Mailleux, P. and Vanderhaeghen, J.-J. (1992b) Localization of cannabinoid receptor in the human developing and adult basal ganglia. Higher levels in the striatonigral neurons. *Neurosci. Lett.* **148**, 173–176.

Mailleux, P., Verslijpe, M. and Vanderhaeghen, J.-J. (1992) Initial observations on the distribution of cannabinoid receptor binding sites in the human adult basal ganglia using autoradiography. *Neurosci. Lett.* **139**, 7–9.

Maneuf, Y.P., Nash, J.E., Crossman, A.R. and Brotchie, J.M. (1996) Activation of the cannabinoid receptor by Δ^9-tetrahydrocannabinol reduces γ-aminobutyric acid uptake in the globus pallidus. *Eur. J. Pharmacol.* **308**, 161–164.

Mansbach, R.S., Rovetti, C.C., Winston, E.N. and Lowe, J.A. (1996) Effects of the cannabinoid CB1 receptor antagonist SR141716A on the behavior of pigeons and rats. *Psychopharmacology* **124**, 315–322.

Martin, B.R. (1986) Cellular effects of cannabinoids. *Pharmacol. Rev.* **38**, 45–74.

Martin, B.R., Thomas, B.F. and Razdan, R.K. (1995) Structural requirements for cannabinoid receptor probes. In *Cannabinoid Receptors* (ed. Pertwee, R.G.), pp. 35–85. London, Academic Press.

Matsuda, L.A. and Bonner, T.I. (1995) Molecular biology of the cannabinoid receptor. In *Cannabinoid Receptors* (ed. Pertwee, R.G.), pp. 117–143. London, Academic Press.

Matsuda, L.A., Lolait, S.J., Brownstein, M.J., Young, A.C. and Bonner, T.I. (1990) Structure of a cannabinoid receptor and functional expression of the cloned cDNA. *Nature* **346**, 561–564.

Maurelli, S., Bisogno, T., De Petrocellis, L., Di Luccia, A., Marino, G. and Di Marzo, V. (1995) Two novel classes of neuroactive fatty acid amides are substrates for mouse neuroblastoma 'anandamide amidohydrolase'. *FEBS Lett.* **377**, 82–86.

Mechoulam, R., Ben-Shabat, S., Hanuš, L., Ligumsky, M., Kaminski, N.E., Schatz, A.R., Gopher, A., Almog, S., Martin, B.R., Compton, D.R., Pertwee, R.G., Griffin, G., Bayewitch, M., Barg, J. and Vogel, Z. (1995) Identification of an endogenous 2-monoglyceride, present in canine gut, that binds to cannabinoid receptors. *Biochem. Pharmacol.* **50**, 83–90.

Mechoulam, R. and Fride, E. (1995) The unpaved road to the endogenous brain cannabinoid ligands, the anandamides. In *Cannabinoid Receptors* (ed. Pertwee, R.G.), pp. 233–258. London, Academic Press.

Munro, S., Thomas, K.L. and Abu-Shaar, M. (1993) Molecular characterization of a peripheral receptor for cannabinoids. *Nature* **365**, 61–65.

Nowicky, A.V., Teyler, T.J. and Vardaris, R.M. (1987) The modulation of long-term potentiation by delta-9-tetrahydrocannabinol in the rat hippocampus, *in vitro*. *Brain Res. Bull.* **19**, 663–672.

Omeir, R.L., Chin, S., Hong, Y., Ahern, D.G. and Deutsch, D.G. (1995) Arachidonoyl ethanolamide-[1,2-14C] as a substrate for anandamide amidase. *Life Sci.* **56**, 1999–2005.

Onaivi, E.S., Chakrabarti, A. and Chaudhuri, G. (1996a) Cannabinoid receptor genes. *Prog. Neurobiol.* **48**, 275–305.

Onaivi, E.S., Chakrabarti, A., Gwebu, E.T. and Chaudhuri, G. (1996b) Neurobehavioral effects of Δ^9-THC and cannabinoid (CB1) receptor gene expression in mice. *Behav. Brain Res.* **72**, 115–125.

Pacheco, M., Childers, S.R., Arnold, R., Casiano, F. and Ward, S.J. (1991) Aminoalkylindoles: actions on specific G-protein-linked receptors. *J. Pharmacol. Exp. Ther.* **257**, 170–183.

Pacheco, M.A., Ward, S.J. and Childers, S.R. (1993) Identification of cannabinoid receptors in cultures of rat cerebellar granule cells. *Brain Res.* **603**, 102–110.

Pacheco, M.A., Ward, S.J. and Childers, S.R. (1994) Differential requirements of sodium for coupling of cannabinoid receptors to adenylyl cyclase in rat brain membranes. *J. Neurochem.* **62**, 1773–1782.

Pan, X., Ikeda, S.R. and Lewis, D.L. (1996) Rat brain cannabinoid receptor modulates N-type Ca^{2+} channels in a neuronal expression system. *Mol. Pharmacol.* **49**, 707–714.

Paria, B.C., Das, S.K. and Dey, S.K. (1995) The preimplantation mouse embryo is a target for cannabinoid ligand-receptor signaling. *Proc. Natl. Acad. Sci. USA* **92**, 9460–9464.

Paria, B.C., Deutsch, D.D. and Dey, S.K. (1996) The uterus is a potential site for anandamide synthesis and hydrolysis: differential profiles of anandamide synthase and hydrolase activities in the mouse uterus during the periimplantation period. *Mol. Reproduction and Development* **45**, 183–192.

Paterson, S.J. and Pertwee, R.G. (1993) Characterization of the cannabinoid binding site in the guinea-pig. *Br. J. Pharmacol. (Proc. Suppl.)* **109**, 82P.

Paton, W.D.M., and Pertwee, R.G. (1973a) The pharmacology of cannabis in animals. In *Marijuana* (ed. Mechoulam, R.), pp. 191–285. New York, Academic Press.

Paton, W.D.M., and Pertwee, R.G. (1973b) The actions of cannabis in man. In *Marijuana* (ed. Mechoulam, R.), pp. 287–333. New York, Academic Press.

Pério, A., Rinaldi-Carmona, M., Maruani, J., Barth, F., Le Fur, G. and Soubrié, P. (1996) Central mediation of the cannabinoid cue: activity of a selective CB1 antagonist, SR 141716A. *Behav. Pharmacol.* **7**, 65–71.

Pertwee, R.G. (1988) The central neuropharmacology of psychotropic cannabinoids. *Pharmacol. Ther.* **36**, 189–261.

Pertwee, R. (1993) The evidence for the existence of cannabinoid receptors. *Gen. Pharmacol.* **24**, 811–824.

Pertwee, R.G. (1995) Pharmacological, physiological and clinical implications of the discovery of cannabinoid receptors: an overview. In *Cannabinoid Receptors* (ed. Pertwee, R.G.), pp. 1–34. London, Academic Press.

Pertwee, R.G. (1997) Pharmacology of cannabinoid CB_1 and CB_2 receptors. *Pharmacol. Ther.* **74**, 129–180.

Pertwee, R.G. and Fernando, S.R. (1996) Evidence for the presence of cannabinoid CB_1 receptors in mouse urinary bladder. *Br. J. Pharmacol.* **118**, 2053–2058.

Pertwee, R.G., Fernando, S.R., Griffin, G., Abadji, V. and Makriyannis, A. (1995a) Effect of phenylmethylsulphonyl fluoride on the potency of anandamide as an inhibitor of electrically evoked contractions in two isolated tissue preparations. *Eur. J. Pharmacol.* **272**, 73–78.

Pertwee, R.G., Fernando, S.R., Griffin, G., Ryan, W., Razdan, R.K., Compton, D.R. and Martin, B.R. (1996a) Agonist-antagonist characterization of 6′-cyanohex-2′-yne-Δ^8-tetra-hydrocannabinol in two isolated tissue preparations. *Eur. J. Pharmacol.* **315**, 195–201.

Pertwee, R.G., Fernando, S.R., Nash, J.E. and Coutts, A.A. (1996b) Further evidence for the presence of cannabinoid CB_1 receptors in guinea-pig small intestine. *Br. J. Pharmacol.* **118**, 2199–2205.

Pertwee, R.G. and Griffin, G. (1995) A preliminary investigation of the mechanisms underlying cannabinoid tolerance in the mouse vas deferens. *Eur. J. Pharmacol.* **272**, 67–72.

Pertwee, R., Griffin, G., Fernando, S., Li, X., Hill, A. and Makriyannis, A. (1995b) AM630, a competitive cannabinoid receptor antagonist. *Life Sci.* **56**, 1949–1955.

Pertwee, R., Griffin, G., Hanuš, L. and Mechoulam, R. (1994) Effects of two endogenous fatty acid ethanolamides on mouse vasa deferentia. *Eur. J. Pharmacol.* **259**, 115–120.

Pertwee, R.G., Griffin, G., Lainton, J.A.H. and Huffman, J.W. (1995c) Pharmacological characterization of three novel cannabinoid receptor agonists in the mouse isolated vas deferens. *Eur. J. Pharmacol.* **284**, 241–247.

Pertwee, R.G., Joe-Adigwe, G. and Hawksworth, G.M. (1996c) Further evidence for the presence of cannabinoid CB_1 receptors in mouse vas deferens. *Eur. J. Pharmacol.* **296**, 169–172.

Pertwee, R.G., Stevenson, L.A., Elrick, D.B., Mechoulam, R. and Corbett, A.D. (1992) Inhibitory effects of certain enantiomeric cannabinoids in the mouse vas deferens and the myenteric plexus preparation of guinea-pig small intestine. *Br. J. Pharmacol.* **105**, 980–984.

Pertwee, R.G., Stevenson, L.A., Fernando, S.R. and Corbett, A.D. (1993a) *In vitro* effects of the cannabinoid, CP 55,940, and of its (+)-enantiomer, CP 56,667. In *Problems of Drug Dependence, 1992: Proceeding of the 54th Annual Scientific Meeting of the College on Problems of Drug Dependence, Inc.:*

National Institute on Drug Abuse Research Monograph 132 (ed. Harris, L.), p. 374. Rockville, MD., National Institute on Drug Abuse.

Pertwee, R.G., Stevenson, L.A. and Griffin, G. (1993b) Cross-tolerance between delta-9-tetrahydrocannabinol and the cannabimimetic agents, CP-55,940, WIN-55,212-2 and anandamide. *Br. J. Pharmacol.* **110**, 1483–1490.

Petitet, F., Marin, L. and Doble, A. (1996) Biochemical and pharmacological characterization of cannabinoid binding sites using [^{3}H]SR141716A. *Neuroreport* **7**, 789–792.

Pinto, J.C., Potié, F., Rice, K.C., Boring, D., Johnson, M.R., Evans, D.M., Wilken, G.H., Cantrell, C.H. and Howlett, A.C. (1994) Cannabinoid receptor binding and agonist activity of amides and esters of arachidonic acid. *Mol. Pharmacol.* **46**, 516–522.

Poling, J.S., Rogawski, M.A., Salem, N. and Vicini, S. (1996) Anandamide, an endogenous cannabinoid, inhibits *Shaker*-related voltage-gated K^+ channels. *Neuropharmacology* **35**, 983–991.

Reche, I., Fuentes, J.A. and Ruiz-Gayo, M. (1996) A role for central cannabinoid and opioid systems in peripheral Δ^9-tetrahydrocannabinol-induced analgesia in mice. *Eur. J. Pharmacol.* **301**, 75–81.

Reichman, M., Nen, W. and Hokin, L.E. (1991) Δ^9-tetrahydrocannabinol inhibits arachidonic acid acylation of phospholipids and triacylglycerols in guinea pig cerebral cortex slices. *Mol. Pharmacol.* **40**, 547–555.

Rinaldi-Carmona, M., Barth, F., Héaulme, M., Shire, D., Calandra, B., Congy, C., Martinez, S., Maruani, J., Néliat, G., Caput, D., Ferrara, P., Soubrié, P., Brelière, J.C. and Le Fur, G. (1994) SR141716A, a potent and selective antagonist of the brain cannabinoid receptor. *FEBS Lett.* **350**, 240–244.

Rinaldi-Carmona, M., Calandra, B., Shire, D., Bouaboula, M., Oustric, D., Barth, F., Casellas, P., Ferrara, P., and Le Fur, G. (1996a) Characterization of two cloned human CB1 cannabinoid receptor isoforms. *J. Pharmacol. Exp. Ther.* **278**, 871–878.

Rinaldi-Carmona, M., Pialot, F., Congy, C., Redon, E., Barth, F., Bachy, A., Brelière, J.-C., Soubrié, P. and Le Fur, G. (1996b) Characterization and distribution of binding sites for [^{3}H]-SR141716A, a selective brain (CB1) cannabinoid receptor antagonist, in rodent brain. *Life Sci.* **58**, 1239–1247.

Rinaldi-Carmona, M., Barth, F., Millan, J., Derocq, J.-M., Casellas, P., Congy, C., Oustric, D., Sarran, M., Bouaboula, M., Calandra, B., Portier, M., Shire, D., Brelière, J.-C. and Le Fur, G. (1998) SR144528, the first potent and selective antagonist of the CB2 cannabinoid receptor. *J. Pharmacol. Exp. Ther.* In Press

Rowley, J.T. and Rowley, P.T. (1990) Tetrahydrocannabinol inhibits adenyl cyclase in human leukemia cells. *Life Sci.* **46**, 217–222.

Rubino, T., Massi, P., Patrini, G., Venier, I., Giagnoni, G. and Parolaro, D. (1994) Chronic CP-55,940 alters cannabinoid receptor mRNA in the rat brain: an *in situ* hybridization study. *Neuroreport* **5**, 2493–2496.

Santucci, V., Storme, J.-J., Soubrié, P. and Le Fur, G. (1996) Arousal-enhancing properties of the CB1 cannabinoid receptor antagonist SR141716A in rats as assessed by electroencephalographic spectral and sleep-waking cycle analysis. *Life Sci.* **58**, PL103–PL110.

Schatz, A.R., Kessler, F.K. and Kaminski, N.E. (1992) Inhibition of adenylate cyclase by Δ^9-tetrahydrocannabinol in mouse spleen cells: a potential mechanism for cannabinoid-mediated immunosuppression. *Life Sci.* **51**, 25–30.

Schlicker, E., Timm, J. and Göthert, M. (1996) Cannabinoid receptor-mediated inhibition of dopamine release in the retina. *Naunyn-Schmeideberg's Arch. Pharmacol.* **354**, 791–795.

Schmid, P.C., Krebsbach, R.J., Perry, S.R., Dettmer, T.M., Maasson, J.L. and Schmid, H.H.O. (1995) Occurrence and postmortem generation of anandamide and other long-chain N-acylethanolamines in mammalian brain. *FEBS Lett.* **375**, 117–120.

Shen, M., Piser, T.M., Seybold, V.S. and Thayer, S.A. (1996) Cannabinoid receptor agonists inhibit glutamatergic synaptic transmission in rat hippocampal cultures. *J. Neurosci.* **16**, 4322–4334.

Shire, D., Calandra, B., Rinaldi-Carmona, M., Oustric, D., Pessègue, B., Bonnin-Cabanne, O., Le Fur, G., Caput, D. and Ferrara, P. (1996) Molecular cloning, expression and function of murine CB2 peripheral cannabinoid receptor. *Biochim. Biophys. Acta* **1307**, 132–136.

Shire, D., Carillon, C., Kaghad, M., Calandra, B., Rinaldi-Carmona, M., Le Fur, G., Caput, D. and Ferrara, P. (1995) An amino-terminal variant of the central cannabinoid receptor resulting from alternative splicing. *J. Biol. Chem.* **270**, 3726–3731.

Shivachar, A.C., Martin, B.R. and Ellis, E.F. (1996) Anandamide- and Δ^9-tetrahydrocannabinol-evoked arachidonic acid mobilization and blockade by SR141716A [*N*-(piperidin-1-yl)-5-(4-chlorophenyl)-1-(2,4-dichlorophenyl)-4-methyl-1*H*-pyrazole-3-carboximide hydrochloride]. *Biochem. Pharmacol.* **51**, 669–676.

Showalter, V.M., Compton, D.R., Martin, B.R. and Abood, M.E. (1996) Evaluation of binding in a transfected cell line expressing a peripheral cannabinoid receptor (CB2): identification of cannabinoid receptor subtype selective ligands. *J. Pharmacol. Exp. Ther.* **278**, 989–999.

Skaper, S.D., Buriani, A., Dal Toso, R., Petrelli, L., Romanello, S., Facci, L. and Leon, A. (1996) The ALIAmide palymitoylethanolamide and cannabinoids, but not anandamide, are protective in a delayed postglutamate paradigm of excitotoxic death in cerebellar granule neurons. *Proc. Nat. Acad. Sci. USA* **93**, 3984–3989.

Sim, L.J., Selley, D.E. and Childers, S.R. (1995) *In vitro* autoradiography of receptor-activated G proteins in rat brain by agonist-stimulated guanylyl 5'-[γ-[^{35}S]thio]-triphosphate binding. *Proc. Natl. Acad. Sci. USA* **92**, 7242–7246.

Slipetz, D.M., O'Neill, G.P., Favreau, L., Dufresne, C., Gallant, M., Gareau, Y., Guay, D., Labelle, M. and Metters, K.M. (1995) Activation of the human peripheral cannabinoid receptor results in inhibition of adenylyl cyclase. *Mol. Pharmacol.* **48**, 352–361.

Song, Z.-H. and Bonner, T.I. (1996) A lysine residue of the cannabinoid receptor is critical for receptor recognition by several agonists but not WIN55212-2. *Mol. Pharmacol.* **49**, 891–896.

Stjärne, L., Åstrand, P. (1985) Relative pre- and postjunctional roles of noradrenaline and adenosine 5'-triphosphate as neurotransmitters of the sympathetic nerves of guinea-pig and mouse vas deferens. *Neuroscience* **14**, 929–946.

Stubbs, L., Chittenden, L., Chakrabarti, A. and Onaivi, E. (1996) The gene encoding the central cannabinoid receptor is located in proximal mouse Chromosome 4. *Mammalian Genome* **7**, 165–166.

Sugiura, T., Kondo, S., Sukagawa, A., Nakane, S., Shinoda, A., Itoh, K., Yamashita, A. and Waku, K. (1995) 2-Arachidonoylglycerol: a possible endogenous cannabinoid receptor ligand in brain. *Biochem. Biophys. Res. Commun.* **215**, 89–97.

Sugiura, T., Kodaka, T., Kondo, S., Tonegawa, T., Nakane, S., Kishimoto, S., Yamashita, A. and Waku, K. (1996a) 2-arachidonoylglycerol, a putative endogenous cannabinoid receptor ligand, induces rapid, transient elevation of intracellular free Ca^{2+} in neuroblastoma x glioma hybrid NG108-15 cells. *Biochem. Biophys. Res. Commun.* **229**, 58–64.

Sugiura, T., Kondo, S., Sukagawa, A., Tonegawa, T., Nakane, S., Yamashita, A., Ishima, Y. and Waku, K. (1996b) Transacylase-mediated and phosphodiesterase-mediated synthesis of *N*-arachidonoylethanolamine, an endogenous cannabinoid-receptor ligand, in rat brain microsomes. Comparison with synthesis from free arachidonic acid and ethanolamine. *Eur. J. Biochem.* **240**, 53–62.

Sugiura, T., Kondo, S., Sukagawa, A., Tonegawa, T., Nakane, S., Yamashita, A. and Waku, K. (1996c) Enzymatic synthesis of anandamide, an endogenous cannabinoid receptor ligand, through N-acylphosphatidylethanolamine pathway in testis: involvement of Ca^{2+}-dependent transacylase and phosphodiesterase activities. *Biochem. Biophys. Res. Commun.* **218**, 113–117.

Terranova, J.-P., Michaud, J.-C., Le Fur, G. and Soubrié, P. (1995) Inhibition of long-term potentiation in rat hippocampal slices by anandamide and WIN55212-2: reversal by SR141716 A, a selective antagonist of CB1 cannabinoid receptors. *Naunyn Schmiedebergs Arch. Pharmacol.* **352**, 576–579.

Terranova, J.-P., Storme, J.-J., Lafon, N., Pério, A., Rinaldi-Carmona, M., Le Fur, G. and Soubrié, P. (1996) Improvement of memory in rodents by the selective CB1 cannabinoid receptor antagonist, SR 141716. *Psychopharmacology* **126**, 165–172.

Thomas, B.F., Adams, I.B., Mascarella, S.W., Martin, B.R. and Razdan, R.K. (1996) Structure-activity analysis of anandamide analogs: relationship to a cannabinoid pharmacophore. *J. Med. Chem.* **39**, 471–479.

Thomas, B.F., Wei, X. and Martin, B.R. (1992) Characterization and autoradiographic localization of the cannabinoid binding site in rat brain using [^{3}H]11-OH-Δ^{9}-THC-DMH. *J. Pharmacol. Exp. Ther.* **263**, 1383–1390.

Turkanis, S.A., Karler, R. and Partlow, L.M. (1991a) Differential effects of delta-9-tetrahydrocannabinol and its 11-hydroxy metabolite on sodium current in neuroblastoma cells. *Brain Res.* **560**, 245–250.

Turkanis, S.A., Partlow, L.M. and Karler, R. (1991b) Delta-9-tetrahydrocannabinol depresses inward sodium current in mouse neuroblastoma cells. *Neuropharmacology* **30**, 73–77.

Twitchell, W. and Mackie, K. (1996) Cannabinoids inhibit voltage-dependent calcium channels in cultured hippocampal neurons. *Proc. Int. Cannabinoid Res. Soc.*, p. 17.

Ueda, N., Kurahashi, Y., Yamamoto, S. and Tokunaga, T. (1995a) Partial purification and characterization of the porcine brain enzyme hydrolyzing and synthesizing anandamide. *J. Biol. Chem.* **270**, 23823–23827.

Ueda, N., Yamamoto, K., Yamamoto, S., Tokunaga, T., Shirakawa, E., Shinkai, H., Ogawa, M., Sato, T., Kudo, I., Inoue, K., Takizawa, H., Nagano, T., Hirobe, M., Matsuki, N. and Saito, H. (1995b) Lipoxygenase-catalyzed oxygenation of arachidonylethanolamide, a cannabinoid receptor agonist. *Biochim. Biophys. Acta* **1254**, 127–134.

Varga, K., Lake, K., Martin, B.R. and Kunos, G. (1995) Novel antagonist implicates the CB_1 cannabinoid receptor in the hypotensive action of anandamide. *Eur. J. Pharmacol.* **278**, 279–283.

Vogel, Z., Barg, J., Levy, R., Saya, D., Heldman, E. and Mechoulam, R. (1993) Anandamide, a brain endogenous compound, interacts specifically with cannabinoid receptors and inhibits adenylate cyclase. *J. Neurochem.* **61**, 352–355.

Von Kügelgen, I., Schöffel, E., Starke, K. (1989) Inhibition by nucleotides acting at presynaptic P_2-receptors of sympathetic neuro-effector transmission in the mouse isolated vas deferens. *Naunyn-Schmiedeberg's Arch. Pharmacol.* **340**, 522–532.

Wartmann, M., Campbell, D., Subramanian, A., Burstein, S.H. and Davis, R.J. (1995) The MAP kinase signal transduction pathway is activated by the endogenous cannabinoid anandamide. *FEBS Lett.* **359**, 133–136.

Watanabe, K., Kayano, Y., Matsunaga, T., Yamamoto, I. and Yoshimura, H. (1996) Inhibition of anandamide amidase activity in mouse brain microsomes by cannabinoids. *Biol. Pharmaceut. Bull.* **19**, 1109–1111.

Welch, S.P., Dunlow, L.D., Patrick, G.S. and Razdan, R.K. (1995) Characterization of anandamide- and fluoroanandamide-induced antinociception and cross-tolerance to Δ^{9}-THC after intrathecal administration to mice: blockade of Δ^{9}-THC-induced antinociception. *J. Pharmacol. Exp. Ther.* **273**, 1235–1244.

Westlake, T.M., Howlett, A.C., Bonner, T.I., Matsuda, L.A. and Herkenham, M. (1994) Cannabinoid receptor binding and messenger RNA expression in human brain: an *in vitro* receptor autoradiography and *in situ* hybridization histochemistry study of normal aged and Alzheimer's brains. *Neuroscience* **63**, 637–652.

Wiley, J.L., Lowe, J.A., Balster, R.L. and Martin, B.R. (1995) Antagonism of the discriminative stimulus effects of Δ^9-tetrahydrocannabinol in rats and rhesus monkeys. *J. Pharmacol. Exp. Ther.* **275**, 1–6.

Yamaguchi, F., Macrae, A.D. and Brenner, S. (1996) Molecular cloning of two cannabinoid type 1-like receptor genes from the puffer fish *Fugu rubripes*. *Genomics* **35**, 603–605.

7. THE THERAPEUTIC POTENTIAL FOR CANNABIS AND ITS DERIVATIVES

DAVID T. BROWN

School of Pharmacy, University of Portsmouth, Portsmouth, Hampshire, UK

1 INTRODUCTION

Chapter 1 of this book provides a fascinating, historical account of the use of cannabis across many cultures and centuries. Suffice it to say here that any natural substance with over 5000 years of medical history will have attached to it a heritage of hearsay and legend through which one must sift to identify areas of true therapeutic potential for us in the late twentieth century and beyond. A summary of conditions for which cannabis has been used, ranging through various shades of rationality, appears in Table 1.

Many areas of use are shrouded, maybe totally obscured, in anecdotal reports, merging with folklore and tradition, which come nowhere near providing the minimum of scientific evidence required to show that cannabis has any real beneficial effect at all. In other areas, a body of knowledge is expanding rapidly to reveal indications in which the use of cannabis or its derivatives is justified, or at least, where research effort into new synthetic analogues of the cannabinoids holds most promise.

As mentioned in Chapter 3, cannabis contains over thirty cannabinoids, some of which may be harmful; others may have beneficial effects (see also Chapters 4 and 6). The cannabinoid content of raw cannabis can vary considerably, both in quantity and proportion, depending on cultivation conditions of the plant and the methods used to extract them. It is little wonder then that one of the major problems one encounters when trying to compare clinical trials involving crude extracts of cannabis lies in determining the standard of the preparation used.

After the isolation of delta-9-tetrahydrocannabinol (THC) in 1964 (Gaoni and Mechoulam, 1964), efforts to assess its therapeutic potential proceeded apace in a diverse range of conditions, including: hypertension, asthma, endogenous depression, glaucoma, bacterial infection, epilepsy, emesis, pain and anxiety. Additional compounds were also synthesised in attempts to create useful therapeutic agents based on the chemistry of THC and other cannabinoids. The availability of these purified compounds allowed firstly, an objective assessment of their pharmacology in animals and in some cases, subsequently in man; secondly, it was possible to achieve some standardisation of dosing in clinical investigations. One further advantage was the recognition that a single agent, although derived from cannabis, might not appear so attractive to potential abusers as the parent material, simplifying the path to obtaining an investigator's licence.

The legislative environment in which these experiments were carried out, in academia, government laboratories and the pharmaceutical industry, was not a friendly one (see Chapter 4). In the US, the Marihuana Tax Act of 1937 prompted the removal of

Table 1 Medicinal and quasi-medicinal uses for cannabis and its derivatives

Indications for which only anecdote or reports of traditional use exist:

aphrodisiac	muscular spasm in rabies/ tetanus	Huntingdon's chorea	jaundice	toothache
earache	tumour growth	cough	hysteria	insanity
menstrual cramps	rheumatism	movement disorders	gut spasm	pyrexia
inflammed tonsils	migraine	headache	increasing uterine contractions in childbirth	urinary retention/ bladder spasm
parasite infection	fatigue	allergy	fever	herpetic pain
hypertension	joint inflammation	diarrhoea	malaria	forgetfulness

Indications for which there are at least some trial data:

depression	anxiety	insomnia	infected wounds	muscle spasm
asthma	muscle spasticity	pain of various aetiologies	sedation	alcoholism
epilepsy		phantom limb pain	Parkinsonian tremor	Tourette's syndrome/ torsion dystonia
postoperative pain				

Indications for which there is general acceptance of efficacy, at least in some patients, at some stage of their disease:

glaucoma	chemotherapy induced nausea and vomiting	relief of pain and muscle spasm in multiple sclerosis	lost appetite and anorexia particularly in AIDS patients

The major applications mentioned in this table are discussed in the text. The use of cannabis in the relief of pain is discussed in the following chapter.

cannabis from the US Pharmacopoeia in 1941. Cannabis had already been removed from the British Pharmacopoeia in 1932.

To this day in the US, in spite of stiff lobbying, cannabis remains a Schedule 1 substance under the Controlled Substance Act, as a drug which has a high abuse potential, lacks an "accepted" medical use and is unsafe for use under medical supervision. Similarly in the UK, the listing of cannabis under Schedule 1 of the Misuse of Drugs Act 1971, makes it illegal to prescribe cannabis for therapeutic purposes or to research the plant without a special licence from the Home Office.

The term "accepted" is an interesting one when applied to cannabis. For specific indications such as the control of cancer chemotherapy-induced nausea and vomiting, a growing number of patients and their doctors, reported relief after smoking cannabis when other therapies had proven ineffective. Clinicians could refer to limited clinical trial evidence which did provide some proof in this area. This lead, albeit temporarily, to a number of patients receiving cannabis legally, for therapeutic purposes, in the US. From 1978, legislation allowing patients to use cannabis (as marihuana), for therapeutic purposes, under their doctor's supervision, was enacted in 36 states. Ten states went as far as to establish, with the necessary government approval, formal research programs with cannabis. These have since foundered because of the bureaucracy still associated with handling the drug.

The growing demand did however persuade the Food and Drug Administration (FDA) to investigate Compassionate Investigational New Drug (CIND) status for cannabis, for physicians whose patients needed cannabis where no other drug was effective. Uptake of this CIND was small initially; however, it increased dramatically in 1989, in parallel with an upturn in the number of patients diagnosed as having AIDS where the drug was used for pain relief and as an appetite stimulant.

In 1991, the US Public Health Service suspended the CIND programme on the grounds that it was contrary to the administration's opposition to the use of illegal drugs, and it withered on the vine, to the extent that no new CINDs were granted in 1992. A few patients remained who were receiving cannabis legally; for all others, use was illegal.

As Chapter 4 in this book indicates, the legal status of cannabis and its derivatives is in a state of flux in many countries. In Australia, where the cultivation and possession of cannabis is permitted, the authorities have judged the evidence, such as it is, and have concluded that the benefits outweigh the risks. In others, such as the UK and Holland, legislators remain unconvinced, citing the available evidence as being deficient in scope, size and other methodology. However, two cannabis derivatives – nabilone and dronabinol ('synthetic' THC) – are available on prescription for extremely limited indications.

It is logical, but not necessarily likely, that advocates of the use of cannabis and its derivatives should indeed focus their pleas for legalisation for prescription use in a limited number of identifiable, serious conditions for which there is convincing evidence of efficacy, allied to acceptable risk, if they are to be successful.

Most recently, legislation (proposition 215) was passed to allow doctors in California to recommend cannabis verbally but not by written prescription, for treating conditions such as cancer, AIDS, anorexia, chronic pain, spasticity and glaucoma. Similar laws have been passed in Arizona, Florida, Idaho, Ohio and Washington. These new state laws conflict with federal law however, as it is still a crime to sell or possess cannabis and therefore in theory, the drug cannot be purchased legally. The law can be invoked as a legitimate defence by someone who is arrested for possession and can persuade the authorities that it was being used for a medical condition on their Doctor's recommendation. There is considerable inter-state variation; for example, the Arizona law requires the prescriber to write a scientific opinion, which must be corroborated by a second physician, on why a particular patient needs cannabis, before the patient can obtain the drug.

The remainder of this chapter discusses the available evidence on a disease state basis. An analysis of the efficacy of the crude drug (cannabis) is followed by a discussion of trials with THC or other cannabinoid derivatives and where possible, a brief summary of theories on the mode of action is given.

Pain features as a component of many of the illnesses discussed and a separate chapter by Dr Bill Notcutt and Mario Price (Chapter 8), describes the clinical evidence available which supports this use.

The chapter concludes with a discussion of where the real therapeutic and developmental potential lies for the derivatives of cannabis.

2   SPECIFIC MEDICINAL USES OF CANNABIS

The historical and contemporary, medicinal uses of cannabis have been reviewed on several occasions (Cohen and Andrysiak, 1982; Hollister, 1986; Ashton, 1987; Formukong *et al.*, 1989; Grinspoon and Bakalar, 1995).

Perhaps the earliest published report to contain at least some objectivity on the subject was that of O'Shaughnessy (1842), an Irish surgeon, working in India, who described the analgesic, anticonvulsant and muscle relaxant properties of the drug. This report triggered the appearance of over 100 publications on the medicinal use of cannabis in American and European medical journals over the next 60 years. One such use was to treat nausea and vomiting; but it was not until the advent of potent cancer chemotherapeutic drugs that the antiemetic properties of cannabis became more widely investigated and then employed.

One can argue that the available clinical evidence of efficacy is stronger here than for any other application and that proponents of its use are most likely to be successful in arguing that cannabis should be re-scheduled (to permit its use as a medicine) because it has a "currently accepted medical use".

2.1   Use as an Antiemetic

Many agents used in cancer chemotherapy produce severe nausea and vomiting in most patients. Symptoms can last for hours or days and have a major impact on patient nutrition and electrolyte status, body weight and physical and mental resilience to both the disease and its treatment. The current choice of available anti-emetics is limited and most are only partially effective, which may lead patients to refuse therapy all together, or for clinicians to use chemotherapeutic regimens which are less than optimum. For these reasons, the search for more effective antiemetics continues.

2.1.1   Cannabis

In the late 1960s and early 1970s, patients receiving various cancer chemotherapy regimes (including mustine, vincristine, prednisone and procarbazine) noted that smoking cannabis from illicit sources, before and during chemotherapy, reduced the incidence of nausea and vomiting to a variable degree. Only since the isolation of THC have formal clinical trials on the safety and efficacy of cannabis derivatives been conducted. As far as crude cannabis is concerned, we have only anecdotal evidence that inhaling its smoke is effective in a variable percentage of patients who vomit, despite supposedly adequate doses of standard antiemetics (Poster *et al.*, 1981a,b).

There is a growing population who, with or without the approval of their doctors, smoke cannabis to combat the emetogenic effects of cancer chemotherapy. The results of questionnaire surveys of oncologists' attitudes to this use have been ambiguous. In a survey of US oncologists (Doblin and Kleiman, 1991), 44% said that they had recommended cannabis to at least one of their patients and 48% said that they would prescribe cannabis if it were legal to do so. Many of the respondents believed that smoking cannabis was more effective than using orally administered 'synthetic' cannabinoids. However, these findings were subject to response bias and were certainly not supported by a later and more credible US study (Schwartz and Beveridge, 1994) where, in a survey of clinical oncologists' prescribing habits, although 65% reported having prescribed marijuana or oral THC, prescribing rates were extremely low, relative to other antiemetics.

Cannabis ranked ninth behind established antiemetics for treating mild/moderate chemotherapy induced nausea and vomiting and sixth, behind metoclopramide, lorazepam, corticosteroids, prochlorperazine and promazine, for the treatment of severe vomiting; 3.5% had prescribed marijuana more than 100 times. Those who had prescribed cannabis thought that it had been effective in 50% of patients but that 25% of patients had experienced unpleasant side effects. Just 6% of respondents said that they would prescribe cannabis much more frequently if there were no legal barriers to its medicinal use; 76% said that they would not prescribe cannabis more frequently if legal restrictions were eased.

2.1.2   THC

Clinical trials with cannabinoids were encouraged by anecdotal reports of decreased emesis in younger patients who smoked marijuana when receiving their chemotherapy and North American states had by 1985, enacted legislation to allow the creation of medical research programs on cannabis and its constituents.

On September 10, 1980, the US Surgeon General announced that 'synthetic' THC capsules were approved by the FDA, permitting use by up to 4000 cancer specialists to treat cancer chemotherapy-induced nausea and vomiting.

Most studies investigating the antiemetic efficacy of cannabinoids have involved THC or nabilone (see below). Detailed reviews of these studies appear elsewhere (Poster et al., 1981a; Carey et al., 1983; Vincent et al., 1983; Gralla and Tyson, 1985). A summary of the salient features of trials with THC, involving a total of 1001 evaluable patients appears in Table 2.

The earliest study involving THC was that reported by Sallan et al. (1975). Oral THC was compared to placebo in 22 patients. No antiemetic effect was noted for the placebo, but several complete and partial responses were noted in those given THC, which was found to be statistically superior. The most frequent side effect was somnolence in two thirds of patients; ten suffered brief periods of dysphoria. The characteristic cannabis 'high' was seen in patients taking THC, but was positively associated with its antiemetic effect.

Chang et al. (1979) compared THC with placebo in 15 patients receiving high dose methotrexate for osteogenic sarcoma. The problem of subjects vomiting their orally

Table 2   Summary of clinical trials investigating THC as an antiemetic for cancer chemotherapy-induced nausea and vomiting

Reference	n*	Chemotherapy	Antiemetic	Outcome+
2.1 Open Trials				
Lucas and Laszlo (1980)[2]	20	adriamycin, vinblastine cisplatin, methyl lomustine	THC 15 mg/m², every 6 hours, starting one hour before chemotherapy and continuing for 4 doses. Reduced to 5 mg/m² every 4 hours starting 8–12 hours before chemotherapy and continuing for 24 hours.	10 had complete response, 28 had partial response. All responders experienced a 'high'.
Sweet et al. (1981)[2]	25	Multiple combinations.	THC 5 mg/m², q8h for 6 doses.	72% partial responses; 8% complete responses.
2.2 Placebo Controlled Trials				
Sallan et al. (1975)[1]	20	Adriamycin, 5-azacytidine nitrogen mustard, dacarbazine procarbazine, 6-thioguanine, cytosine arabinoside	THC, 10 mg/m² PO 2 hours prior to chemotherapy, then 2 and 6 hours after chemotherapy vs placebo.	THC response partial in 9, complete in 5. No response with placebo.
Chang et al. (1979)[1]	15	High dose methotrexate	THC 10 mg/m² PO every 3 hours for 5 doses. First dose given 2 hours before chemotherapy, vs placebo.	Incidence of nausea and vomiting was 0–6% with THC levels >.10 ng/ml,
	5	Adriamycin + cyclophosphamide		
				Nausea and vomiting significantly reduced compared to placebo. Response associated with a 'high'; THC less effective in patients having adriamycin and cyclophosphamide.

Table 2 (*Continued*)

Reference	n[*]	Chemotherapy	Antiemetic	Outcome[+]
Kluin-Neleman et al. (1979)[1]	11	nitrogen mustard, vincristine procarbazine, prednisone	THC 10 mg/m^2 PO or placebo, 2 hours prior to chemotherapy; repeated 4 and 8 hours later.	THC was antiemetic, but most patients preferred the nausea and vomiting to the side effects of THC.
Chang et al. (1981)[1]	8	Doxorubicin and cyclophosphamide.	THC 10 mg/m^2, q3 hours for 5 doses vs placebo.	THC no better than placebo.

2.3 Comparative Studies

THC vs metoclopramide

Ekert et al. (1979)[1]	33	Multiple combinations including high dose methotrexate, anthracyclines, nitrosoureas, asparaginase nitrogen mustard. No cisplatin.	THC 10 mg/m^2 PO vs metoclopramide 5–10 mg PO, each given 2 hours prior to chemotherapy; Then at 8, 16 and 24 hours after chemotherapy (19 children). Also THC 10 mg/m^2 PO vs prochlorperazine 5–10 mg PO, given as a bove (14 children).	THC had a lower incidence of nausea and vomiting compared to both comparators.
Colls et al. (1980)[1]	35	Multiple combinations. including cyclophosphamide and mustine.	THC 12 mg/m^2 PO vs thiethylperazine 6 mg/m^2 vs metoclopramide 15 mg/m^2.	THC equivalent to both comparators.
Gralla et al. (1982 and 1984)[4]	31	Cisplatin	THC 10 mg/m^2, q3h for 5 doses; vs metoclopramide 2 mg/kg.	Metoclopramide superior to THC.

THC vs prochlorperazine/chlorperazine/haloperidol

Frytak et al. (1979)[3]	116	5-fluorouracil plus lomustine with or without vincristine, adriamycin, triazinate, ICRF-159.	THC 15 mg PO; prochlorperazine 10 mg PO or placebo; given 2 hours before chemotherapy, then 2 and 8 hours after initiation of chemotherapy. Then daily for 3 days, half an hour before each meal, 3 times a day.	THC was superior to placebo and equal in efficacy to prochlorperazine

Table 2 (*Continued*)

Reference	n[*]	Chemotherapy	Antiemetic	Outcome[+]
See also Ekert *et al.* (1979) above				
Sallan *et al.* (1980)[1]	83	Adriamycin, cyclophosphamide high dose methotrexate cisplatin, bleomycin, vincristine dacarbazine, actinomycin	THC 10 mg/m^2 PO vs prochlorperazine 10 mg PO; each given as one dose, 1 hour before chemotherapy another dose 3 hours after chemotherapy, then a third dose, four hours later.	Complete response in 36 of 79 THC courses and 16 of 78 prochlorperazine. 32 courses of THC associated with a high, but were complete responders.
Orr *et al.* (1980)[1]	55	Adriamycin, cyclophosphamide 5-fluorouracil, methotrexate, nitrogen mustard, dacarbazine nitrosoureas, cytosine arabinoside	THC, 7 mg/m^2 PO, every 4 hours for 4 doses vs prochlorperazine, 12 mg/m^2 PO, as above vs placebo. First dose given 1 hour before chemotherapy.	THC supressed nausea in 36; no emesis in 52; prochlorperazine suppressed nausea in 8, with no emesis in 37. Placebo suppressed nausea in 5, with no emesis in 26 patients.
Garb *et al.* (1980)[1]	34	Cyclophosphamide, adriamycin, cisplatin, dacarbazine actinomycin, nitrogen mustard.	THC 10 mg PO plus prochlorperazine or thiethylperazine 10 mg PO, given for four doses prior to chemotherapy and on day of chemotherapy vs placebo plus phenothiazine. THC plus prochlorperazine or thiethylperazine PO in escalating doses given 24 and 48 hours before chemotherapy and on day of chemotherapy.	THC plus phenothiazine superior to placebo combination. Higher doses of THC superior to lower doses.
Levitt *et al.* (1981)[3]	120	Multiple combinations	THC, 5, 10, and 15 mg q4h for 4 doses; vs prochlorperazine 10 mg, vs placebo, vs none.	THC superior vs vomiting; prochlorperazine better vs nausea.

Table 2   (*Continued*)

Reference	n^*	Chemotherapy	Antiemetic	Outcome[+]
McCabe *et al.* (1981)[1]	36	Multiple combinations	THC 15 mg, q4h for 6 doses; prochlorperazine 10 mg.	THC superior to prochlorperazine.
Neidhart *et al.* (1981)[1]	52	Cisplatin, nitrogen mustard doxorubicin.	THC 10 mg q4h for 8 doses; vs haloperidol 2 mg.	THC equivalent to haloperidol.
Ungerleider *et al.* (1982)[1]	214	Multiple combinations.	THC 7.5–12.5 mg/ m^2, q4h vs prochlorperazine 10 mg.	THC equivalent to prochlorperazine.
Lane *et al.* (1989)[4]	55	Multiple combinations except high-dose cisplatin.	THC 10 mg q6h vs prochlorperazine 10 mg q6h vs combination. Given 24h prior to and continued 24h after chemotherapy	Reduction in nausea and vomiting in all groups. Significant reduction in severity with combination compared to either drug alone. Higher incidence of side effects with THC.

* = number of evaluable patients.
$^+$ Therapeutic efficacy determined qualitatively or by measuring vomiting frequency.
Trial design: 1 = randomised, double blind, crossover trial; 2 = single group, before and after comparison; 3 = randomised, non-crossover; 4 = randomised, double blind, non-crossover.

administered treatment was circumvented by supplementing it with cigarettes containing THC or placebo. THC was found to be superior to placebo, providing "substantial therapeutic benefit and minimal toxicity". Efficacy could be related to THC blood levels – at THC levels of 10 ng/ml, the incidence of nausea and vomiting was 6%. THC was not so effective in five patients who received adriamycin plus cylophosphamide. Somnolence, tachycardia and orthostatic hypotension were observed, together with a low incidence (2% of all THC doses) of dysphoric reactions.

It is of significance that the authors were able to demonstrate superiority of THC over placebo in patients receiving methotrexate; however they could not demonstrate a difference in a later study where more emetogenic agents (cyclophosphamide, doxorubicin) were used (Chang *et al.*, 1981). A reduction in antiemetic response was observed in several patients who continued to receive THC after the trial had ended; but it was not possible to determine if this was due to tolerance, normal variation in response to chemotherapy or the development of anticipatory nausea and vomiting.

Sallan *et al.* (1980) compared THC with prochlorperazine in 84 cancer patients, 82 of which were refractory to standard antiemetic drugs. THC was described as completely effective in 36 of 79 courses; in comparison, prochlorperazine was only as effective in 16 of 78 courses. Younger patients appeared to respond better than older patients. Of

25 patients who were treated with both drugs and who expressed a preference, 20 preferred THC. Increased food intake occurred more frequently with THC. The only reactions to THC were somnolence and a high associated with the beneficial effects of THC; four patients developed highs described as "excessive".

Lucas and Laszlo (1980) tested oral THC in 53 patients, refractory to standard antiemetics. A response rate of 72% was obtained (19% complete and 53% partial). A dose of $15\,mg/m^2$ produced dysphoric reactions in 3 of 9 patients. It was found that this effect could be avoided, whilst maintaining antiemetic efficacy at a dose of $5\,mg/m^2$; somnolence and dry mouth were observed at this dose.

It is clear from these studies that the effect of THC was reproducible in individual patients upon repeat courses of the same chemotherapy and that there was no tachyphylaxis. The dependency of the degree of response to THC on the nature of the chemotherapy used was illustrated by the fact that of 21 patients who failed to derive relief from THC, 11 were receiving the highly emetogenic cisplatin in combination with other agents. In contrast, THC was effective in patients given high-dose intravenous infusions of carmustine or cyclophosphamide – both experimental regimes, characteristically associated with a high incidence of nausea and vomiting.

Orr et al. (1980) compared the antiemetic efficacy of THC to prochlorperazine and placebo in 55 patients, described as refractory to conventional antiemetic therapy. Of the 55 patients, only 3 actually vomited whilst taking THC and 12 felt nauseous. With prochlorperazine, the corresponding figures were 18 and 29 respectively. Both treatments were considered to be superior to placebo. In the THC group, reported 'highs' were described as being favourably related to the antiemetic effect. Loss of physical control, possibly ataxia, was reported in an unspecified number of patients and dysphoric reactions were observed in two patients; these responded to tranquilliser therapy.

Ungerleider et al. (1982) found THC to be as effective as prochlorperazine in the reduction of nausea and vomiting associated with cancer chemotherapy in a randomised, double-blind, crossover study in 214 patients. The same group (Ungerleider et al., 1985) gathered data on patient preferences in 139 of these patients in order to determine the relative influence of perceived efficacy and side effect profile. Nausea reduction was the main determinant of preference. Suprisingly, preference for THC was associated with an increased level of side effects, notably sedation; the latter may reflect the patients' desire for sedation during chemotherapy. Subjects who reported being anxious or depressed prior to therapy did not experience accentuation as a result of either regime and there was no difference in the numbers of patients preferring THC in age groups above or below 50 years. The common assumptions that THC was contraindicated in older, cannabis-naive patients or indeed, those who are anxious and depressed, were not supported by this study.

Ekert et al. (1979) compared the efficacy of THC versus metoclopramide in a double-blind, parallel group, trial in 19 children, 5–19 years old, receiving a wide range of chemotherapies, including high-dose methotrexate. The incidence of both nausea and vomiting was significantly reduced with THC compared with metoclopramide. Similar results were obtained in a subgroup of eight children where crossover of antiemetic drugs was possible. THC was also shown to be superior to prochlorperazine in a paediatric population, both when the drugs were given double-blind to parallel groups (18 episodes each) or in crossover fashion in seven children. By no means all children obtained relief with THC in these trials. Somnolence was observed after THC; one patient had a dysphoric reaction and another reported agitation, anxiety and bad dreams. This is interesting; although it has been argued that the brain receptor(s) responsible for these effects are less well-developed in children and therefore that THC may be used in higher doses to prevent vomiting, it appears that the central side effects cannot be avoided completely.

Combinations of standard antiemetics are commonly used in refractory patients and it seems logical that THC should also be investigated in this respect. Garb et al. (1980) combined THC with prochlorperazine or thiethylperazine. Eight of 10 patients responded better to the combination of actives compared with a placebo-phenothiazine combination. In a second study, the dose of THC was escalated to a maximum of 120 mg/day, in combination with phenothiazine according to the following schedule: THC up to 40 mg/day combined with equal doses of phenothiazine; 7–9 mg increases of phenothiazine were given for each step increase of 5 mg THC. A mean protection rate of 83% was achieved among 24 evaluable patients. At lower doses of THC, the degree of protection was dependent on the nature of the cancer chemotherapy; THC was effective against cyclophosphamide and adriamycin but higher doses were needed against nitrogen mustard, dacarbazine, actinomycin D and cisplatin. It was suggested that the phenothiazine component protected against the development of a THC 'high'. Two patients did develop dysphoria, managed with additional phenothiazine. Drowsiness was a major side effect; other reactions were confusion, orthostatic hypotension, decreased concentration, anxiety and hallucinations.

Lane et al. (1989) studied 55 patients with a variety of tumors who were randomised in a placebo-controlled, parallel-group trial, to receive THC 10 mg, prochlorperazine 10 mg, or a combination of the two, four times a day when receiving a range of cancer chemotherapies, with the exception of high-dose cisplatin. Side effects, primarily related to the CNS, were more common in the THC-only group. The antipsychotic effect of prochlorperazine may have decreased the incidence and severity of the psychrotropic effects of THC. The THC–prochlorperazine combination was superior to each agent alone in terms of reducing the median duration and severity of nausea and in comparison with prochlorperazine, the mean duration of vomiting.

Not every trial has demonstrated a favourable profile for THC as an antiemetic. Frytak et al. (1979) conducted a placebo controlled comparison of THC and prochlorperazine in 116 patients with gastrointestinal carcinoma, treated with either 5-fluorouracil plus methyl-lomustine or vincristine, adriamycin and an investigational compound, ICRF-159. Nausea and vomiting were suppressed in 42% of patients in both the THC and prochlorperazine groups; the corresponding figure for the placebo group was 19%. Thirty-two percent of patients in the THC group had dose-limiting toxicity, which included ataxia, hypotension, visual hallucinations, blurred vision, muddled thinking, paresthesias, faecal incontinence and dysphoria. In this study of generally older patients, fewer subjective side effects were seen with the phenothiazine and it was the preferred treatment over THC. Elderly patients may be less willing to accept the adverse effects associated with THC or may metabolise the drug differently.

In other trials, THC was found to be as effective as metoclopramide or thiethylperazine, but with a significantly greater incidence of side effects (Colls et al., 1980); to the extent that the authors could not recommend the drug for routine use as an adjunct to cancer chemotherapy. Neidhart *et al.* (1981) reported similar antiemetic efficacy for THC and haloperidol; but again, more side effects were noted with THC.

Gralla *et al.* (1982, 1984) conducted a randomised clinical trial of THC versus high-dose metoclopramide in patients receiving their first dose of cisplatin (120 mg/m² over 20 minutes). The first cisplatin dose was chosen to avoid the possibility of conditioned emesis. Metoclopramide was significantly superior to THC in reducing the number of emetic episodes and the percentage of patients receiving major emesis support over the 24-hour observation period. Also the volume of vomit and duration of nausea and vomiting were less with metoclopramide. However, over one quarter of the patients taking THC showed a major antiemetic effect for the drug. Interestingly, THC appeared to be effective in combating cisplatin-induced diarrhoea – a traditional indication for cannabis!

Kluin-Neleman *et al.* (1979) compared the use of THC with placebo in a small group of patients receiving a combination of nitrogen mustard, vincristine, procarbazine and prednisone. THC had appreciable antiemetic efficacy, but there was a high incidence of serious side effects, including somnolence, dizziness, depersonalisation and derealisation; mania was triggered in one patient. Although a direct relationship between either antiemetic effect or side effects and THC blood level was not established, the increased incidence of side effects was explained by the high blood levels of THC measured in this trial – several hundred ng/ml, compared with an accepted therapeutic level of 10 ng/ml (Chang *et al.*, 1979). These may have been produced by allowing patients a second dose of THC if they vomited after the first.

Data from an uncontrolled, open-label trial by Stanton (1983) suggest that 5.0 and 7.5 mg/m² doses of THC given every 4 hours, during a variety of cancer chemotherapy regimens, were equipotent in terms of antiemetic efficacy but that side effects (mainly sedation) were less with the lower dose.

In an interesting footnote study, Ungerleider *et al.* (1984) reported a randomised, double-blind, crossover comparison of THC (7.5–12.5 mg, every 4 hours, three times a day) with prochlorperazine, 10 mg, on the same schedule, in 11 patients with various cancers requiring radiotherapy of the abdomen. Drug administration coincided with the five day per-week radiotherapy schedule. Four patients withdrew from the trial, including two who had dizziness and depersonalisation after THC and one who experienced excessive nausea and vomiting. Five of the remaining 7 patients had previous experience with marijuana. Patients were asked to rate the severity of their illness, as well as the extent of their subsequent moods, their level of concentration, their amount of physical activity and their desire for social interaction. They chose the drug they preferred and recorded its side effects. Improved alleviation of nausea and vomiting was noted in 4 patients taking THC and 3 taking prochlorperazine; but the difference between drugs was not significant. Significant differences in favour of THC were a reduction in appetite suppression and ability to concentrate. Side effects noted with THC included somnolence, dizziness, dry mouth, increased heart rate and dysphoria. The authors concluded that based on all the psychological and physiological parameters assessed, THC was slightly more advantageous than prochlorperazine.

Plasse *et al.* (1991) conducted a meta-analysis of 454 patients from published clinical trials reporting the efficacy and side effects of THC as a function of dose. Drowsiness and other, non-psychotropic symptoms were as common in patients receiving oral doses of less than or equal to $7\,mg/m^2$ as in those receiving a greater dose. The incidence of dysphoric effects was only 12% in the low dose group compared to 28% in the high dose group; a corresponding reduction of efficacy with dose was not observed. The authors concluded that a relatively low dose of THC could minimise side effects while preserving efficacy.

In all of the studies, patients were receiving chemotherapy for a variety of solid tumours and haematological dyscrasias. Usual doses of THC ranged from $5\text{--}15\,mg/m^2$, given at fixed time intervals, prior to, during and after chemotherapy. When given with a phenothiazine, it was possible to give higher doses without severe toxicity, including the characteristic high (Garb et al., 1980). Of the comparative studies carried out, there was only one study in which THC did not have an effect which was superior to prochlorperazine (Frytak et al., 1979). Regelson et al. (1976) have pointed out that the drug not only reduces nausea and vomiting, but can improve appetite and mood in nutritionally depleted cancer patients.

All of the studies in Table 2 reported some degree of efficacy for oral THC as an anti-emetic, even where the level of toxicity was considered to be unacceptable (Frytak et al., 1979; Kluin-Neleman et al., 1979; Colls et al., 1980). The incidence and severity of dysphoria observed in some cases is worrying although in those patients with refractory nausea and vomiting, attempts to manage dysphoria, either by psycho- or pharmaco-therapy may reduce this reaction to an acceptable risk level (Chang et al., 1979; Garb et al., 1980). Hypotension, ataxia and tachycardia may be manageable by THC dose reduction.

Penta et al. (1981) have reviewed the side effect profile of THC in a collection of the clinical trials summarised in Table 2; the effects they describe as frequent, include somnolence (31%), dry mouth (9.1%), ataxia (8.2%), dizziness (6.1%), dysphorias (5.8%) and orthostatic hypotension (3.6%). Infrequent toxicities associated with the use of THC were: visual distortions/hallucinations (1.8%), confusion (0.9%), muddled thinking (0.6%), paresthesias (0.6%), amnesia (0.3%), syncope (0.3%), slurred speech (0.3%) and faecal incontinence (0.3%). The nature of this side effect profile is not dissimilar to that of other cannabinoids when used under the same conditions (Penta et al., 1981).

It has been suggested that younger patients are less susceptible to developing side effects compared with older subjects (Frytak et al., 1979); this may be because older subjects are not first-time users and have developed some tolerance to the side effects. Another reason may be that brain cannabinoid receptors associated with CNS side effects are not so well-developed in children and therefore the effects are less pronounced.

So although the side effect profile of THC is notable, it is manageable, and it may still be an attractive alternative to intractable nausea and vomiting; however there are additional problems with its use.

Firstly, response does appear to be related to the emetogenicity of the chemother-apeutic agents used, with high dose methotrexate (Chang et al., 1979; Orr et al., 1980), doxorubicin and cyclophosphamide/fluorouracil combinations responding better than nitrosourea, mustine or cisplatin (Orr et al., 1980; Garb et al., 1980; Gralla et al., 1984).

Secondly, THC is usually given, formulated in sesame seed oil, in gelatin capsules. Oral bioavailability is known to be low (5–10% only) and unpredictable (Nahas, 1979;

Perez-Reyes *et al.*, 1972). This may have led to the wide range of efficacies observed in clinical trials. Indeed, this may be the reason why patients have been reported to derive faster and more predictable relief from smoking cannabis (Doblin and Kleiman, 1991; Grinspoon and Bakalar, 1995). Thirdly, because of the abuse potential of this agent, secure storage, prescription and dispensing procedures are required. Finally, it is clear that there are certain groups of patients in whom THC is contraindicated. Those patients with epilepsy; cardiovascular disorders; mental disorders (Treffert, 1978); and children – where THC can cause neurohormonal regulatory disorders (Chang *et al.*, 1979; Nahas, 1979).

In conclusion, THC appears to be a useful drug in sub-sets of patients where the only alternative is the misery of intractable chemotherapy-induced nausea and vomiting and where the stakes of non-compliance with chemotherapy are high indeed. Many anti-emetics are available but none is entirely satisfactory in all patients. Even the serotonin antagonists such as ondansetron, can have failure rates as high as 40%, depending on dose and the nature of the chemotherapy – once again, cisplatin appears to be the most difficult agent in this respect (Khojasteh *et al.*, 1990; Markham and Sorkin, 1993). Insufficient research has been conducted on the combination with established anti-emetics and on producing a stable, oral formulation which has a satisfactory and reproducible bioavailability.

In an outstanding review of the use of THC as an antiemetic in clinical trials, Carey *et al.* (1983) noted that the often conflicting and confusing results obtained owe much to the inadequacy of the study designs and methods. While acknowledging that a study of antiemetic activity in cancer patients on a wide variety of cancer chemotherapies holds unavoidable difficulties, there are many variables, such as chemotherapy regimen, age, drug dose regimen, drug tolerance, route of administration, toxicity and drug interactions which have been analysed inadequately. More research is needed into the management of nausea and vomiting which is conditioned rather than organic. Environment variables, such as in- or out-patient setting, and the attitudes of carers to the use of THC also need closer scrutiny. The authors argue that only through carefully designed and controlled trials, can the efficacy of THC be identified, its limits defined and its effectiveness relative to other treatments established.

From September 1995, THC was made available in the UK, to be prescribed on a named patient basis, as an antiemetic in patients receiving cancer chemotherapy. This change followed advice from the World health Organisation, accepted by the UN Commission on Narcotic Drugs, that THC has a recognised therapeutic use in cancer patients. This situation now mirrors that in the US where THC formulated as an oral capsule in sesame seed oil, and called dronabinol, is already marketed as a second-line antiemetic and to treat anorexia in AIDS patients.

2.1.3 Nabilone

This synthetic derivative of THC is licenced in a number of countries, solely for the prevention of chemotherapy-induced nausea and vomiting. It is the result of research on a range of THC analogues, driven by the desire to produce a drug which is effective, but lacking the often severe, central side effects of THC.

It appears that with nabilone, some separation between euphoric and antiemetic effects has been achieved (Herman *et al.*, 1977; Lemberger, 1976). Key investigations featuring nabilone, involving 579 evaluable patients, are shown in Table 3.

Like THC, nabilone appears to have significant antiemetic activity and has been shown to be superior to prochlorperazine in both animals (McCarthy and Borison, 1981; Ward and Holmes, 1985) and man (Nagy *et al.*, 1978; Herman *et al.*, 1979; Steele *et al.*, 1980; Einhorn *et al.*, 1981; Cone *et al.*, 1982; Einhorn, 1982; Johansson *et al.*, 1982; Long *et al.*, 1982; Ahmedzai *et al.*, 1983; Niiranen *et al.*, 1983) and in one small study, equivalent to intramuscular chlorpromazine (George *et al.*, 1983). It appears to be useful in cases of nausea and vomiting refractory to other antiemetics. It is also interesting to note that in the studies reported by Einhorn *et al.* (1981) and Vincent *et al.* (1983) nabilone was judged to be effective from the first day of treatment when nausea and vomiting were most severe.

Archer *et al.* (1986) have reviewed phase 2 and phase 3 clinical trials of nabilone in the control of nausea and vomiting associated with cancer chemotherapy. As with THC, nabilone is more effective against lower doses of cisplatin compared with higher doses, and against regimes which do not contain this drug (Herman *et al.*, 1979; Steele *et al.*, 1980; George *et al.*, 1983). Nabilone is by no means free of side effects however: drowsiness, dry mouth, divided co-ordination, blurred vision, postural hypotension, and dizziness occur in significant proportions of patients (Nagy *et al.*, 1978; Herman *et al.*, 1979; Steele *et al.*, 1980; Wada *et al.*, 1982; Levitt, 1982); rare cases of depersonalisation and hallucinations have been reported (Herman *et al.*, 1979) and psychotic reactions have also been noted (Niiranen *et al.*, 1983). Reports of other CNS side effects persist, including descriptions of 'highs', which can be euphoric or dysphoric, (Johansson *et al.*, 1982; Cornbleet *et al.*, 1982; Ahmedzai *et al.*, 1983); these have prompted patients to withdraw from clinical trials (Wada *et al.*, 1982; Jones *et al.*, 1982), but are likely to be dose-related (Einhorn *et al.*, 1981; Cornbleet *et al.*, 1982). There is no agreement on whether the 'high' is an essential component of the antiemetic activity. These reports are of concern, especially as nabilone has been shown to be effective in children, where the side effect profile was considered acceptable (Patel *et al.*, 1983; Chan *et al.*, 1984).

Ahmedzai *et al.* (1983) could find no statistical association between age, sex and nabilone toxicity and recommended that if the dose of nabilone was restricted to 2 mg, 12-hourly, for anti-emetic control in regimens which did not contain platinum-based drugs, its use would generally be associated with a moderate but overall, acceptable incidence of side effects.

Other (unlicenced) areas where nabilone has been shown to be effective are in combating nausea and vomiting in patients undergoing radiotherapy (Priestman and Priestman, 1984); total abdominal hysterectomy (Lewis *et al.*, 1994) and in a patient experiencing nausea and vomiting due to an AIDS related cryptosporydial infection (Green *et al.*, 1989).

2.1.4   *Levonantradol*

Cronin and Sallan (1981) described the intramuscular use of this synthetic analogue in patients receiving a wide range of cancer chemotherapies, who were refractory to

Table 3   Summary of clinical trials investigating nabilone as an antiemetic for cancer chemotherapy-induced nausea and vomiting

Reference	n^*	Chemotherapy	Antiemetic	Outcome[+]
Nagy et al. (1978)[2]	33	Cisplatin in combination	Nabilone, 2 mg PO, q6h as required for nausea or vomiting, vs prochlorperazine, 10 mg PO, q6h, started 30 mins prior to treatment.	Nabilone superior to prochlorperazine. 84% of subjects had less nausea and vomiting with nabilone than with prochlorperazine. 21 of 28 patients wanted nabilone for future chemotherapy.
Herman et al. (1979)[2]	113	1. Cisplatin combination $n=70$ 2. Doxorubicin plus cyclophosphamide. $n=12$ 3. Mustine combination. $n=11$ 4. various combinations. $n=20$	Nabilone, 2 mg PO, q8h vs prochlorperazine, 10 mg PO q8h, 2 doses before treatment. Nabilone, 2 mg PO, q6h vs prochlorperazine, 10 mg PO q6h, started 30 mins before therapy.	In both trials, nausea and vomiting significantly less with nabilone. 75% patient preference for nabilone; 15% for prochlorperazine. 80% response to nabilone, 32% response to prochlorperazine.
Steele et al. (1980)[1]	37	1. High dose cisplatin 120 mg/m^2 2. Low dose cisplatin 45–70 mg/m^2 3. Streptazosin, actinomycin, dacarbazine, mustine.	Nabilone 2 mg PO, q12h for 3–5 doses vs, prochlorperazine 10 mg PO, q12h for 3–5 doses, begun 12 hours prior to treatment,	1. Nabilone = prochlorperazine 2. Nabilone > prochlorperazine 3. Nabilone > prochlorperazine
Einhorn et al. (1981)[2]	80	Cisplatin in combination, with various agents including vinblastine, bleomycin and adriamycin, $n=77$. Other combinations, $n=3$.	Nabilone 2 mg PO, q6h, as required. Prochlorperazine 10 mg PO, q6h, as required.	Nabilone preferred by 75% of patients. 21% preferred, prochlorperazine. Nausea and vomiting reduced significantly with nabilone compared to prochlorperazine. Efficacy of nabilone most pronounced after the first day of therapy.

Table 3 *(Continued)*

Reference	n[*]	Chemotherapy	Antiemetic	Outcome[+]
Einhorn et al. (1982)[1]	50	Various combinations.	Nabilone 2 mg PO, on evening prior to chemotherapy then twice daily vs placebo.	Nabilone reduced nausea and vomiting significantly compared to placebo. 90% preferred nabilone, 4% placebo.
Johansson et al. (1982)[2]	18	Various combinations including cisplatin.	Nbilone 2 mg PO, twice daily vs prochlorperazine 10 mg three times a day.	Significantly less nausea severity and vomiting with nabilone than prochlorperazine. 72% of patients preferred nabilone. Side effects more frequent and severe with nabilone.
Wada et al. (1982)[1]	92	Various combinations including cisplatin and adriamycin.	Nabilone 2 mg PO twice daily vs placebo. Given before and for at least 24 hours after chemotherapy.	Significant reductions in nausea and vomiting with nabilone vs placebo. 35% experienced complete relief vs 11% on placebo. 70% patient preference for nabilone vs 22% for placebo.
Jones et al. (1982)[1]	24	Various combinations including cisplatin and adriamycin.	Nabilone 2 mg PO, twice daily vs placebo. Given before and for at least 24 hours after chemotherapy.	Nausea and vomiting significantly reduced with nabilone vs placebo. 67% patient preference for nabilone vs 8% for placebo.
Levitt (1982)[1]	36	Various combinations including cisplatin and adriamycin.	Nabilone 2 mg PO, twice daily vs placebo. Given before and for at least 24 hours after chemotherapy.	Nausea and vomiting significantly reduced with nabilone vs placebo. 78% preferred nabilone vs 8% for placebo. Nabilone also increased food intake significantly.
Vincent et al. (1983)[1]	42	Various combinations, including cisplatin. Also doxorubicin, cyclophosphamide and fluorouracil.	Nabilone 2 mg PO, twice daily vs placebo, two doses prior to treatment.	Nabilone superior to placebo; preferred by 91% of patients. Nabilone superior to placebo on first day.

Table 3   (*Continued*)

Reference	n[*]	Chemotherapy	Antiemetic	Outcome[+]
Ahmedzai et al. (1983)[2]	34	Combination of cyclophosphamide adriamycin, etoposide.	Nabilone 2 mg PO, twice daily. Prochlorperazine 10 mg PO three times a day.	Significant reductions in vomiting and nausea symptoms with nabilone. 16 preferred nabilone, 3 preferred prochlorperazine.
George et al. (1983)[3]	20	Combinations with cisplatin or cisplatin alone.	Nabilone 3 mg PO, three times a day, chlorpromazine, 12.5 mg IM, 15 minutes before chemotherapy.	Nabilone = chlorpromazine in reducing vomiting. 10 patients preferred nabilone, 5 preferred chlorpromazine.

[*] = number of evaluable patients. [+] Therapeutic efficacy determined qualitatively or by recording changes in vomiting frequency.
Trial design: 1 = randomised, double blind, placebo-controlled, crossover trial; 2 = randomised, double blind, crossover trial; 3 = randomised, double blind, crossover trial.

treatment with prochloperazine or thiethylperazine. Of the 18 chemotherapy regimes used, 11 contained cisplatin. After initial dose titration, 15 of 16 patients showed improvement in the extent of nausea and in the number of vomiting episodes. Two patients experienced dysphoria. Other side effects were somnolence, drowsiness and dizziness; these were described as mild and acceptable.

Tyson et al. (1985) described a phase 1 trial of intramuscular levonantradol in 34 patients who received a total of 52 courses of cancer chemotherapy. Chemotherapy included high dose cisplatin alone or in combination with other drugs. Major (0–2 emetic episodes) or minor (3–5 episodes) antiemetic effects were observed in 23% of patients receiving cisplatin and in 53% receiving non-cisplatin regimens, at levonantradol doses ranging from 0.5 to 4 mg, every four hours. Centrally-mediated side effects (sedation: 44%; dysphoria: 29%) were observed, together with dizziness (65%), postural hypotension (37%) pain at the injection site (48%) dry mouth (67%) and urinary retention (10%). Levonantradol was acceptable to most patients at doses of 3 mg or less, although side effects were common. Marked toxicity, including urinary retention, occurred at higher doses. While antiemetic efficacy is undoubtedly dose – related (Stuart et al., 1982), Laszlo et al. (1981) have recommended a maximum dose of 1 mg to mini-mise the development of euphoria and that both oral and intramuscular routes might be combined to good effect to combat persistent nausea and vomiting.

Further randomised, double blind studies have confirmed that levonantradol has antiemetic activity equal to (Scheidler et al., 1984) or superior to placebo or other antiemetic agents including prochlorperazine (Stambaugh et al., 1982; Long et al., 1982; Staurt-Harris et al., 1983; Heim et al., 1984); but that its use is frequently associated with a greater incidence of side effects than these comparators. Levonantradol has also been shown to be as effective as THC, with the advantage that it may be given intramus-cularly (Citron et al., 1985).

2.1.5   Other Derivatives

In further attempts to dissociate dysphoric reactions from antiemetic activity and to develop agents which are more potent at combating nausea and vomiting caused by cisplatin, additional chemical modifications of cannabinoids have been made and at least partial efficacy has been demonstrated for some of these agents in man (Bron *et al.*, 1981; Howes, 1981). In the study by Bron *et al.* (1981), the analogue, BRL-4664 was more effective at combating cisplatin-induced nausea and vomiting in patients who had not received cisplatin before. In cisplatin-experienced individuals, the reduction in vomiting episodes was not nearly so marked. This emphasises the role of conditioned reflexes in trials of this nature. The authors make the point that a stratified, parallel-group design might be better than a crossover study in future trials.

In summary, the effectiveness of at least two derivatives of cannabis (THC and nabilone) in treating chemotherapy-induced nausea and vomiting is now firmly established. Acceptance has come only after gathering clear clinical evidence of efficacy. The side effect profiles of the synthetic derivatives of cannabis appear to be qualitatively similar and in the absence of head to head clinical trial data, it is difficult to favour one compound over another. Suffice it to say that the level and nature of side effects with any of them is far from ideal. If the development of additional agents from the cannabinoids is to succeed, then trials must avoid the pitfalls of earlier work which have hampered direct comparison of trial data; these include the use of:

(i)   Several different dosing schedules, doses and routes of administration;
(ii)   differing chemotherapy regimens;
(iii) differing antiemetic combinations;
(iv)   differing patient populations, in terms of pre-treatment history, age, sex, psychological status and premorbid personality structure;
(v)   different assessment techniques;
(vi)   different response criteria;
(vii) differing environmental factors associated with the administration setting.

By the same token, there is no denying that cannabinoids are extremely effective antiemetics, given particular combinations of the above conditions. To date, the combinations, have not been clearly defined.

One observation, supported by observations when cannabis has been used in other conditions (Grinspoon and Bakalar, 1995), is that patients derive equal, if not greater relief after smoking cannabis compared with the oral ingestion of a single cannabinoid; this may be due to a number of reasons:

(i)   THC is more readily absorbed via the lung than through the gut;
(ii)   enteral absorption is slowed in patients with gastrointestinal hypermotility;
(iii) the impressive cocktail of actives in crude cannabis may modify the antiemetic activity of certain cannabinoids in a synergistic or additive way;
(iv)   smoking allows self-titration which is not possible with a fixed oral dose of a cannabinoid.

To-date, there have been no trials comparing cannabis or its derivatives with newer antiemetics such as ondansetron. One can speculate from accumulated wisdom that the latter would prove to be more effective; however, because of their unique pharmaco-logical profile, rather than in spite of it, the cannabinoids still retain a place in the antiemetic armamentarium, particularly for intractable nausea and vomiting, perhaps in combination with more established agents.

2.1.6   Mode of Action

How cannabis and its isolated cannabinoids achieve their antiemetic effects is far from clear (Formukong et al., 1989). Bhargava (1978) has suggested that, as THC is known to decrease brain dopamine levels, the drug may be inactivating central dopaminergic activity associated with the vomiting reflex. Studies with nabilone in the cat suggest that it may act in the forebrain, inhibiting the vomiting control mechanism in the medulla oblongata through descending pathways as well as the chemoreceptor trigger zone (London et al., 1979). The intravenous administration of naloxone blocks the antiemetic effect of nabilone in the cat (McCarthy and Borison, 1977) indicating an action on central opiate receptors. The mode of action of cannabinoids is likely to be different from that of the phenothiazines, which are known to block dopaminergic receptors responsible for apomorphine-induced nausea and vomiting, as they do not significantly inhibit apomorphine-induced vomiting in dogs (Steele et al., 1980) and cats (McCarthy and Borison, 1981).

2.2   Glaucoma

Cannabis smoking and the oral ingestion of several of its derivatives have been shown to cause an appreciable drop in intraocular pressure (Hepler and Frank, 1971; Cooler and Gregg, 1976; West and Lockhart, 1978); and it is known that patients with open angle closure glaucoma smoke cannabis for this purpose.

2.2.1   Cannabis

When smoked, cannabis containing the equivalent of 20–30 mg of THC has been shown to lower intraocular pressure in an heterogeneous group of glaucoma patients (Crawford and Merritt, 1979) and more specifically, patients with open angle glaucoma (Merritt et al., 1979, 1981a). However, the treatment was not without side effects: six of 32 patients developed severe systemic hypotension; this was significantly greater in hypertensive glaucoma patients (Crawford and Merritt, 1979). Cannabis caused a dose-related, clinically significant, reduction in intraocular pressure of 25–30%, occur-ring at 1 hour and lasting 5–6 hours, which was discrete from the sedative effects of the drug. Orthostatic hypotension was observed mainly in cannabis-naive patients.

Cannabis does not cure glaucoma but has been shown to slow progressive sight loss when conventional medicines have failed or where the risks of surgery are too great (Hepler et al., 1976). Tolerance to this effect of cannabis has not been observed; but the degree of reduction in intraocular pressure seen in cannabis-naive patients may not be observed in experienced users (Flom et al., 1975; Dawson et al., 1977).

2.2.2   THC

THC has a comparable effect on intraocular pressure to smoking cannabis at comparable doses (Cooler and Gregg, 1976; Merritt *et al.*, 1980) and is effective when administered orally, encapsulated in sesame seed oil and also by the intravenous route. The latter may however be accompanied by severe dysphoria.

Initial studies in animals suggested that formulated in a light mineral oil, THC eye drops might also be effective (Cool *et al.*, 1974; Green and Bowman, 1976; Merritt *et al.*, 1981c); this route was also shown to be effective in hypertensive glaucoma sufferers (Merritt *et al.*, 1981b) but not in their normotensive counterparts (Merritt *et al.*, 1981a). Studies where patients could not be taken off their existing medication, showed an additive effect for THC.

2.2.3   Nabilone

Nabilone has proven to be effective at lowering intraocular pressure in animal models (D'Ermo *et al.*, 1980; Elsohly *et al.*, 1981, 1984) and in man, where an oral dose of 1–2 mg produced an average drop of 34% (range 10–54%), in 9 patients with open angle glaucoma (Newell *et al.*, 1979). Failure to find a suitable vehicle for an eye drop formulation of nabilone has thwarted studies of topical application.

2.2.4   Other Derivatives

Attempts to develop other cannabinoid derivatives for glaucoma have not met with great success. A water soluble extract of cannabis with no demonstrable central effects, had an additive effect when administered with timolol, a beta adrenoreceptor blocking agent, in patients with refractory raised intraocular pressure (West and Lockhart, 1978); but the active ingredient(s) in this cocktail were not identified. Green *et al.* (1978) were unable to show that cannabidiol had any effect on raised intraocular pressure. This has been confirmed by Waller *et al.* (1984) in both rabbit and unanaesthetised monkey models.

The compound BW 146Y, a synthetic anologue of THC, lowered intraocular pressure in glaucoma patients after oral administration, but this was accompanied by mild hypotension and subjective central side effects (Tiedmann *et al.*, 1981). The water soluble maleate salt of 1,2-dimethylheptyl, delta-6-THC has been shown to have significant antiglaucoma activity in rabbits without significant CNS activity in monkeys (Mechoulam *et al.*, 1978). Another synthetic THC derivative, naboctate, reduced intraocular pressure when administered orally to human volunteers without subjective CNS effects (Razdan *et al.*, 1982).

Summarising, there is little doubt that certain glaucoma patients may benefit from cannabis or its derivatives. From a practical viewpoint, it is estimated that a daily dose of four marijuana cigarettes may be needed to control intraocular pressure; although this level of consumption may be lower if the drug is used as adjunctive therapy.

Further work is required; there is no evidence that any cannabis derivative would remain effective in controlling intraocular pressure during chronic use; and there is precious little evidence that the use of cannabinoids in glaucoma can preserve visual function. Central side effects noted in clinical trials, the risks of long term consumption, attendant driving restrictions and social stigma attached to the drug may prevent more widespread use. Glaucoma is a disease which needs close monitoring if sight is to be preserved; to this should be added the monitoring of casual use to make sure this does not get out of hand or diverted for abuse and the development of side effects to cannabis, especially postural hypotension. This advice also applies to THC, particularly as there is no feasible way of topical administration at present.

There is a need for an agent with a more specific action on the eye or a delivery system which minimises systemic effects. In the meantime, it seems reasonable to suggest that selected patients, i.e. those failing to respond to conventional medication or who cannot tolerate side effects, or where established therapies are contraindicated, should receive inhaled cannabis to augment their present therapy in the hope of preventing visual impairment.

2.2.4 Mode of Action

The way in which cannabis reduces intraocular pressure in glaucoma is unclear. A central mechanism has been proposed (Perez-Reyes et al., 1976; Colasanti and Powell, 1985) although the observation that the effect outlasts the subjective cannabis 'high' suggests a peripheral action (Purrell and Gregg, 1975). THC can decrease aqueous humor production in rabbits; although the mechanism is not known (Green and Peterson, 1973). Inhibition of prostaglandin synthesis may be at least partially responsible, as high doses of prostaglandins are known to cause intraocular hypertension in animals (Green et al., 1987; Korczyn, 1980). Another theory is that cannabinoids act on adrenergic receptors, constricting afferent blood vessels to the ciliary epithelium, causing a fall in pressure and flow to the region of aqueous humor formation; as phentolamine, a specific alpha blocker, can inhibit the fall in intraocular pressure induced by THC (Green et al., 1977; Green and Kim, 1976).

2.3 Multiple Sclerosis (MS)

Like so many other applications, there are numerous anecdotal reports from both patients and their carers who say that cannabis has proffered relief from a range of symptoms associated with MS, including tremor, spasticity and muscle pain (Petro, 1980; Meinck et al., 1989; Waters, 1993). Evidence for the efficacy of cannabis in the relief of spasticity other than that found in MS is discussed in Section 2.4.

2.3.1 Cannabis

Meinck et al. (1989) describe a case where the benefits of smoking cannabis reported by the patient – improvement in muscle tone, reflexes, spasticity, tremor and walking ability – were quantifiable in the laboratory and deteriorated on withdrawal. In a double-blind, placebo controlled trial of cannabis in 10 ambulant patients with MS (Greenberg et al., 1994), the drug impaired posture and balance although several patients reported an improvement in subjective feelings of well-being; a formal psychological assessment was not carried out.

Anecdotal evidence gathered from the testimonials of MS sufferers indicates that a considerable proportion obtain at least partial relief from night-time spasticity, and reduced muscle pain, tremor and depression (Anon, 1997). See also Chapter 8.

2.3.2   THC

Petro and Ellenburger (1981) reported a placebo-controlled trial of oral THC in 9, cannabis-naive patients with MS. Each patient was given a single dose of 5 or 10 mg of THC or placebo on 3 successive days. Both doses of THC were reported to improve reflexes, muscle tone, mobility and spasticity, compared to placebo in some patients; but only 4 patients showed a 'substantial', objectively measured, improvement and just two of these claimed subjective improvement of their symptoms. One additional patient claimed subjective improvement only. Where an effect was observed, it reached a maximum at three hours and persisted for a further two hours.

In a similar trial using 5–15 mg doses of THC, two of eight patients showed objective as well as subjective improvement in tremor. An additional five patients claimed mild subjective improvements in tremor and general well-being, but this was not substantiated objectively (Clifford, 1983). Interestingly, one of the patients had treated his tremor to his own satisfaction for 1 year prior to the study, without appreciating any decrease in response. The THC dosage form and any concurrent antitremor medication were not stated.

Ungerleider et al. (1987) gave THC, 2.5–15 mg daily, to 13 patients with MS, refractory to treatment with standard muscle relaxants, in a double-blind, placebo-controlled, crossover trial. The patients reported subjective improvement when taking THC but the assessing neurologist could not distinguish between treatment and placebo groups. Reduction in spasticity, as rated by the patients, started at the 7.5 mg dose; all but one patient reported side effects at this dose and at higher doses, including dry mouth, weakness, dizziness and psychoactive effects; although the difference in mean side effect scores for THC and placebo did not reach statistical significance, few patients wished to continue THC therapy after the trial.

Most recently, Brenneisen et al. (1996) demonstrated that THC might be useful in ms, given by both oral and rectal routes. The greater bioavailability of the rectal route led to greater efficacy.

2.3.3   Nabilone

Martyn et al. (1995) report a case study, where a single MS patient was given nabilone, 1 mg every other day for 4 weeks, followed by a placebo administered in the same fashion over the same period. The trial was double-blind, and continued for a second eight-week cycle. Muscle spasm, nocturia and general feeling of well being were all improved during the active phases but disappeared rapidly when the placebo was given.

Overall, there is little data on which to make an assessment of the utility of cannabis or its derivatives in MS. Reports to date have been sparse, of poor quality, subjective, anecdotal and have taken little account of confounding factors such as the duration or severity of the illness and concurrent medication. Cases where improvement has been observed are more often than not heavily biased towards the subjective feelings of the patients and may simply be confirming the known mood altering, anxiolytic, sedative or analgesic properties of the drug.

Large scale studies are required, preferably with a cannabinoid derivative where the CNS effects have been dissociated from the peripheral actions and the formulation and dosing can be more closely controlled. It is interesting that the call for further investigation has not come just from the medical community, but also from MS sufferers themselves, in the form of the organisation 'Alliance for Cannabis Thera-peutics' and in the UK, the MS Society. Even a small improvement may be of value in this distressing and multifaceted disorder; however, caution should be exercised in the use of any agent known to cause cognitive impairment, excessive muscle relaxation or central sedation in patients with organic CNS disease.

2.3.4   Mode of Action

Dystonias are known to be largely associated with raised levels of acetylcholine, dopamine and decreased GABA in the basal ganglia and cannabinoids appear to be capable of affecting all three neurotransmitters, possible by an inhibitory effect on the polysynaptic process: reducing acetylcholine turnover (Revuelta *et al.*, 1978), decreasing ligand binding to dopamine receptors (Bloom, 1985) and increasing GABA turnover (Revuelta *et al.*, 1979). It is possible that some, if not all of these effects could be operating in MS to improve motor co-ordination and relieve spasticity although the picture is far from clear.

Most recently, it has been suggested that cannabinoids may act in MS through an effect on specific cannabinoid (CB_1) receptors in the brain and spinal chord to control movement and improve motor function (see Dr Roger Pertwee, Chapter 6 this volume).

2.4   Spastic Conditions

A discussion of the efficacy of cannabis in the relief of spasticity associated with multiple sclerosis appears in Section 2.3. This should not be viewed in isolation, as it is possible that the underlying neuropharmacology and thus approach to treatment of a range of disorders, may be similar.

2.4.1   Cannabis

Two studies have noted improvement of spinal trauma symptoms secondary to cannabis use (Dunn and Davis, 1974; Petro and Ellenburger, 1981). In the former study, 5 of 10 male patients who admitted using cannabis after injury reported improve-ments in spasticity; but 3 had no such effect. Two patients had no initial spasticity anyway.

Despite anecdotal reports of improvement (Malee *et al.*, 1982), cannabis smoking has not stood up to objective, laboratory assessment of controlling Parkinsonian tremor (Frankel *et al.*, 1990). Similarly, early results in Huntingdon's disease were not confirmed in placebo controlled trials (Consroe *et al.*, 1991).

However, cannabis has been used successfully to treat motor tics in Tourette's syndrome and torsion dystonia (Sandyk *et al.*, 1985). Idiopathic dystonia (a group of disorders characterised by abnormal movements and postures produced by prolonged muscle spasms) has also responded to smoking cannabis (Marsden, 1981).

2.4.2   Cannabidiol (CBD)

CBD has been shown to have significant muscle relaxant effects and to reduce muscular spasm in humans (Petro, 1980). This derivative has been used with reported success to treat a variety of dystonias (Sandyk *et al.*, 1985; Consroe *et al.*, 1986) at oral doses ranging from 100 to 600 mg daily. Consroe *et al.* (1986) observed dose related improvement in dystonia in each of five patients, ranging from 20–50%. Side effects were described as mild and included hypotension, dry mouth, psychomotor slowing, light-headedness and sedation. At doses of over 300 mg per day, hypokinesia and tremor were exacerbated in two patients with coexisting Parkinsonian features.

Consroe *et al.* (1991) reported the use of CBD in 15, neuroleptic-free patients with Huntingdon's disease. CBD, in doses up to 700 mg daily, and placebo were given orally, in double blind, randomised, crossover fashion for 6 weeks. No clinically significant advantages were demonstrated for CBD compared to treatment with placebo. The level of side effects did not differ between groups either; although serum levels of CBD, demonstrating adequate absorption throughout the study, were measured in this careful trial.

In conclusion, evidence that cannabis can relieve muscle spasm in a range of dystonias is largely anecdotal and larger, properly controlled trials are required. An unconfirmed observation that cannabinoids can effect the action of neuroleptic drugs beneficially, is interesting (Moss *et al.*, 1989); experiments with individual cannabinoids may hold clues as to how cannabis works in this respect.

2.4.3   Mode of Action

See Section 2.3.4.

2.5   Anticonvulsant Activity

As early as 1890, the use of cannabis was advocated for the suppression of convulsions in man (Reynolds, 1890). While there are some reports of the effects of smoking cannabis in this area, most effort has focused on synthetic cannabinoid derivatives.

2.5.1   Cannabis

Two reports indicate that regular cannabis smoking contributes to better control of seizures in epileptic patients whose disease is refractory to conventional therapy (Consroe *et al.*, 1975; Cartwright, 1983). These studies were not controlled and contained few patients.

2.5.2   THC

There have been no reports of the use of THC in epileptics; this compound has induced or exacerbated seizure activity in animal studies (see Section 2.5.4).

2.5.3   Cannabidiol (CBD)

Carlini and Cunha (1981) added CBD, 200–300 mg daily, to the anticonvulsant regime of eight patients with generalised secondary seizures, refractory to antiepileptic drugs.

Over the 5-month observation period, improvement was noted in seven. Very similar results were reported in an earlier trial by Karler and Turkanis (1976). In a placebo controlled study in patients with grand mal seizures (Cunha *et al.*, 1980), CBD produced improvement in seizure control in 7 patients, three of these showing great improvement. One patient showed some improvement on placebo however.

Summarising, there seems to be little to recommend the regular smoking of cannabis as an adjunct to conventional anticonvulsant medication. Only a few, poorly controlled studies in a limited number of patients are available and these have not demonstrated efficacy to the required standard. Indeed, in the light of the unpredictable tendency of at least one of the active constituents, THC, to cause seizures, this custom may put susceptible individuals at risk.

There is some evidence that other derivatives, notably CBD, have anticonvulsant activity in man. The precise pharmacological basis for this effect is far from clear. In experiments where CBD was shown to enhance the effect of other anticonvulsants, it is possible that this was achieved not by a direct action on receptors, but rather via a pharmacokinetic interaction in which the metabolism of the anticonvulsant was inhibited by CBD; serum levels were not measured in these experiments.

Because CBD lacks central activity but retains anticonvulsant activity and has a lower potential for tolerance, it has been suggested that this compound, or at least, other cannabinoids devoid of psychotropic activity, might be used as the basis for developing analogues with anticonvulsant activity (Formukong *et al.*, 1989).

2.5.4   Mode of Action

A deal of complex, neuropharmacological research has been undertaken in attempts to discover the reasons why CBD, a compound apparently devoid of central activity, has anticonvulsant properties, whereas THC, which undoubtedly has a central effect, has a tendency to produce, rather than prevent seizures.

THC produced seizures in rodents (Karler and Turkanis, 1979) and rabbits (Consroe and Fish, 1980). The effect was dose-related and elicited at doses comparable to those consumed by man. Metabolites of THC appeared to be more epileptogenic than the parent compound (Karler *et al.*, 1974). Interestingly, CBD has been shown to antagonise the epileptogenic activity of THC (Consroe and Wolkin, 1977) – a state of affairs which may prevail when crude cannabis is smoked.

Animal studies have demonstrated anticonvulsant activity for cannabis resin and a range of synthetic analogues using artificial means to induce seizures such as electroshocks (Consroe and Man, 1973); and chemicals (Sofia *et al.*, 1971). Data from comparative studies shows that CBD has anticonvulsant properties which are comparable with a range of anticonvulsants including phenytoin and ethosuximide (Karler and Turkanis, 1981). CBD has also been shown to potentiate the anticonvulsant activity of diazepam and valproic acid (Ehlers *et al.*, 1981) and phenytoin (Chesher and Jackson, 1974) against electronically induced seizures in mice. In the latter study THC had a similar effect; when given with CBD, the effect was additive. The activity of CBD in suppressing hippocampal discharges induced by afferent stimulation was comparable with other anticonvulsants including mysoline and phenytoin at doses which produced little or no behavioural impairment (Izquierdo *et al.*, 1973).

It has been suggested that the variable response of THC is related to its central activity, whereas CBD acts by altering post-synaptic membrane conductance (Turkanis and Karler, 1985). Centrally, cannabinoids have been shown to reduce cortical-evoked responses and spinal reflexes, which is consistent with a neurotransmission theory for the efficacy of CBD. Other mechanisms for which there is some evidence include altered neurotransmitter release, altered transmitter equilibrium potential and altered drug/receptor interactions.

For example, brain monoamine activity may be altered. The anticonvulsant activity of cannabis resin has been decreased after pre-treatment with drugs which decrease brain serotonin levels (Ghosh and Bhattacharya, 1978). GABAergic pathways may also be involved, as CBD inhibited convulsions caused by GABA-inhibiting drugs such as 3-mercaptopropionic acid and hydrazine (Consroe et al., 1982). One interesting observation by Mechoulam (1981) is that CBD and the established anticonvulsant, phenytoin share striking molecular similarities, particularly in terms of the spacing of electron-dense centres; hence they may have a common or similar mode of anticonvulsant action.

2.6   Anti-Asthmatic Activity

2.6.1   Cannabis

Early reports indicated that acute inhalation increased airway patency in healthy individuals (Tashkin et al., 1973; Vachon et al., 1973; Shapiro and Tashkin, 1976) and improved specific airway conductance in asthmatics (Tashkin et al., 1974, 1975b; Vachon et al., 1976). However, there is good evidence that chronic smoking of cannabis restricts the airways and that, as with tobacco, unspecified irritants in the smoke can cause bronchospasm (Henderson et al., 1972).

2.6.2   THC

Aerosolised preparations of THC have produced bronchodilation in both healthy individuals and asthmatics (Tashkin et al., 1974, 1975a; Abboud and Sanders, 1976; Williams et al., 1976); but in asthmatic subjects, bronchoconstriction has also been observed. For example, Abboud and Sanders (1976) administered aerosolised THC to six asthmatic patients, three of whom showed increases in specific airway conductance indicating bronchodilation; the remaining three subjects showed a decrease in conductance and the overall mean, excluding one of these patients who experienced severe bronchoconstriction which had to be reversed with salbutamol, was not significant compared to placebo. Similar experiments have found no evidence of tolerance to the bronchodilating effect of THC for up to 20 days of use (Gong et al., 1984). Williams et al. (1976) showed that a combination of salbutamol and THC produced a more rapid improvement in ventilatory function in asthmatic subjects than salbutamol alone.

Problems with aerosolised THC may be at least in part due to the failure to develop a satisfactory formulation of the drug (Graham et al., 1976). Oral THC does produce some bronchodilation (Tashkin et al., 1976; Shapiro and Tashkin, 1976), but due to unpredictable absorption, onset is delayed and the degree of bronchodilation uncertain. Both delta-8- and delta-9-THC have bronchodilating effects whereas cannabidiol and cannabinol do not; it therefore seems that the bronchodilatory property resides in psychoactive material.

2.6.3 Nabilone

Oral nabilone has been shown to inhibit bronchospasm induced in healthy volunteers by methacholine but was, in contrast to terbutyline, ineffective in patients with chronic stable asthma (Gong et al., 1983). No further data on nabilone are available.

In conclusion, problems with bronchoconstriction after smoking or aerosolised administration, slow onset of action of orally administered cannabinoids and the existence of a range of effective and relatively safe alternatives, preclude the use of cannabis or its derivatives as adjuvants in the management of asthma at present. This is reflected in the low level of research activity in this area.

Indeed, asthmatics should be warned against smoking marihuana for two reasons: firstly, the irritant effects could lead to bronchospasm; secondly, components of marihuana smoke are known to inhibit the metabolism of at least one drug used in asthma – theophylline (Brown, 1994).

2.6.4 Mode of Action

The pharmacology of established anti-asthmatic drugs is well-characterised. If any of the cannabinoids or their derivatives are to be useful in asthma, development of a lead compound is only likely when based on a good understanding of the way in which it promotes bronchodilation and how side effects can be minimised.

The bronchodilator effect of THC does not appear to be related to beta-adrenergic stimulation or muscarinic inhibition (Shapiro and Tashkin, 1976). Orzalek et al. (1979) showed that the bronchodilator effects of the cannabinoids are due to mechanisms which are different to those of the more familiar anti-asthmatic drugs. THC and nabilone did not appear to interfere with cholinergic or histaminergic responses; nor did they alter prostaglandin F_2-alpha-induced contractile responses. The authors suggest that the effects may be centrally mediated. Laviolette and Belanger (1986) showed that when THC was smoked by healthy volunteers, significant increases in specific airway conductance and forced expiratory volume were produced which were not inhibited by bronchoconstricting prostaglandin administration. This is in contrast to in vitro experiments which suggest that many of the physiological effects of marijuana may be mediated via prostaglandins.

2.7 Effects on Anxiety and Insomnia

Cannabis smoking produces a relaxant effect which most users value and it has been suggested that the beneficial effects of cannabis and THC observed in neurological disorders such as motor tics, dystonias and Huntingdon's chorea are due to sedative and anxiolytic actions. In addition, sedation is by far the most common side effect of cannabis, and in particular THC, observed in clinical trials against a range of disorders. This has lead to the suggestion that cannabis and some cannabinoids may be useful in disorders accompanied by anxiety and/or insomnia.

2.7.1   Cannabis

Sethi *et al.* (1986) noted a reduction of anxiety in 50 chronic cannabis users compared to controls, in terms of scores on the Taylor Manifest Anxiety Scale. Oral preparations of cannabis have a sedative or tranquillising effect in man, accompanied by diminished anxiety at doses much lower than those producing psychoactivity (Graham and Li, 1976). However, anxiety and panic, possibly due to depersonalisation, intoxication and loss of control, can also feature as side effects. These symptoms have been observed after smoking or oral ingestion of cannabis, but particularly after intravenous administration of aqueous extracts. This may be due to the rapid onset of altered mental state induced under these conditions. Which effect is produced appears to be independent of age, previous cannabis use or setting, but may be associated with the degree of interpersonal support (Cohen and Andrysiak, 1982).

2.7.2   THC

Initial studies in man showed THC to be not much different from conventional hypnotics. Single doses decreased REM sleep and increased stage 4 sleep; some REM rebound was observed on discontinuation of the drug (Pivik *et al.*, 1972). When THC was given to healthy insomniacs in doses of 10, 20 or 30 mg sleep latency was reduced but subjects experienced dose-related dysphoria prior to falling asleep and similar symptoms the following day (Cousens and DiMascio, 1973). When given as a dose of 20 mg for several consecutive nights, THC reduced REM sleep but upon discontinuation, mild insomnia ensued, with a slight degree of REM rebound (Freemon, 1974).

Marijuana extract, given chronically in daily equivalent THC doses ranging from 70 to 200 mg, reduced REM activity, but with the development of some degree of tolerance. Abrupt withdrawal led to marked REM rebound (Feinberg *et al.*, 1976).

2.7.3   Nabilone

While single dose studies in anxious volunteers have shown little therapeutic benefit (Glass *et al.*, 1981) three trials have demonstrated some anxiolytic effect after repeated doses. In the first (Ilaria *et al.*, 1981), 11 anxious patients received 1–2.5 mg nabilone or placebo, twice daily, for one week in a double-blind, randomised, cross-over study. Nabilone was superior to placebo as judged by scores on the Hamilton Anxiety Rating and the Clinical Global Impression scales. In the second study (Fabre and McLendon, 1981), 25 psychoneurotic anxiety patients were given either 1 mg nabilone or placebo, three times a day for 28 days. Nabilone was significantly superior to placebo after four days of treatment, as judged by scores on the Patient and Clinical Global Impression and Hamilton Anxiety Rating scales. Nabilone also appeared to improve other psychosomatic symptoms and concomitant depression suffered by some patients. The third trial was placebo-controlled and double-blind; 2 mg nabilone was compared to 5 mg diazepam in experimentally induced anxiety in 36 volunteers with high levels of trait anxiety. Both drugs relieved experimentally-induced anxiety but diazepam was superior. Nabilone was associated with more side effects than diazepam and as nabilone was tested at the highest practicable dose, the drug was deemed unlikely to be able to compete with existing anxiolytic drugs (Nakano *et al.*, 1978).

2.7.4   Cannabidiol

In an early study (Carlini and Cunha, 1981) CBD doses of 40, 80, and 160 mg were compared to placebo and 5 mg nitrazepam in 15 insomniac volunteers. Only subjects receiving 160 mg CBD reported sleeping significantly longer than those taking the placebo.

Animal and human experiments have demonstrated that this compound has only weak anxiolytic properties (approximately one thirtieth to one sixtieth of the activity of diazepam); it is therefore unlikely that cannabidiol will ever be considered for clinical use (Mistry *et al.*, 1985).

In summary, its seems unlikely that cannabis or the cannabinoids will ever be accepted as satisfactory anxiolytics or hypnotics. The changes in sleep produced by these compounds (decreased sleep latency, decreased REM sleep, increased slow wave sleep) are not unique and use is accompanied by tolerance, rebound insomnia and dysphoric reactions experienced both before falling asleep and as hangover the next day. Occasional reports of excitement and panic and the inability to identify susceptible patients is problematic.

2.7.5   Mode of Action

The mode of action of the cannabinoids, both as anxiolytics and hypnotics is unclear. Results from animal experiments suggest that both the prostaglandin system and central benzodiazepine receptors are involved (Formukung *et al.*, 1989).

2.8   Depression

It is well recognised that smoking marijuana can produce a euphoric high. When psychological tests have been performed on patients receiving cannabis or THC for nausea and vomiting associated with cancer chemotherapy, mood elevations have been observed (Regelson *et al.*, 1976); however, results are far from consistent. Eight psychiatric patients, hospitalised for moderate or severe depression, received low doses of THC each day for a week; but no mood elevation was observed (Kotin *et al.*, 1973).

In the absence of rigorously controlled trials of these agents, it is impossible to predict how useful the mild, mood elevating properties, noted in trials of THC when used for other purposes, would be in specifically depressed patients; further objective measurement is required.

2.9   Appetite Loss and Anorexia

Users of cannabis often experience hunger and a desire to eat sweet foods, leading to the proposal that its use may increase food intake and therefore slow weight loss in cancer patients (Regelson *et al.*, 1976).

2.9.1   Cannabis

Hollister (1971) described a double-blind study using cannabis or placebo cigarettes. Increased appetite and caloric consumption was demonstrated in those using the drug but the degree to which this took place was highly variable between patients.

2.9.2   THC

Patients receiving THC for nausea and vomiting associated with cancer chemotherapy have often shown improved appetite after receiving THC, but the effect is unpredictable (Regelson *et al.*, 1976). In a small study comparing THC with diazepam in patients with anorexia nervosa, THC did not improve caloric intake and three of the eleven patients who took the drug developed paranoia (Gross *et al.*, 1980). The Food and Drug Administration approved the use of synthetic THC (dronabinol) for anorexia associated with weight loss in patients with AIDS, based on clinical studies in which the effect of THC was sustained for up to 5 months: a placebo-controlled trial involving 139 patients with AIDS in the US and Puerto Rico was carried out for an initial period of 6 weeks. The patients in the active arm of the trial received 2.5 mg THC twice daily before meals. THC produced a significant increase in appetite compared to the placebo but no corresponding increase in weight gain. Side effects associated with THC included palpitations, tachycardia, confusion, dizziness, euphoria and ataxia. THC did not interact with drugs commonly used to treat AIDS patients, such as zidovudine, but the side effects mentioned above were accentuated when drugs with similar side effect profiles were used, eg: increased drowsiness with benzodiazepines (Nightingale, 1993).

Plasse *et al.* (1991) described a study in ten AIDS patients in which doses of THC (as dronabinol) which were sufficiently tolerable for chronic administration up to 5 months, were effective in stabilising or improving weight and enhancing appetite. The dose used was a maximum of 2.5 mg, three times a day, but patients were allowed to adjust the dose downward to avoid side effects. The same group subsequently conducted a multicenter, randomised, double-blind, placebo-controlled trial in 88 evaluable patients with AIDS-related anorexia and weight loss (Beale *et al.*, 1994). Patients received 2.5 mg THC twice daily or placebo. Significant improvements above baseline and compared to placebo were observed after 6 weeks, in appetite and reduced nausea; mood was also improved but did not reach statistical significance ($p = 0.06$). Subjects taking the placebo had a mean loss of 0.4 kg, while the weight of the patients taking THC remained stable. Side effects were mostly mild and there was no statistical difference between the number of discontinuations in either group. THC was adjudged to be safe and effective for anorexia and associated weight loss in patients with AIDS, making a significant contribution to quality of life.

Summarising, use of cannabis or its derivatives as specific appetite stimulants does not look promising, because of the wide variability in response. Based on the evidence discussed above, synthetic THC (dronabinol) has been approved in the US as an appetite stimulant in AIDS patients; but because of its putative abuse potential, the drug remains a schedule 2 controlled substance.

2.9.3   Mode of Action

The way in which appetite stimulation occurs in debilitated AIDS or cancer patients is probably multifactorial. Relief of pain, nausea, anxiety and depression may all play a part. Regelson *et al.* (1976) have suggested a specific effect on the mechanisms producing cachexia in cancer patients.

2.10   Alcoholism

The acute and chronic toxicity profile of cannabis is described in Chapter 10 of this volume. Liver damage is not a prominent feature and it has been suggested that alcoholics might be encouraged to shift their dependence from alcohol to cannabis on grounds of increased safety, in a parallel with the substitution of methadone for heroin in addiction with the latter. There are no reports of individual cannabinoids being investigated in this respect.

In spite of early reports of success in weaning alcoholics from their primary addiction (Mikuriya, 1970; Scher, 1971), subsequent findings are far from encouraging. Rosenberg et al. (1978) reported a trial in which 56 alcoholics were given disulfiram or cannabis, alone or in combination, in an attempt to wean them from alcohol. Cannabis proved ineffective at reducing alcohol consumption.

Cannabis and alcohol are commonly used and abused together and surveys have shown that it is the most prevalent drug combination among adolescents and young adults; however, substitution of one addictive substance with another whose toxicity has not been fully characterised, whose use is still illegal in many countries and whose effectiveness in reducing alcohol consumption is in question, would appear to be unjustified.

2.11   Hypertension

Hypotension and tachycardia are well-recognised effects of cannabis and cannabinoid use in man (Benowitz and Jones, 1981). The antihypertensive effects of the cannabinoids appear to be independent of their psychotropic effects; for example CBD, a non-psychotropic derivative, is active (Adams et al., 1977). There is no clinical trial evidence that cannabis or any of its derivatives would be satisfactory alternatives to established agents in the long-term control of elevated blood pressure. While these agents have a hypotensive effect, the pharmacology of this process remains unclear and is complex, possibly involving stimulation of vasodilatory prostaglandins (Burstein and Hunter, 1984).

Interestingly, CBD appears to have a bradycardic effect in animal models where THC produces tachycardia (Smiley et al., 1976); therefore if cannabinoid analogues are to be developed as antihypertensive agents, it may be more useful to explore derivatives based on the structure of the former rather than the latter.

2.11.1   Mode of Action

While it has been suggested that atropine-like material in raw cannabis is responsible for the hypotension and tachycardic effect (Gill et al., 1970), this is difficult to reconcile with the fact that supposedly pure cannabinoids have similar activity. Certainly the effects do seem to be mediated through the autonomic nervous system (Bloom, 1982). Beta-adreno-receptor blockers, such as propranolol, and muscarinic antagonists, like atropine have been shown to antagonise the vascular effects of THC in man (Benowitz et al., 1979).

Animal studies have shown the situation to be complex. THC can reduce myocardial contractility without decreasing heart rate (Benmoyal et al., 1971) although the mechanism remains obscure. Burstein and Ossman (1982) showed that aspirin inhibited the hypotensive effect of THC, suggesting the prostaglandins are involved. However this does not appear to be the case with tachycardia (Laviolette, 1986). A dimethylheptyl side-chain derivative of THC has been shown to be a potent hypotensive agent with less psychoactivity in man; however the vascular and psychological activities have by no means been divorced (Lemberger et al., 1974).

2.12   Inflammation

Extracts of cannabis were administered during the 19th century for fever before the introduction of aspirin and cannabinoids have been shown to possess antipyretic and analgesic properties. While there are no studies of cannabis or the cannabinoids in acute or chronic inflammatory conditions in man, a study of their actions in animals may provide an impression of the potential usefulness of the cannabinoids in this area. Assessments using models other than those involving pain production and heat have often produced contradictory results. Koserky et al. (1973) could not demonstrate any anti-inflammatory action for THC in carrageenan-induced oedema in the rat paw model. However, Sofia et al. (1973) showed that, in the same model, oral THC was 20 times more potent than aspirin and twice as potent as hydrocortisone. The selective inhibitory action of various cannabinoids at various stages of the inflammatory process has been demonstrated in animals. Formukong et al. (1989) reviewed the in vitro and in vivo anti-inflammatory activity of a range of cannabinoids, including THC and CBD; both compounds were active, but CBD was the most effective compound in terms of inhibiting the erythema produced by applying tetradecanoylphorbol acetate (TPA) to the skin of mice and inhibition of rabbit blood platelet aggregation. CBD was also more potent then THC in inhibiting soyabean lipo-oxygenase; but THC was more effective than CBD in inhibiting cyclooxygenase in microsomal preparations. The authors concluded that the effects of the cannabinoids are complex, probably involving several membrane-associated enzymes.

The beneficial effects of cannabis in several disease processes have been attributed, at least in part, to its ability to suppress the production or subsequent action of prostaglandins and other inflammatory mediators. Recent evidence suggests that additional compounds in cannabis, which are structurally and biologically distinct from the cannabinoids, have a significant inhibitory effect on prostaglandin release in vitro (Barrett et al., 1985). In the absence of trials in man, it is impossible to tell if cannabis or its derivatives have any potential as anti-inflammatory drugs. It is clear that several may find a role in relief of pain, as indicated in Chapter 8 and perhaps evidence of anti-inflammatory efficacy will result from trials in this area. It has been pointed out earlier that chronic cannabis smoking may affect lung function, not least because the compounds in the smoke may suppress the immune response (Hollister, 1986). This may be encouraging from the viewpoint of inflammatory lung disease, but worrying if the patient has a lung infection or is already immunocompromised.

2.13   Anti-Tumour Activity

There have been no studies of cannabis or its derivatives in man and studies in animal models have yielded inconclusive results. THC was half as potent as standard anti-cancer chemotherapy in reducing tumour growth in mice innoculated with a murine lung cancer strain; cannabidiol was ineffective (Harris, 1976). Inhibition of tumour growth and improved animal survival following treatment with THC may be in part due to the ability of cannabinoids to inhibit macromolecular synthesis (Harris *et al.*, 1976). In another study, cannabinol accelerated tumour growth in tissue culture (White *et al.*, 1976).

The importance of these findings is unclear; however, it could be surmised that when cannabis is used to alleviate nausea and vomiting associated with cancer chemotherapy, certain of its components could promote tumour growth. It would be reassuring to know that this is not the case and this potential problem should be investigated.

2.14   Anti-infective Effect

Traditional uses of cannabis include inhalation for a range of systemic disorders which may or may not have an underlying infectious pathology and local application of lotions or ointments for skin disorders (Krejci, 1961). While *in vitro* studies have shown cannabis to have some antibacterial and antifungal activity (Turner and Elsohly, 1981) and cannabidiol to possess activity against Gram positive bacteria (Radosevic *et al.*, 1962), there are no reports of efficacy *in vivo* and no comparisons with standard antibiotics. It therefore seems unlikely that cannabis derivatives will be developed for their antibiotic effects.

Blevins and Dumic (1984) were able to show that *in vitro*, THC was capable of decreasing Herpes simplex virus replication and/or infectivity in human cell cultures; further developments of this research have not been published.

3   SIDE EFFECTS ASSOCIATED WITH THE THERAPEUTIC USE OF CANNABINOIDS

The unwanted effects associated with the therapeutic use of cannabis and its derivatives have been evaluated in numerous clinical trials in a variety of disorders. Table 4 provides a summary of the findings and provides a broad analysis by route of administration and severity.

The vast majority of these effects fit the description of a Type A adverse drug reaction, i.e., the severity is related to dose and the reaction resolves upon stopping treatment. Indeed, as with any drug, early phase clinical trials have focused on achieving an acceptable balance between therapeutic dosing and side effects. One argument against the broader acceptance of the cannabinoids as therapeutic agents has been the inability to achieve a satisfactory trade-off, particularly with respect to CNS effects and in particular, the ability to stimulate disphoria, hallucinations and depersonalisation in susceptible individuals.

The safety of cannabis has been described as remarkable. There is no known case of lethal overdose and in animal models, the ratio of lethal to 'effective' dose has been estimated to be 20,000–40,000 to 1 (compared to 4–10 to 1 for alcohol) (Grinspoon and Bakalar, 1995).

Table 4 Acute side effects of cannabis and its derivatives observed in clinical trials

Effects observed after inhaling cannabinoids in smoke (including marijuana):

panic	depersonalisation	paranoia	tachycardia	palpitations
cough	postural hypotension	bronchocon- striction	reddening of the eyes	

Effects after ingesting cannabinoids at 'therapeutic' doses by the oral route (common, >10% of subjects):

postural hypotension	dizziness	dry mouth	blurred vision	hangover
confusion	disorientation	space–time distortion	tremor	muscle weakness
drowsiness	euphoria	dysphoria	ataxia	decreased concentration
discoordination	increased appetite/hunger	nausea	restlessness	

Effects after ingesting cannabinoids at 'therapeutic' doses by the oral route (rare, <10% of subjects):

hallucinations (visual and auditory)	asthenia	paraesthesia (facial and extremities)	amnesia	syncope
slurred speech	faecal incontinence	tachycardia	mania	nightmares
lethargy	headache	psychosis	mood elevation	urinary retention
hypertension	excess sweating	dysphagia		

Effects after Parenteral administration of cannabinoids (intramuscular or intravenous). In addition to those mentioned above for the enteral route.

Pain at injection site	abdominal pain	leg cramps	facial neuralgia	rash
flushing				

See text for side effects noted when cannabinoids were administered by the ophthalmic route.
The results of laboratory tests on the blood of patients treated with cannabinoids have been performed rarely. Available results indicate that changes are inconsistent and rarely serious (Consroe *et al.*, 1991).

4 DRUG INTERACTIONS WITH CANNABINOIDS

Most clinical trials have administered cannabinoids alone to treat a particular disorder rather than in combination with other drugs. There appear have been no adverse interactions of any consequence between a wide range of cancer chemotherapies, when cannabinoids have been given to counteract the nausea and vomiting associated with the latter or indeed during specific studies in man (Riggs *et al.*, 1981). The beneficial effects of administering THC with another antiemetic, prochlorperazine, have been referred to above. Considering the central nervous system depressant action of these agents, it would be wise to be cautious when administering them with other, CNS depressants, such as benzodiazepines, barbiturates, antidepressants and alcohol. The interaction between marijuana smoking and the enhanced metabolism of theophylline has been well-documented. The ability of cannabinoids to produce significant hypoten-sion, particularly postural hypotension, should be remembered if the patient is receiving antihypertensive medication.

The potential for more spectacular drug interactions to occur seems greatest when cannabinoids are consumed for pleasure, often in high doses and frequently as a cocktail with other psychoactive substances, over the counter medicines and alcohol. One would anticipate a much lower threat during the more controlled, measured use of these agents to treat a disease under medical supervision.

5   FUTURE PROSPECTS

Polls and voter referenda have repeatedly indicated that the vast majority of Americans think cannabis should be medically available and recent surveys in the UK have also reflected this attitude.

In reviewing the potential for medicinal application of cannabis and its derivatives, one is hard-pressed to conclude that any of these related compounds would withstand rigorous, clinical trial and turn out to be effective alternatives to contemporary, established therapy. Two things are clear: cannabis is not the Dr Jekyll which some devotees would have us believe; neither is it without redemption and the Mr Hyde others assume. Evaluation of clinical trial data, where it exists, is made problematic by differences in material used (e.g.: cannabis or THC), the route of administration (inhaled smoke, oral or parenteral), dosage regimen and the study populations.

Therapeutic research with cannabis in the fields of depression, anxiety and insomnia has ceased. There is objective evidence that in several disorders, such as nausea and vomiting associated with cancer chemotherapy, glaucoma, pain relief and relief of muscle spasm, adjunctive use will provide benefit in some patients at certain stages of their disease. In some cases, e.g.: cancer, multiple sclerosis, AIDS, the nature of the illness is such that to deny access to this potentially valuable drug without first obtaining the necessary proof of efficacy from conventional clinical trials, seems unreasonably pedantic. There is considerable evidence that selected patients would and should obtain relief from the symptoms of their disease using cannabis or derivatives thereof, when no other substance will suffice. Reasons for this may be that conventional therapy is contraindicated in terms of route, dose, side effect profile or hypersensitivity or that the cannabinoid in question has a mechanism of action which makes it uniquely effective.

The potential for cannabis and its constituents to be used as medicines is still, largely unrealised. Modern investigation of cannabis spans less than a few decades and this seems hardly sufficient time to settle the issue. Pharmacologically, crude cannabis, as opposed to THC or other specific cannabinoids is a dirty drug and has little future as a therapeutic agent for any illness. It contains many substances with multiple, sometimes antagonising, effects and unknown modes of action. It should be remembered that some of the studies mentioned in this review are no more than initial explorations of serendipitous hunches (e.g., asthma and glaucoma) and they should not be interpreted as proven therapeutic uses of cannabis. They are reflective of Phase One or early Phase Two developmental trials which address issues such as dose ranging to achieve an effect without disturbing side effects. The results of Phase 3 – type, double blind, placebo-controlled studies against accepted therapies in the disease concerned, must be generated before any firm (evidence based) conclusions can be drawn. Further information is also required on the pharmacokinetics of the cannabinoids when administered by different routes and in the presence of comorbid pathology, for example kidney disease and their potential for disturbing other biochemical and physiological processes.

As a therapeutic agent, THC itself has problems because of its central activity and disadvantages of poor water solubility, long half-life and molecular instability. It will be clear from Section 2 that several THC analogues are at an advanced stage of investigation with regard to their therapeutic potential. Much research has been directed at divorcing CNS activity, which is largely non-beneficial, from the peripheral, beneficial, effects. For example, derivatives of THC having a C-5 hydroxy moiety in the cannabinoid structure have no central activity while having potent activity against a range of enzymes such as lipooxygenase or cyclooxygenase, in the periphery. Such compounds may prove useful in asthma and other inflammatory disorders and may make useful analgesics. Analogues of CBD, which is not active centrally also hold promise in this area.

The recent discovery and cloning of THC receptors from rat brain may prove helpful in designing cannabinoid derivatives which interact with these to elicit specific, therapeutic responses (Matsuda et al., 1990). Cannabinoids are able to modulate a range of systems – dopaminergic, GABAergic, serotonergic and cholinergic; but now that a specific receptor has been defined, the search is on for receptor sub-types, the existence of which is suggested by this diversity of pharmacological activity (Snyder, 1990). Most recently, a peripheral cannabinoid receptor has been identified, with distinct structural differences from brain THC receptors (Munro, 1993). If endogenous substances which act at this site can be identified, it should be possible to define structural features which could help in synthetic cannabinoid design.

If single cannabinoid analogues find their way into routine medical practice, it must be after rigorous proof of relative safety and efficacy and then will probably be in a modest role, providing symptomatic rather than curative relief, as adjuncts to conventional therapy or used as a last resort where such treatment has failed; unless new and specific indications emerge. Although cannabinoids do fill a gap in helping a small minority of refractory patients, for the majority, effective therapies with fewer side effects already exist.

The current legislative climate is slowing both private and publicly funded research aimed at providing useful, synthetic cannabinoid derivatives and has in the past caused it to cease altogether. Pharmaceutical companies fear that even if a useful compound were discovered, it would inherit the unsavoury background of cannabis with the possibility of having prohibitive prescribing restrictions enforced upon it, which would at the very least pose marketing problems. This situation might be eased if regulating bodies were to review individual cannabinoid derivatives on the basis of their pharmacological profiles and not legislate solely on the political profile of the class.

The possibility that cannabis may be used as a medicine raises moral, ethical, and political issues in addition to scientific ones. Advocates of cannabis may seek respectability for general use by having it approved as a medicine; however, it is difficult to fathom how this approval would make non-medicinal use acceptable as the methods of use are so different. Therapeutic use would be low-dose and closely controlled by prescription-only status; in addition, the availability of the oral route avoids the problems associated with smoking. Whereas recreational use would be associated largely with the latter in an uncontrolled (in terms of dose, setting and combination with other drugs of abuse) and hedonistic way.

One only has to browse the Internet to see that cannabis still arouses considerable international interest. At least 23 international symposia covering a wide range of scientific aspects have been organised in the last 27 years (Musty *et al.*, 1995) and interest shows no sign of abating. Our increasing understanding of this fascinating phytopharmaceutical may yield a few therapeutic surprises yet.

REFERENCES

Abboud, R.T. and Sanders, H.D. (1976) Effect of oral administration of delta-9-tetrahydrocannabinol on airway mechanics in normal and asthmatic subjects. *Chest*, **70**, 480–485.

Adams, M.D., Earnhardt, J.T., Martin, B.R., Harris, I.S., Dewey, W.L. and Raazdan, K.K. (1977) A cannabinoid with cardiovascular activity but no overt behavioural effects. *Experientia*, **33**, 1204–1205.

Ahmedzai, S., Carlyle, D.L., Calder, I.T. and Moran, F. (1983) Anti-emetic efficacy and toxicity of nabilone, a synthetic cannabinoid, in lung cancer chemotherapy. *Br. J. Cancer*, **48**, 657–663.

Anon (1997) Current treatment of multiple sclerosis and new research into the disease. *Pharm. J.*, **258**, 645–646.

Archer, R.A., Stark, P. and Lemberger, L. (1986) Nabilone. In: R. Mechoulam, (ed.), *Cannabinoids as therapeutic agents*, CRC Press, Boca Raton, pp. 85–103.

Ashton, C.H. (1987) Cannabis: dangers and possible uses. *Br. Med. J.*, **294**, 141–142.

Barrett, M.L., Gordon, D. and Evans, F.J. (1985) Isolation form Cannabis Sativa L, of cannflavin – a novel inhibitor of prostaglandin production. *Biochem. Pharmacol.*, **34**(11), 2019–2024.

Beal, J.E., Olson, R., Laubenstein, L., Morales, J.O., Bellman, P., Yangco, B., Lefkowitz, L., Plasse, T.F. and Shepard, K.V. (1995) Dronabinol as a treatment for anorexia associated with weight loss in patients with AIDS. *J. Pain. Symptom Manage.*, **10**, 89–97.

Benmoyal, E., Corte, E. and Morin, Y. (1971) A direct action of delta-9-THC on myocardial contractility. *Clin. Res.*, **19**, 758.

Benowitz, N.R. and Jones, R.T. (1981) Cardiovascular and metabolic considerations in prolonged cannabinoid administration in man. *J. Clin. Pharmacol.*, **21**, 214–235.

Benowitz, N.R., Rosenberg, J., Rogers, W., Bachman, J.T. and Jones, R.T. (1979) Cardiovascular effects of delta-1-THC: autonomic nervous mechanisms. *Clin. Pharmacol. Ther.*, **25**, 440–446.

Bhargava, H.N. (1978) Potential therapeutic applications of naturally occurring and synthetic cannabinoids. *General Pharmacol.*, **9**, 195–213.

Blevins R.D. and Dumic, M.P. (1984) Delta-9-Tetrahydrocannabinol and herpes simplex virus replication. In: S. Agurell, W.L. Dewey, R.E. Willette (eds.) *The cannabinoids: chemical, pharmacologic and therapeutic aspects*. Academic Press, Orlando, pp. 891–899.

Bloom, A.S. (1982) Cannabinoids and neurotransmitter receptors. *Brain Res.*, **235**, 370–375.

Bloom, A.S. (1985) Effect of the cannabinoids on neurotransmitter receptors and brain membranes In: D.J. Harvey (ed.), *Marihuana '84*, IRL Press, Oxford, pp. 217–231.

Brenneisen, R., Egli, A., Elsohly, M.A., Henn, V. and Spiess, Y. (1996) The effect of orally and rectally administered delta-9-tetrahydrocannabinol on spasticity: A pilot study with 2 patients. *Int. J. Clin. Pharmacol. Ther.*, **34**(10), 446–452.

Bron, D., Staquet, M., Rozensweig, M. and Kenis, Y. (1981) Phase I-II study of a tetrahydrocannabinol analogue (BRL-4664) in the prevention of cisplatin-induced vomiting. In: D.S. Poster, J.S. Penta, and S. Bruno, (eds.), *Treatment of cancer chemotherapy-induced nausea and vomiting*, Masson, New York, pp. 147–151.

Brown, D.T. (1994) Drug abuse-influence on theophylline clearance. *Pharm. J.*, **253**, 595.

Burstein, S.H. and Hunter, S.A. (1984) The role of prostaglandins in the actions of the cannabinoids. In: S. Agurell, W.L. Dewey, and R.E. Willette, (eds.), *The cannabinoids: chemical, pharmacologic and therapeutic aspects*, Academic Press, Orlando, pp. 729–738.

Burstein. S.H. and Ozman, K. (1982) Prostaglandins and cannabis XI: inhibition of delta-1-THC induced hypotension by aspirin. *Biochem. Pharmacol.*, **31**, 591–592.

Carey, M.P., Burish, T.G. and Brenner, D.E. (1983) Delta-9-tetrahydrocannabinol in cancer chemotherapy: research problems and issues. *Ann. Intern. Med.*, **99**, 106–114.

Carlini, E.A. and Cunha, J.A. (1981) Hypnotic and antiepileptic effects of cannabidiol. *J. Clin. Pharmacol.*, **21**, 417S–427S.

Cartwright, L. (1983) Marihuana *Curr. Aff. Bull.*, **59**(10), 19–31.

Chan, H.S.L., Macleod, S.M. and Correia, J.A. (1984) Nabilone vs prochlorperazine for control of cancer chemotherapy-induced emesis in children. *Proc. Am. Soc. Clin. Oncol.*, **c-108**, 47.

Chang, A.E., Shilling, D.J. and Stillman, R.C. (1979) Delta-9-tetrahydrocannabinol as an emetic in cancer patients receiving high-dose methotrexate. *Ann. Intern. Med.*, **91**, 891–824.

Chang, E.A., Shilling, D.J., Stillman, R.C., Goldberg, N.H., Seipp, C.A., Barofsky, I. and Rosenberg, S.A. (1981) A prospective evaluation of delta-9-tetrahydrocannabinol as an antiemetic in patients receiving adriamycin and cytoxan therapy. *Cancer.*, **47**, 1746–1751.

Chesher, G.B. and Jackson, D.M. (1974) Anticonvulsant effects of cannabinoids and cannabinoid interactions with phenytoin. *Psychopharmacol.*, **37**, 255–264.

Citron, M., Herman, T.S., Vreeland, F., Krasnow, S.H., Fossieck, B.E., Harwood, S., Franklin, R. and Cohen, M.H. (1985) Antiemetic efficacy of levonantradol compared to delta-9-tetrahydrocannabinol for chemotherapy-induced nausea and vomiting. *Cancer Treat. Rep.*, **69**(1), 109–112.

Clifford, D.B. (1983) Tetrahydrocannabinol for tremor in multiple sclerosis. *Ann. Neurol.*, **13**(6), 669–671.

Cohen, S. and Andrysiak, T. (1982) *The therapeutic potential of marijuana's components.* American Council on marijuana and other psychoactive substances, New York.

Colasanti, B.K. and Powell, S.R. (1985) Effects of delta-9-tetrahydrocannabinol on intra-ocular pressure after removal of autonomic input. *J. Ocular. Pharmacol.*, **1**, 47–57.

Colls, B.M., Ferry, D.G., Gray, A.J., Harvey, A.J. and McQueen, E.G. (1980) The antiemetic activity of THC versus metoclopramide and thiethylperazine in patients undergoing chemotherapy. *N. Z. Med. J.*, **91**, 449–451.

Cone, L.A., Greene, D.S. and Helm, N.A. (1982) Use of nabilone in treatment of chemotherapy-induced vomiting in an out-patient setting. *Cancer Treat. Rev.*, **9**(suppl. B), 55.

Consroe, P. and Fish, B.S. 1980 Behavioural pharmacology of tetrahydrocannabinol convulsions in rabbits. *Comm. Psychopharmacol.*, **4**, 287–291.

Consroe, P., Laguna, J., Allender, J., Snider, S., Sern, S., Stern, L., Sandyk, R., Kennedy, K. and Schram, K.I. (1991) Controlled clinical trial of cannabidiol in Huntingdon's disease. *Pharmacol. Biochem. Behav.*, **40**, 701–708.

Consroe, P. and Man, D.P. (1971) Effect of delta-8 and delta-9-THC on experimentally induced seizures. *Life Sci.*, **13**, 429–439.

Consroe, P., Martin, A.R. and Fish, B.S. (1982) Use of a potential rabbit model for structural-behavioural activity studies of cannabinoids. *J. Med. Chem.*, **25**, 569–599.

Consroe, P., Sandyk, K.R. and Snider, S.R. (1986) Open label evaluation of cannabidiol in dystonic movement disorder. *Int. J. Neurosci.*, **30**(4), 277–282.

Consroe, P. and Wolkin, A.L. (1977) Cannabidiol – antiepileptic drug comparisons and interactions in experimentally-induced seizures in rats. *J. Pharmacol. Exp. Ther.*, **201**, 26–32.

Consroe, P.F., Wood, G.C. and Buchsbaum, H. (1975) Anticonvulsant nature of marijuana smoking. *J. Am. Med. Assoc.*, **234**, 306–307.

Cool, S.J., Kaye, S. and Cullen, A.P. (1974) Topical and intravenous dosage of delta-9-THC. Effects on blood pressure and intraocular pressure. *Nat. Assoc. Res. Vision Ophthalmol.*, April 25th, Florida.

Cooler, P. and Gregg, J. (1976) The effect of delta-9-tetrahydrocannabinol on intraocular pressure in humans. In: S. Cohen and R.C. Stillman (eds.), *The therapeutic potential for marijuana*, Plenum Press, New York, pp. 77–94.

Cornbleet, M.A., Hamilton, D.A., Christian, P. and Smyth, J.F. (1982) Evaluation of nabilone as an anti-emetic. *Cancer Treat. Rev.*, **49**, 492–493.

Cousens, K. and DiMascio, A. (1973) Delta-9-THC as an hypnotic. An experimental study at three dose levels. *Psychopharmacol.*, **33**, 355–364.

Crawford, W.J. and Merritt, J.C. (1979) Effect of tetrahydrocannabinol on arterial and intraocular hypertension. *Int. J. Clin. Pharmacol. Biopharm.*, **17**, 191–196.

Cronin, C.M. and Sallan, S.E. (1981) Early results of the antiemetic activity of intramuscular levonantradol. In: D.S. Poster, J.S. Penta, S. Bruno, (eds.), *Treatment of cancer chemotherapy-induced nausea and vomiting*. Masson, New York, pp. 137–141.

Cunha, J.M., Carlini, E.A., Pereira, A.E., Ramos, O.L., Pimentel, C., Gagliardi, R., Sanvito, W.L., Lander, N. and Mechoulam, R. (1980) Chronic administration of cannabidiol to healthy volunteers and epileptic patients. *Pharmacology*, **21**(3), 175–185.

Dawson, W.W., Jinsenez-Antillen, C.F., Perez, J.M. and Zeskind, J.A. (1977) Marijuana and vision 10 years use in Costs Rica. *Invest. Ophthalmol. Vis. Sci.*, **16**, 689.

D'Ermo, F., Tomazzoli-Gerola, L. and DeLiberato, P., (1980) Effects of nabilone on experimentally-induced glaucoma in animals. *Boll. Ocul.*, **59**, 639.

Doblin, R.E. and Kleiman, A.R. (1991) Marijuana as an antiemetic medicine: a survey of oncologists' experiences and attitudes. *J. Clin. Oncol.*, **9**(7), 1314–1319.

Dunn, M. and Davis, R. (1974) The perceived effects of marijuana on spinal cord-injured males. *Paraplegia*, **12**, 175.

Ehlers, C.L., Henriken, S.J. and Bloom, F.E. (1981) Levonantradol potentiates the anticonvulsant effects of diazepam and valproic acid in the kindling model of epilepsy. *J. Clin. Pharmacol.*, **21**(Suppl.), 413S–416S.

Einhorn, L. (1982) Nabilone: an effective antiemetic agent in patients receiving cancer chemotherapy. *Cancer Treat. Rev.*, **9**(Suppl.B), 55–61.

Einhorn, L.H., Nagy, C., Furnas, B. and Williams, S.D. (1981) Nabilone: an effective antiemetic in patients receiving cancer chemotherapy. *J. Clin. Pharmacol.*, **21**, 64S–69S.

Ekert, H., Waters, K.D., Jurk, I.H., Mobilia, J. and Loughnan, P. (1979) Amelioration of cancer chemotherapy induced nausea and vomiting by delta-9-tetrahydrocannabinol. *Med. J. Aust.*, **2**, 657–659.

Elsohly, M.A., Harland, E.C., Benigni, D.A. and Waller, C.W. (1984) Cannabinoids in glaucoma II: the effect of different cannabinoids on intraocular pressure of the rabbit. *Curr. Eye Res.*, **3**(6), 841–850.

Elsohly, M.A., Harland, E., Murphy, J.C., Wirth, P. and Waller C.W. (1981) Cannabinoids in glaucoma: a primary screening procedure. *J. Clin. Pharmacol.*, **21**, 472S–478S.

Fabre, L.F. and McLendon, D.M. (1981) Efficacy and safety of nabilone (a synthetic cannabinoid) in the treatment of anxiety. *J. Clin. Pharmacol.*, **21**, 377S–382S.

Feinberg, I., Jones, R., Walker, J., Cavness, C. and Floyd, T. (1976) Effects of marijuana extract and tetrahydrocannabinol on electroencephalographic sleep patterns. *Clin. Pharmacol. Ther.*, **19**(6), 782–794.

Flom, M.C., Adams, A.J. and Jones, R.T. (1975) Marijuana smoking and reduced pressure in human eyes: drug actions or epiphenomena? *Invest. Ophthalmol. Vis. Sci.*, **14**, 869.

Formukong, E.A., Evans, A.T. and Evans, F.J. (1989) The medicinal uses of cannabis and its constituents. *Phytother. Res.*, **3**(6), 219–231.

Frankel, J.P., Hughes, A., Lees, A.J. and Stern, G.M. (1990) Marijuana for Parkinsonian tremor. *J. Neurol. Neurosurg. Psychiat.*, **53**, 436–442.

Freemon, F.R. (1974) The effect of delta-9-tetrahydrocannabinol on sleep. *Psychopharmacol.*, **35**, 39–44.

Frytak, S., Moertel, C.G., O'Fallon, J.R., Rubin, J., Creagan, E.T., O'Connel, M.J., Schut, A.J. and Schwartau, N.W. (1979) Delta-9-tetrahydrocannabinol as an antiemetic for patients receiving cancer chemotherapy. A comparison with prochlorperazine and a placebo. *Ann. Int. Med.*, **91**, 825–830.

Gaoni, Y. and Mechoulam, R. (1964) Isolation, structure and partial synthesis of active constituent of hashish. *J. Am. Chem. Soc.*, **86**, 1646–1647.

Garb, S., Beers, A.L., Bograd, M., McMahon, R.T., Mangalik, A., Ashman, A.C. and Levine, S. I. (1980) Two-pronged study of tetrahydrocannabinol (THC) prevention of vomiting from cancer chemotherapy. *IRCS J. Med. Soc.*, **8**, 203–204.

George, M., Pejovic, M.H., Thuaire, M., Kramar, A. and Wolff, J.P. (1983) Randomized comparative trial of a new anti-emetic, nabilone, in cancer patients treated with cisplatin. *Biomed. Pharmacother.*, **37**, 24–27.

Ghosh, P. and Bhattacharya, S.R. (1978) Anticonvulsant action of cannabis in the rat-role of monoamines. *Psychopharmacol.*, **59**, 293–297.

Gill, E.W., Paton, W.D.M. and Pertwee, R.G. (1970) Preliminary experiments on the chemistry and pharmacology of Cannabis. *Nature*, **228**, 134–136.

Glass, R.M., Uhlenhuth, E.H. and Hartel, F.W. (1981) Single dose study of nabilone in anxious volunteers. *J. Clin. Pharmacol.*, **21**, 383S–396S.

Gong, H., Tashkin, D.P. and Calvarese, B. (1983) Comparison of bronchial effects of nabilone and terbutyline in healthy and asthmatic subjects. *J. Clin. Pharmacol.*, **23**, 127–133.

Gong, H., Tashkin, D.P., Simmons, M.S., Calvarese, B. and Shapiro, B.J. (1984) Acute and subacute bronchial effects of oral cannabinoids. *Clin. Pharmacol. Ther.*, **35**, 26–32.

Graham, J.D.P., Davies, B.H., Seaton, A. and Weatherstone, R.M. (1976) Bronchodilator action of an extract of cannabis and delta-1-tetrahydrocannabinol. In: M.C. Braude and C. Szara, (eds.), *The pharmacology of marijuana*. Raven Press, New York, pp. 269–276.

Graham, J.D.P. and Li, D.M.F. (1976) The pharmacology of cannabis and the cannabinoids. In: J.D.P. Graham (ed.) *Cannabis and health*. Academic Press, London, pp. 142–265.

Gralla, R.J. and Tyson, L.B. (1985) Delta-9-tetrahydrocannabinol as an antiemetic. In: D.J. Harvey (ed.), *Marihuana '84* – Proceedings of an Oxford Symposium on Cannabis. IRL Press, Washington, pp. 721–727.

Gralla, R.J., Tyson, L.B., Bordin, L.A., Clark, R.A., Kelsen, D.P., Kris, M.G., Kalman, L.B. and Groshen, S. (1984) Antiemetic therapy: a review of recent studies and a report of a random-assignment trial comparing metoclopramide with delta-9-tetrahydrocannabinol. *Cancer Treat. Rep.*, **68**(1), 163–172.

Gralla, R.J., Tyson, L.B., Clarke, R.A., Bordin, L.A., Kelsen, D.P. and Kalman, L.B. (1982) Antiemetic trials with high dose metoclopramide: superiority over THC and preservation of efficacy in subsequent chemotherapy courses. *Pro. Am. Assoc. Cancer Res./Am. Soc. Clin. Oncol.*, **23**, 58.

Green, K. and Bowman, K. (1976) Effect of marihuana and derivatives on aqueous humor dynamics in the rabbit. In: M.C. Braude and S. Szara, (eds.), *The Pharmacology of Marijuana.* Raven Press, New York, pp. 803–814.

Green, S.T., Nathwani, D., Goldberg, D.J. and Kennedy, D.H. Nabilone as effective therapy for intractable nausea and vomiting in AIDS. *Br. J. Clin. Pharmacol.*, **28**, 494–495.

Green, K., Bigger, J.F., Kim, K. and Bowman, K. (1977) Cannabinoid action in the eye as mediated through central nervous system and local adrenergic activity. *Expl. Eye Res.*, **24**, 189–196.

Green, K., Checks, K.E., Watkins, L., Bowman, K. and McDonald, T.F. (1987) Prostaglandin involvement in the responses of the rabbit eye to water soluble marijuana derived material. *Curr. Eye Res.*, **6**, 337–344.

Green, K. and Kim, K. (1976) Mediation of ocular tetrahydrocannabinol effects by adrenergic nervous system. *Expl. Eye. Res.*, **23**, 443–448.

Green, K. and Peterson, J.E. (1973) Effect of delta-1-THC on aqueous humour dynamics and ciliary body permeability in rabbit eye. *Expl. Eye Res.*, **15**, 499–507.

Green, K., Wynn, H. and Bowman, K.A. (1978) A comparison of topical cannabinoids on intraocular pressure. *Expl. Eye Res.*, **24**, 189–196.

Greenberg, H.S., Werness, S.A.S., Pugh, J.E., Andrus, R.O., Anderson, D.J. and Domino, E.F. (1994) Short-term effect of smoking marijuana on balance in patients with multiple sclerosis and normal volunteers. *Clin. Pharmacol. Ther.*, **55**(3), 324–328.

Grinspoon, L. and Bakalar, J.B. (1995) Marijuana as a medicine – a plea for reconsideration. *J. Am. Med. Assoc.*, **273**(23), 1875–1876.

Gross, H.A., Ebert, M. and Goldberg, S. (1980) A trial of delta-9-THC in primary anorexia nervosa. *Proc. Am. Psychiat. Assoc. San Francisco*, 1980.

Harris, L.S. (1976) Analgesic and antitumor potential of cannabinoids. In: S.A. Cohen and R.C. Stillman (eds.), *The Therapeutic Potential of Marijuana.* Plenum Press, New York, pp. 299–312.

Harris, L.S., Munson, A.E. and Carchman, R.A. (1976) Antitumor properties of cannabinoids. In: M.C. Braude and S. Szara (eds.), *Pharmacology of Marihuana Vol. 2*, Raven Press, New York, pp. 749 – 762.

Heim, M.E., Queisser, W. and Altenberg, H. (1984) Randomized crossover study of the antiemetic activity of levonantradol and metoclopramide in cancer patients receiving chemotherapy. *Cancer Chemother. Pharmacol.*, **13**, 123–125.

Henderson, R.L., Tennant, F.S. and Guernay, R. (1972) Respiratory manifestations of hashish smoking. *Arch. Otolaryngol.*, **95**, 248–251.

Hepler, R.S. and Frank, I.M. (1971) Marijuana smoking and intraocular pressure. *J. Am. Med. Assoc.*, **217**, 1392.

Hepler, R.S., Frank, I.M. and Petrus, R. (1976) Ocular effect of marihuana smoking. In M.C. Braude and S. Szara (eds.), *Pharmacology of Marihuana.* Raven Press, New York, pp. 815–824.

Herman, T.S., Einhorn, L.H., Jones, S.E., Nagy, C., Chester, A.B., Dean, J., Furnas, B., Williams, S.D., Leigh, S.A., Dorr, R.T. and Moon, T.E. (1979) Superiority of nabilone over prochlorperazine as an antiemetic in patients receiving cancer chemotherapy. *N. Engl. J. Med.*, **300**, 1295–1297.

Hollister, L.E. (1971) Hunger and appetite after single doses of marihuana, alcohol and dextroamphetamine. *Clin. Pharmacol. Ther.*, **12**, 44–49.

Hollister, L.E. (1986) Health aspects of cannabis. *Pharmacological Rev.*, **38**(1), 1–20.

Howes, J.F. (1981) SP-106 (Nabitan): Preliminary clinical results. In: D.S. Poster, J.S. Penta and S. Bruno (eds.), *Treatment of Cancer Chemotherapy-Induced Nausea and Vomiting.* Masson, New York, pp. 143–146.

Ilaria, R.L., Thornby, J.I. and Fann, W.E. (1981) Nabilone, a cannabinol derivative, in the treatment of anxiety neurosis. *Curr. Ther. Res. Clin. Exp.*, **29**, 243.

Izquierdo, I., Orsingher, O.A. and Berardi, A.C. (1973) Effect of cannabidiol and other Cannabis sativa compounds in hippocampal seizure discharges. *Psychopharmacol.*, **28**, 95–102.

Johansson, R., Kilkku, P. and Groenroos, M. A. (1982) A double-blind, controlled trial of nabilone vs prochloperazine for refractory emesis induced by cancer chemotherapy. *Cancer Treat. Rev.*, **9** (Suppl. B), 25–33.

Jones, S.E., Durant, J.R., Greco , A. and Robertone, A. (1982) A multi-institutional Phase III study of nabilone vs placebo in chemotherapy-induced nausea and vomiting. *Cancer Treat. Rep.*, **9** (Suppl. B), 45–48.

Karler, R., Cely, W. and Turkanis, S.A. (1974) Anticonvulsant properties of delta-9-THC and other cannabinoids. *Life Sci.*, **15**, 9131–9147.

Karler, R. and Turkanis, S.A. (1976) The antiepileptic potential of the cannabinoids. In: S. Cohen and R.C. Stillman (eds.), *The therapeutic potential of Marijuana.* Plenum Press, New York, pp. 393–397.

Karler, R. and Turkanis, S.A. (1979) Cannabis and epilepsy. In Marihuana: biological effects. G.G. Nahas and W. Paton. (eds.), Pergamon Press, New York, p. 619.

Karler, R. and Turkanis, S.A. (1981) The cannabinoids as potential antiepileptics. *J. Pharm. Pharmacol.*, **21**, 437S–448S.

Khojasteh, A., Sartiano, G., Tapazoglou, E., Lester, E., Gandera, D., Bernard, S. and Finn, A. (1990) Ondansetron for the prevention of emesis induced by high-dose cisplatin. *Cancer*, **66**, 1101–1105.

Kluin-Neleman, J.C., Neleman, F.A., Meuwissen, O.J.A. and Maess, R.A.A. (1979) Delta-9-tetrahydrocannabinol (THC) as an antiemetic in patients treated with cancer chemotherapy: a double blind cross-over trial against placebo. *Vet. Hum. Toxicol.*, **21**(5), 338–340.

Korczyn, A.D. (1980) The ocular effect of the cannabinoids. *Gen. Pharmacol.*, **11**, 419–423.

Koserky, D.S., Dewey, W.L. and Harris, L.S. (1973) Antipyretic, analgesic and anti-inflammatory effects of delta-9-THC in the rat. *Eur. J. Pharmacol.*, **24**, 1–7.

Kotin, J., Post, R.M. and Goodwin, F.K. (1973) Delta-9-tetrahydrocannabinol in depressed patients. *Arch. Gen. Psychiat.*, **28**, 345–348.

Krejci, Z. (1961) To the problem of substances with anti-bacterial action: Cannabis effect. *Casopis. Lekaru. Ceskych.*, **43**, 1351–1354.

Lane, M., Vogel, C.L., Ferguson, J., Krasnov, S., Saiers, J.H., Hamm, J., Fuks, J., Wiernik, T., Holroyde, C.P., Blossom, M., Shepard, K. and Plasse, T. (1989) Dronabinol and prochlorperazine in combination are better than either agent alone for treatment of chemotherapy-induced nausea and vomiting. *Proc. Am. Soc. Clin. Oncol.*, **8**, 326.

Laszlo, J., Lucas, U.S., Hanson, D.C., Cronin, C.M. and Sallan, S.E. (1981) Levonantradol for chemotherapy-induced emesis: Phase 1-2 oral administration. *J. Clin. Pharmacol.*, **21**, 51S–56S.

Laviolette, M. and Belanger, J. (1986) Role of prostaglandins in marihuana-induced bronchodilation. *Respiration*, **49**, 10–15.

Lemberger, L. (1976) Clinical pharmacology of nabilone, a cannabinol derivative. *Clin. Pharmacol. Ther.*, **18**, 720–726.

Lemberger, L., McMahon, R., Archer, R., Matsumoto, K. and Rowe, H. (1974) Pharmacological effects and physiological disposition of delta – 6alpha, 10alpha – dimethyl heptyl tetrahydrocannabinol (HMHP) in man. *Clin. Pharmacol. Ther.*, **15**, 380–386.

Levitt, M. (1982) Nabilone vs placebo in the treatment of chemotherapy-induced nausea and vomiting in cancer patients. *Cancer Treat. Rep.*, **9**(Suppl. B), 49–43.

Levitt, M., Wilson, A. and Bowman, D. (1981) Physiologic observations in a controlled clinical trial of the effectiveness of 5, 10 and 15 mg of delta-9-tetrahydrocannabinol in cancer chemotherapy – ophthalmological implications. *J. Clin. Pharmacol.*, **21**, 103S–109S.

Lewis, I.H., Campbell, D.N. and Barrowcliffe, M.P. (1994) Effect of nabilone on nausea and vomiting after total abdominal hysterectomy. *Br. J. Anaesth.*, **73**, 244–246.

London, S.W., McCarthy, L.E. and Borison, H.L. (1979) Suppression of cancer chemotherapy-induced vomiting in the cat by nabilone, a synthetic cannabinoid (40465). *Proc. Soc. Exp. Biol. Med.*, **160**, 437–440.

Long, A., Mioduszewski, J. and Natale, R. (1982) A randomised double-blind crossover comparison of the antiemetic activity of levonantradol and prochlorperazine. *Proc. Am. Soc. Clin. Oncol.*, **1**, 57.

Lucas, V.S. and Laszlo, J. (1980) Delta-9-tetrahydrocannabinol for refractory vomiting induced by cancer chemotherapy. *J. Am. Med. Assoc.*, **243**, 1241–1243.

Malee, J., Harvey, R.F. and Cayner, J.J. (1982) Cannabis effect on spasticity in spinal cord injury. Arch. *Phys. Med. Rehab.*, **63**, 116–118.

Markham, A. and Sorkin, E.M. (1993) Ondansetron. An update of its therapeutic use in chemotherapy-induced and postoperative nausea and vomiting. *Drugs*, **45**, 6, 931–952.

Marsden, C.D. (1981) Treatment of torsion dystonia. In A. Barbeau (ed.) *Disorders of movement, current status of modern therapy.* Lippencott, Philadelphia, pp. 81–104.

Martyn, C.N., Illis, L.S. and Thom, J. (1995) Nabilone in the treatment of multiple sclerosis. *Lancet*, **345**, 579.

Matsuda, L.A., Lolait, S.J., Brownstein, M.J., Young, A.C. and Bonner, T.I. (1990) Structure of a cannabinoid receptor and functional expression of the clones cDNA. *Nature*, **346**(6284), 561–564.

McCabe, M., Smith, F.P. and Goldberg, D. (1981) Comparative trial of oral delta-9-tetrahydro-cannabinol (THC) and prochlorperazine (PCZ) for cancer chemotherapy-related nausea and vomiting. *Proc. Am. Assoc. Cancer. Res./Am. Soc. Clin. Oncol.*, **22**, 416.

McCarthy, C.E. and Borison, H.L. (1977) Antiemetic activity of nabilone, a cannabinol derivative, reversed by naloxone in awake cats. *Pharmacologist*, **19**, 239.

McCarthy, C.E. and Borison, H.L. (1981) Antiemetic activity of N-methylevonantradol and nabilone in cisplatin-treated cats. *J. Clin. Pharmacol.*, **21**, 305–375.

Mechoulam, R. (1981) Current status of therapeutic opportunities based on cannabinoid research. An overview. *J. Clin. Pharmacol.*, **21**, 2S–7S.

Mechoulam, R., Lander, N. and Distein, S. (1978) Israel Patent 55.274; U.S. Patent 4,282,248 (1981).

Meinck, H.M., Schonle, P.W. and Conrad, B. (1989) Effect of cannabinoids on spasticity and ataxia in multiple sclerosis. *J. Neurol.*, **236**, 120–122.

Merritt, J.C., Crawford, W.J., Alexander, P.C., Anduze, A.L. and Gelbart, S.S. (1979) Effect of marijuana inhalation on the intraocular pressure and blood pressure in open angle glaucoma. *Ophthalmology*, **86**, 45.

Merritt, J.C., McKinnon, S.M., Armstrong, J.R., Hatem, G. and Reid, L.A. (1980) Oral delta-9-tetrahydrocannabinol in heterogeneous glaucomas. *Ann. Ophthalmol.*, **14**, 947–950.

Merritt, J.C., Olson, J.L., Armstrong, J.R. and MacKinnon, S.M. (1981b) Topical delta-9-terahydrocannabinol in hypertensive glaucomas. *J. Pharm. Pharmacol.*, **33**, 40–41.

Merritt, J.C., Peiffer, R.L., McKinnon, S.M., Stapleton, S.S., Goodwin, R.T. and Risco, J.M. (1981c) Effects of topical delta-9-tetrahydrocannabinol in intraocular pressure in dogs. *Glaucoma*, **3**, 13–16.

Merritt, J.C., Perry, D.D., Russell, D.N. and Jones, B.F. (1981a) Topical delta-9-tetrahydrocan-nabinol and aqueous humor dynamics in glaucoma. *J. Clin. Pharmacol.*, **21**, 467S–471S.

Mikurya, T.H. (1970) Cannabis substitution: an adjunctive therapeutic tool in the treatment of alcoholism. *New Physician*, **98**, 187–191.

Mistry, R.E., Conti, L.H. and Mechoulam, R. (1985) Anxiolytic properties of cannabidiol. In D.J. Harvey (ed.), *Marijuana '84* – Proceedings of an Oxford Symposium on Cannabis, IRL Press, Washington, pp. 713–719.

Musty, R.E., Reggio, P. and Consroe, P. (1995) The international cannabis research society and the 1994 international symposium of cannabis and the cannabinoids. *Life Sciences*, **56**(223/24), 1931–1932.

Munro, S., Thomas, K.L. and Abu Shaar, M. (1993) Molecular characterisation of a peripheral receptor for cannabinoids. *Nature*, **365**, 61.

Nagy, C.M., Furnas, B.E., Einhorn, L.H. and Bond, W.H. (1978) Nabilone antiemetic crossover study in cancer chemotherapy patients. *Proc. Am. Assoc. Cancer. Res. and ASCO.*, **19**, 30.

Nahas, G.G. (1979) Current status of marijuana research. *J. Am. Med. Assoc.*, **242**, 2775–2778.

Nakano, S., Gillespie, H.K. and Hollister, L.E. (1978) Model for the evaluation of antianxiety drugs with the use of experimentally induced stress: Comparison of nabilone and diazepam. *Clin. Pharmacol. Ther.*, **23**(1), 54–62.

Neidhart, J.A., Gagen, M.M., Wilson, H.E. and Young, D.C. (1981) Comparative trial of the antiemetic effects of THC and haloperidol. *J. Clin. Pharmacol.*, **21**, 38S–42S.

Newell, F.W., Stark, P., Jay, W.M. and Schanzlin, D.J. (1979) Nabilone: a pressure-reducing synthetic benzopyran in open angle glaucoma. *Ophthalmology*, **86**(1), 156–160.

Nightingale, S.L. (1993) Dronabinol approved for use in anorexia associated with weight loss in patients with AIDS. *J. Am. Med. Assoc.*, **269**(11), 1361.

Niiranen, A., Mattson, K. and Poppius, H. A. (1983) A crossover comparison of nabilone vs prochlorperazine for emesis induced by cancer chemotherapy. *Proc. 2nd Eur. Conf. Clin. Oncol., Amsterdam 1983*, 216.

Orr, L.E., McKernan, J.F. and Lee, P. (1980) Antiemetic effect of delta-9-tetrahydrocannabilnol in chemotherapy-associated nausea and emesis as compared to placebo and compazine. *Arch. Int. Med.*, **140**, 1431–1433.

Orzalek, R.M., Goodman, F.R. and Forney, R.B. (1979) The effects of delta-9-tetrahydrocannabinol and nabilone on the isolated guinea pig bronchus. *Toxicol. Appl. Pharmacol.*, **48**, A67.

O'Shaughnessy, W.B. (1890) On the preparation of Indian hemp, or ganjah (Cannabis indica): the effects on the animal system in health and their utility in the treatment of tetanus and other convulsive diseases. *Trans. Med. Phys. Soc. Bombay.*, **8**, 421.

Patel, N., Hunt, J. and McElwain, T.J. (1983) A comparison of antiemetic efficacy and safety of nabilone (NAB) and prochlorperazine in paediatric patients with cytotoxic-induced nausea and vomiting. *Proc. 2nd Eur. Conf. Clin. Oncol., Amsterdam*, 299.

Penta, J.S., Poster, D.S., Bruno, S. and MacDonald, J.S. (1981) Clinical trials with antiemetic agents in cancer patients receiving chemotherapy. *J. Clin. Pharmacol.*, **21**, 11S–22S.

Perez-Reyes, M., Lipton, M.A. and Timmons, C. (1972) Pharmacology of orally administered delta-9-tetrahydrocannabinol. *Clin. Pharm. Ther.*, **14**, 48–55.

Perez-Reyes, M., Wagner, D., Wall, M.E. and Davis, K.H. (1976) Intravenous administration of cannabinoids and intraocular pressure. In M.C. Braude and S. Szara (eds.), *The pharmacology of marijuana*. Raven Press, New York, pp. 829–832.

Petro, D.J. (1980) Marihuana as a therapeutic agent for muscle spasm or spasticity. *Psychosomatics*, **21**(1), 81–85

Petro, D.J. and Ellenburger, C. (1981) Treatment of human spasticity with delta-9-tetrahydro-cannabinol. *J. Clin. Pharmacol.*, **21**, 413S–416S.

Pivic, R.T., Zarcone, V., Dement, W.C. and Hollister, L.E. (1972) Delt-9-tetrahydrocannabinol and synhexyl: effects on human sleep patterns. *Clin. Pharmacol. Ther.*, **13**, 426–435.

Plasse, T.F., Gorter, R.W., Krasnow, S.H., Lane, M., Shepard, K.V. and Wadleigh, R.G. (1991) Recent clinical experience with dronabinol. *Pharmacol. Biochem. Behaviour*, **40**, 695–700.

Poster, D.S., Penta, J.S., Bruno, S., Vilk, P., Davignon, J.P. and MacDonald, J.S. (1981a) A review of oral delta-9-tetrahydrocannabinol clinical antiemetic studies, 1975–1980. In: D.S. Poster,

J.S. Penta, S. Bruno (eds.), *Treatment of cancer chemotherapy-induced nausea and vomiting*. Masson, New York, pp. 55–60.

Poster, D.S., Penta, J.S., Bruno, S. and Macdonald, J.S. (1981b) Delta-9-tetrahydrocannabinol in clinical oncology. *J. Am. Med. Assoc.*, **245**(20), 2047–2051.

Priestman, J. and Priestman, S.G. (1984) An initial evaluation of nabilone in the control of radiotherapy-induced nausea and vomiting. *Clin. Radiol.*, **35**, 265–266.

Purrell, W.D. and Gregg, J.M. (1975) Delta-9-tetrahydrocannabinol euphoria and intraocular pressure in man. *Ann. Ophthalmol.*, **7**, 1921–1923.

Radosevic, A., Kupinic, M. and Grilic, L. (1962) Antibiotic activity of various types of cannabis resin. *Nature*, **195**, 1007–1009.

Razdan, R.K., Pars, H.G. and Howes, J.F. Development of orally active cannabinoids for the treatment of glaucoma. *Proceedings of the 44th International Annual Scientific Meeting on the Problems of Drug Dependence*, Toronto, 1982.

Regelson, W., Butler, J.R., Schultz, J., Kirk, T., Peck, L. and Green, M.L. (1976) Delta-9-tetrahydrocannabinol as an effective antidepressant and appetite stimulating agent in advanced cancer patients. In: M.C. Braude and S. Szara (eds.), *Pharmacology of marihuana*, Raven Press, New York, pp. 763–776.

Revuelta, A.V., Cheney, D., Wood, P.L. and Costa, E. (1979) GABAergic mediation in the inhibition of hippocampal acetylcholine turnover rate elicited by delt-9-THC. *Neuropharmacol.*, **18**, 525–530.

Revuelta, A.V., Moroni, F., Cheney, D.L. and Costa, E. (1978) Effect of cannabinoids on the turnover of acetylcholine in the rat hippocampus, striatum and cortex. *Arch. Pharmacol.*, **304**, 107–110.

Reynolds, J. (1890) On the therapeutical uses and toxic effects of Cannabis indica. *Lancet*, **1**, 637–638.

Riggs, C.E., Egorin, M.J., Fuks, J.Z., Schnaper, N. and Duffey, P. (1981) Initial observations on the effects of delta-9-tetrahydrocannabinol on the plasma pharmacokinetics of cylophopsphamide and doxorubicin. *J. Clin. Pharmacol.*, **21**, 90S–98S.

Rosenberg, C.M., Gerrin, J.R. and Schnell, C. (1978) Cannabis in the treatment of alcoholism. *J. Stud. Alcohol*, **39**, 1955–1958.

Sallan, S.E., Zinberg, N.E. and Frei, E. (1975) Antiemetic effect of delta – 9 tetrahydrocannabinol in patients receiving cancer chemotherapy. *N. Eng. J. Med.*, **293**, 795–797.

Sallan, S.E., Cronin, C., Zelen, M. and Zinberg, N.E. (1980) Antiemetics in patients receiving chemotherapy for cancer. A randomised comparison of delta-9-tetrahydrocannabinol and prochlorperazine. *N. Eng. J. Med.*, **302**, 135–138.

Sandyk, R. and Awerbuch, G. (1985) Marijuana and Tourette's syndrome. *J. Clin. Psychopharmacol.*, **8**, 444–445.

Sandyk, K.R., Snider, S.R. and Consroe, P. (1985) Cannabidiol in dystonic movement disorders. *Psychiat. Res.*, **18**, 291.

Scher, J. (1971) Marihuana as an agent in the rehabilitation of alcoholics. *Am. J. Psychiat.*, **127**, 147–148.

Schwartz, R.H. and Beveridge, R.A. (1994) Marijuana as an antiemetic drug: how useful is it today? Opinions of clinical oncologists. *J. Addict. Dis.*, **13**(1) 53–65.

Shapiro, B.J. and Tashkin, D.P. (1976) Effect of beta-adrenergic blockade and muscarinic stimulation on cannabis bronchodilation. In: M.C. Braude and S. Szara (eds.), *The pharmacology of marijuana*. Raven Press, New York, pp. 277–289.

Sheidler, V., Ettinger, D.S., Diaiso, R.B., Enterline, J.P. and Brown, D. (1984) Double-blind multiple-dose crossover study of the antiemetic effect of intramuscular levonantradol compared to prochloperazine. *J. Clin. Pharmacol.*, **24**, 155–159.

Sethi, B.B., Trivedi, J.K., Kumar, P., Gulati, A., Agarwal, A.K. and Sethi, N. (1986) Antianxiety effects of cannabis: involvement of central benzodiazepine receptors. *Biol. Psychiat.*, **21**, 3–10.

Smiley, K.A., Karber, R. and Turkanis, S.A. (1976) Effect of cannabinoids on the perfused rat heart. *Res. Comm. Chem. Pathol. Pharmacol.*, **14**, 659–675.

Snyder, S.H. (1990) Planning for serendipity. *Nature*, **346**, 508.

Sofia, R.D., Nalepa, S.D., Harakel, J.J. and Vassar, A.B. (1973) Antiemetic and analgesic properties of delta-9-THC compared with three other drugs. *Eur. J. Pharmacol.*, **35**, 7–16.

Sofia, R.D., Soloman, T.A. and Barry, H. (1971) The anticonvulsant activity of delta-1-THC in mice. *Pharmacologist*, **13**, 246.

Stambaugh, J.E., McAdams, J. and Vreeland, F. (1992) A phase II randomised trail of the antiemetic activity of levonantradol (CP-50,556) in cancer patients receiving chemotherapy. *Proc. Am. Soc. Clin. Oncol.*, **1**, 61.

Stanton, W., (1983) Antiemetic efficacy and safety of delta-9-tetrahydrocannabinol (THC): effect of dose and anticancer regimen. *Proc. Am. Soc. Clin. Oncol.*, **2**, 94.

Steele, N., Gralla, R.J., Braun, D.W. and Young, C.W. (1980) Double blind comparison of the antiemetic effects of nabilone and prochloroperazine on chemotherapy-induced emesis. *Cancer Treat. Rep.*, **64**(2–3), 219–224.

Stuart, J.F.B., Welsh, J., Sangster, G., Scullion, M., Cash, H., Kaye, S.B. and Calman, K.C. (1982) The antiemetic potential of oral levonantradol in patients receiving cancer chemotherapy. *Cancer Treat. Rev.*, **49**, 492.

Stuart-Harris, R.C., Mooney, C.A. and Smith, I.E. (1983) Levonantradol: a synthetic cannabinoid in the treatment of severe chemotherapy-induced nausea and vomiting resistant to conventional anti-emetic therapy. *Clin. Oncol.*, **9**, 143–146.

Sweet, D.L., Miller, N.J., Weddington, W., Senay, E. and Sushelsky, L. (1981) Delta-9-tetrahydrocannabinol as an antiemetic for patients receiving cancer chemotherapy. *J. Clin. Pharmacol.*, **21**, 70S–75S.

Tashkin, D.P., Reiss, S., Shapiro, B.J., Calvarese, B., Olson, J.L. and Lodge, J.W. (1975a) Bronchial effect of aerosolised delta-9-THC in healthy and asthmatic subjects. *Am. Rev. Resp. Dis.*, **115**, 57–65.

Tashkin, D.P., Shapiro, B.J. and Frank, I.M. (1973) Acute pulmonary and physiological effects of smoked marijuana and oral delta-9-THC in healthy young men. *N. Eng. J. Med.*, **289**, 336–341.

Tashkin, D.P., Shapiro, B.J. and Frank, I.M. (1974) Acute effects of smoked marijuana and oral delta-9-tetrahydrocannabinol on specific airway conductance in asthmatic subjects. *Am. Rev. Resp. Dis.*, **109**, 420.

Tashkin, D.P., Shapiro, B.J. and Frank, I.M. (1976) Acute effects of marihuana on airway dynamics in spontaneous and experimentally induced bronchial asthma. In: M.C. Braude and S. Szara (eds.), *Pharmacology of marihuana*. Raven Press, New York, pp. 785 – 802.

Tashkin, D.P., Shapiro, B.J., Lee, E. and Harper, C.E. (1975b) Effects of smoked marijuana in experimentally induced asthma. *Am. Rev. Resp. Dis.*, **112**, 377–386.

Tiedeman, J.S., Shields, M.B., Wever, P.A. and Crow, J.N. (1981) Effect of synthetic cannabinoids on elevated intraocular pressure. *Ophthalomology*, **88**, 270–277.

Turkanis, S.A. and Karler, R. (1985) Electrophysiological mechanism and loci of delta-9-THC-caused CNS depression. In D.J. Harvey (ed.), *Marihuana '84* – Proceedings of an Oxford Symposium on Cannabis. IRL Press, Oxford, p. 233.

Treffer, t D.A. (1978) Marijuana use in schizophrenics: a clear hazard. *Am. J. Psychiat.*, **135**, 1213–1215.

Turner, C.E. and Elsohly, M.A. (1981) Biological activity of cannabichromene, its homologs and isomers. *J. Clin. Pharmacol.*, **21**(Suppl. 283), 91S.

Tyson, L.B., Gralla, R.J., Clark, R., Kris, M.G., Brodin, L.A. and Bosl, G.J. (1985) Phase 1 trial of levonantradol on chemotherapy-induced emesis. *Am. J. Clin. Oncol.* (CCT), **8**, 528–532.

Ungerleider, J.T. and Andrysiak, T. (1985) Therapeutic issues of marijuana and THC (tetrahydro-cannabinol). *Int. J. Addictions.*, **20**(5), 691–699.

Ungerleider, J.T., Andyrsiak, T., Fairbanks, L.A., Ellison, G.W. and Myers, L.W. (1987) Delta-9-THC in the treatment of spasticity associated with multiple sclerosis. *Adv. Alcohol. Substance. Abuse.*, **7**(1), 39–50.

Ungerleider, J.T., Andrysiak, T., Fairbanks, L., Goodnight, J., Sarna, G. and Jamison, K. (1982) Cannabis and cancer chemotherapy – the UCLA study: a comparison of oral delta-9-THC and prochlorperazine. *Cancer*, **50**(4), 636–645.

Ungerleider, J.T., Andrysiak, T., Fairbanks, L., Tesler, A.S. and Parker, R.G.. (1984) Tetrahyd-rocannabinol vs prochlorperazine: the effects of two antiemetics on patients undergoing radiotherapy. *Radiology*, **150**, 598–599.

Ungerleider, J.T., Fairbanks, L.A., Andrysiak, T., Sarna, G., Goodnight, J. and Jamison, K. (1985). THC and Compazine for the cancer chemotherapy patient – the UCLA study. Part II: patient drug preference. *Am. J. Clin. Oncol. (CCT)*, **8**, 142–147.

Vachon, .L, Fitzgerald, M.X., Solliday, N.H., Gould, I.A. and Gaenster, E.A. (1973) Single-dose effect of marihuana smoke. *N. Eng. J. Med.*, **288**, 985–989.

Vachon, L., Mikus, P., Morrisey, W., Fitzgerald, M.X. and Gaensler, E. (1976) Bronchial effect of marihuana smoke in asthma. In: M.C. Braude and S. Szara (eds.), *The pharmacology of marijuana*. Raven Press, New York, pp. 777–784.

Vincent, B.J., McQuiston, D.J., Einhorn, L.H., Nagy, C.M. and Brames, M.J. (1983) Review of cannabinoids and their antiemetic effectiveness. *Drugs*, **25**(Suppl.1), 52–62.

Wada, J.K., Bogden, D.L., Gunnell, J.C., Hum, G.J., Gota, C.H. and Rieth, T.E. (1982) Double-blind, randomized, crossover trial of nabilone vs placebo in cancer chemotherapy. *Cancer Treat. Rev.*, **9**(Suppl. B), 39–44.

Waller, C.W., Beningi, D.A., Harland, E.C., Bedford, J.A., Murphy, J.C. and Elsohly, M.A. (1984) Cannabinoids in glaucoma III: the effects of different cannabinoids on intraocular pressure in the monkey. In: S. Agurell, W.L. Dewey, R.E. Willette (eds.) *The cannabinoids: chemical, pharmacologic and therapeutic aspects*. Academic Press, Orlando, pp. 871–880.

Ward, A. and Holmes, B. (1985) Nabilone: A preliminary review of its pharmacological properties and therapeutic use. *Drugs*, **30**, 127–144.

Waters, J. (1993) Necessity excuses GP use of cannabis. *Gen. Practitioner*, Nov. 5, 60.

West, M.E. and Lockhart, A.B. (1978) The treatment of glaucoma using a non-psychoactive preparation of Cannabis sativa. *West Indian Med. J.*, **27**, 16–25.

White, H.C., Munson, J.A. and Munson, A.E. (1976) Effects of delta-9-THC Lewis lung adenocarcinoma cells in tissue culture. *J. Nat. Cancer Inst.*, **56**(3), 655–658.

Williams, S.J., Hartley, J.P. and Graham, J.D. (1976) Bronchodilator effect of tetrahydro-cannabinol administered by aerosol in asthmatic patients. *Thorax*, **31**, 720–723.

8. CANNABIS AND CANNABINOIDS IN PAIN RELIEF

MARIO A.P. PRICE and WILLIAM G. NOTCUTT

The James Paget Hospital NHS Trust, Norfolk, UK

INTRODUCTION AND HISTORY

Cannabis is a term that describes products derived from the Indian hemp, *Cannabis sativa*. It has its origins probably in India but now grows all over the world. The chemical compounds responsible for intoxication and medicinal effects are found mainly in a sticky golden resin exuded from the flowers of the female plants and surrounding leaves. *Cannabis sativa* contains a wide range of different chemicals including a family of compounds called "cannabinoids". Of the cannabinoids delta-9 tetrahydrocannabinol (THC) is probably the main compound responsible for the psychotropic activities.

Cannabis has been used as a medicine for thousands of years and is mentioned in a Chinese herbal dating back to 2700 BC. There are records of 'its medicinal use in Egyptian papyri of the sixteenth century BC. Much later, the plant is mentioned in Assyrian texts and in Greek and Roman sources as a medicinal agent.

Early Experiences in the 19th Century

Cannabis Tincture was used in the nineteenth century as an analgesic, as well as numerous other conditions and was considered milder and less dangerous than opium. W.B. O'Shaughnessy was the first of the western physicians to take an interest in cannabis as a medicine on account of his observations on its use in India at the time and was the main figure behind its resurgence. He not only meticulously recorded the popular and medical uses of the various preparations in India but also conducted animal and human experiments and applied his knowledge in the clinic (O'Shaugnessy, 1841, 1843). In 1845 Donovan found cannabis to be highly effective in cases of violent neuralgic pain in the arms and fingers, inflammation of the knee, facial neuralgia and sciatica affecting the hip, knee and foot (Donovan, 1845). The most detailed review of the therapeutic uses of cannabis in the mid-19th Century was from Christison who reports uses of the tincture for rheumatic pain, sciatica and toothe ache (Christison, 1851).

An article written by Dr. J. Russell Reynolds (physician to Queen Victoria) noted, "In almost all painful maladies I have found Indian Hemp by far the most useful of drugs." Dr. Reynolds cites neuralgia, facial pain and neuritis as being particularly responsive to cannabis. He also wrote: "Migraine: Very many victims of this malady have for years kept their suffering in abeyance by taking hemp at the moment of threatening, or onset of the attack."

Hundreds of articles were written in European and American journals on the use of cannabis for many types of pain; but around 1890 the use of cannabis started to decline. There are several reasons that probably contributed to this decline: the increasing availability of injectable opiates; the uncontrollable variability in strength and composition of the cannabis preparations; the unpredictable response of patients to cannabis taken orally; and the introduction of aspirin, chloral hydrate and the barbiturates.

The Law Exerts its Control

In 1928 cannabis was banned for non-medicinal purposes in the UK. In America in 1937 a marihuana tax was introduced to discourage recreational use of cannabis and the tight controls in both countries served to discourage its use medicinally. In 1954 and 1957 the World Health Organisation (WHO) affirmed its view that cannabis had no therapeutic value. In 1968, the UN Economic and Social Council adopted a resolution recommending that all countries concerned should intensify enforcement of restrictions on traffic and use, that they promote research and deal effectively with publicity advocating legalisation or tolerance to the non medical use of the drug.

In 1969, the WHO reported cannabis as not physically habit-forming but as a drug of dependence and recommended keeping it under legal control. By 1960 in America and 1971 in the UK, cannabis was made a Schedule 1 drug making possession and medicinal use illegal without a special licence. This severely restricted clinical research.

The Purification, Analysis and Understanding of Cannabis

In 1964 THC was obtained in its pure form and the structure elucidated (Gaoni and Mechoulam, 1964). Lilly Research Laboratories, in 1968 initiated a cannabinoid research program. Early clinical studies investigated the pharmacological actions of THC and synthetic analogues. The objective was to derive a compound with the benefits of cannabis but without the adverse effects. As a result nabilone came to the forefront and was marketed as an anti emetic in Canada in 1982 and the UK in 1983.

On May 13, 1986, the Drug Enforcement Administration (DEA) in America transferred a synthetic form of THC from Schedule 1 to Schedule 2 for use as an antiemetic for cancer patients undergoing chemotherapy. In effect, this action by the DEA resulted in a dual scheduling of an identical molecule. A molecule of THC derived from the cannabis plant is a schedule 1 molecule, since the definition of "cannabis" includes all derivatives of the plant; but an identical molecule when synthetically derived and encapsulated in a gelatin capsule, is a Schedule 2 molecule.

In 1988 in America, an Administrative Law Judge stated cannabis to be "...one of the safest drugs known to man." The DEA overruled this and in 1992 gave its final rejection to the medicinal use of cannabis. In the same year, Howlett in St Louis USA discovered the first cannabinoid receptor in neuronal tissue (Devane et al., 1988) which led to the discovery of the first endogenous ligand anandamide, the body's own natural cannabinoid in 1992 (Devane et al., 1992).

THE PHYSIOLOGY OF ACUTE AND PERSISTENT PAIN

Pain protects the body from external harm and prevents activity after damage due to trauma and surgery, while it heals. It is also the result of many pathological processes.

To most people the mechanism is similar to an electrical alarm system but this is much too simplistic.

Steps in Pain Perception

Peripheral Nerves

Nociceptors (pain receptors) are present in the skin and most other tissues. They respond to mechanical, thermal or chemical stimulation. The chemical stimulation is due to a variety of substances released into damaged tissues, for example prostaglandins and bradykinin. Sensation is carried to the spinal cord either by "fast" fibres, which detect sharp, localised, short-lived pain, or by "slow" fibres, which carry signals of diffuse, ongoing pain.

Wind Up

Wind up is a normal process in which peripheral and central neurones become sensitised, leading to amplification of signals. The painful area may become hypersensitive to touch. The understanding of this process is still being worked out but it involves a complex chain of neurochemical events.

Transmission

Pain is transmitted up the spinal cord into the brain. This invokes an interaction of arousal, perception, emotion, interpretation and memory. It also triggers physiological changes. The transmission of pain signals across millions of neurones is mediated by neuropeptides, including beta-endorphin, enkephalin, dynorphine, serotonin and other catecholamines, these enhance or inhibit transmission.

Descending Inhibition

There is a descending system of nerves through the spinal cord back to the dorsal horn cells which can inhibit or enhance the pain perceived. Various neurotransmitters are involved. Descending inhibition damps down incoming pain impulses, providing analgesia. It operates when, for example, someone is injured but feels no pain until away from the site of danger. Inhibitory signals travel from the brain down the spinal cord and "damp down" incoming pain impulses. Similarly pain may be increased. This is the mechanism by which for example, happiness or distraction will reduce pain, whilst depression, anxiety or sleeplessness will aggravate it.

Gate Theory

The concept of the "gate" was introduced in 1966 by Melzak and Wall to explain the processing of pain in the dorsal horn of the spinal cord. The wider the gate is open, the more signals are transmitted. The most important control of the gate comes from the brain itself, mediated through the descending pathways described above. From the periphery, touch can be used to close the gate (e.g. transcutaneous nerve stimulation, rubbing, massage). However, in the acute situation touch may have the opposite effect and intensify pain perception.

Neuroplasticity

The nervous system is plastic in both acute and chronic pain. Nerve cells can change the quantity and type of transmitter that they release, receptors can change their activity and new synapses can develop.

Cutting or Damaging a Nerve

Damage to a peripheral nerve can cause major changes in cell function at all levels through to the cerebral cortex. These changes may be permanent, so the idea that a nerve can simply be cut to control pain is incorrect.

ANALGESICS AND HOW THEY WORK

Analgesics affect the transmission of pain in a wide variety of ways and places and are categorised in the following way:

NSAIDs (non-steroidal anti-inflammatory drugs) such as aspirin, ibuprofen and indomethacin, have anti-inflammatory activity inhibiting the formation of prostaglandins from arachidonic acid via the cyclo-oxygenase pathway. Their main site of action is in the periphery at the site of tissue damage but there may also be an effect within the spinal cord. Paracetamol is a para-aminophenol derivative with analgesic and antipyretic activity but no anti-inflammatory activity, with the spinal cord as its site of probable action. Opioid analgesics such as morphine and pethidine act on specific receptors on the descending pathways to inhibit pain. The main mode of action is by pre- and post-synaptic inhibition thereby preventing transmission of neural signals to the brain.

Antidepressants modulate the response to pain within the brain and spinal cord. It has been suggested that the analgesic action of tricyclic antidepressants and monoamine oxidase inhibitors is mediated by their action on central neurotransmitter functions; particularly serotonin and noradrenaline pathways. Anticonvulsants affect the abnormal triggering and transmission of pain along nerve fibres by acting as membrane stabilisers. Carbamazepine is the most often prescribed, although sodium valproate, clonazepam and clobazam are also used.

A wide range of agents are being explored nowadays, targeted on the various steps in the complex pathway of the transmission of pain. Agents such as clonidine, baclofen and ketamine are being used and their roles evaluated.

It is against this background that the pharmacological and clinical investigations of the cannabinoids will now be discussed.

NEUROTRANSMITTERS INVOLVED WITH CANNABINOID ACTION

Cannabis is a complex mixture of cannabinoid molecules (over 61 have been identified) and other chemicals (of which 400 have been identified); with THC as the main active cannabinoid responsible for the psychotropic effects. All these chemicals may have a wide variety of mechanisms of action and that of their metabolites may well be different again. So far, studies have concentrated on THC and a number of synthetic analogues, revealing a number of possible mechanisms of action.

The central nervous system (CNS) transmitters that modulate the perceptions of pain include noradrenaline, serotonin (5HT), acetylcholine, GABA, the opioid peptides and the prostaglandins. Reports suggest that the analgesic effects seen with the cannabinoids involve prostaglandins, noradrenaline, 5HT and the opioid peptides, but not GABA or acetylcholine. The involvement of the prostaglandins is complex. The cannabinoids are stimulators of phospholipase A2, promoting the production of prostaglandins, but also inhibitors of cycloxygenase therefore also inhibiting production. The scene is further complicated by the fact that prostaglandins oppose pain centrally but cause pain at peripheral sites (Bhattacharya, 1986). This may explain why in some tests involving cutaneous electrical pain stimulation to the finger tips in human subjects, cannabis increased sensitivity to both painful and nonpainful stimulation and reduced tolerance to pain (Hill et al., 1974).

The mechanism of the anti-inflammatory effect of THC has been investigated by Burstein et al. (1973). They explain that THC inhibited prostaglandin synthesis in an in-vitro system by reducing the conversion of arachidonic acid to prostaglandin E2. It was also found to be an inhibitor of the formation of prostaglandin E1. Cannabidiol was found to be far more active than THC in this test suggesting a structural relationship between analgesic and anti-inflammatory activity among the cannabinoids. It is also proposed that the cannabinoids interfere with prostaglandin action on adenylate cyclase which is reported to mediate pain perception. Levonantradol, a cannabinoid derivative from Pfizer Laboratories also inhibits prostaglandin induced diarrhoea in animals (Milne et al., 1981).

The involvement of 5HT as a mediator for analgesia with the cannabinoids is debatable. Analgesia is potentiated in the mouse tail flick test by 5-hydroxytryptophan (the precursor of 5HT) and imipramine (a 5HT re-uptake inhibitor) and the cannabinoids are known to affect 5HT. However intrathecally injected methysergide (a 5HT antagonist) has no effect on THC induced analgesia.

The noradrenergic system is a likely mechanism for cannabinoid induced analgesia, as the effects are reduced when yohimbine (an alpha-2 adrenoceptor antagonist) is injected into the lumbar region of the spinal cord. The alpha-1 noradrenergic antagonist, phenoxybenzamine, fails to block cannabinoid induced analgesia. Although the cannabinoids do not act at opiate sites, the effects of both drug classes may be mediated through a common descending noradrenergic mechanism. Analgesia produced by injecting morphine in the periaqueductal grey matter is also blocked by intrathecally injected noradrenergic antagonists (Lichtman and Martin, 1991).

Rats or mice rendered tolerant to the analgesic effects of morphine show a tolerance to cannabinoid induced analgesia (Bloom et al., 1978; Chesher, 1980). Naloxone can decrease the analgesic effects of cannabis in the tail-flick test, the phenylquinone abdominal stretch test, and the hot plate test, but at high doses only. Doses of naloxone known to reverse the analgesic effects of pethidine and morphine in the hot plate and abdominal stretch test do not reverse the analgesic effects of cannabis. After oral administration, THC and morphine produce dose dependent depressions of the passage of a charcoal meal through the gut of mice. THC works out to be about five times less potent than morphine in constipating effect (Chesher et al., 1973). These results tend to suggest that cannabinoids do have an involvement with opioid receptors but that the relationship is not straight forward.

It has been reported (Lichtman et al., 1991) that the kappa opioid antagonist, nor-binaltophimine (nor-BNI) effectively blocks the analgesic effects of the cannabinoids, which is compelling evidence for a link between opioid and cannabinoid analgesic systems. The opioid delta antagonist, ICI 174864 and low doses of naloxone are incapable of blocking cannabinoid induced analgesia and there is evidence of cross tolerance between THC and U50488, a kappa agonist. This suggests that only the opioid kappa receptors are involved. Nor-BNI does not affect the behavioural effects of cannabinoids in mice which raises the possibility of developing a cannabinoid derivative with only the analgesic properties. Both THC and morphine analgesic effects are blocked by potassium channel blockers; however the cannabinoids seem to be blocked by calcium-gated potassium channels via apamin, while morphine interacts with ATP-gated potassium channels. It may be that the potassium channel modulation may explain in part the profound cannabinoid/opioid synergism seen in some pain assessment tests. (International Cannabis Research Society Meeting, Keystone, 1992).

Some synergism must also exist in the mechanisms for mu or delta opioid analgesia with cannabinoid analgesia, because intrathecal pre-treatment of mice with sub-effective doses of THC or several other cannabimimetic compounds was able to shift the dose response curve to the left for intrathecal morphine in the tail-flick test; i.e. increase the potency of the morphine (Welch et al., 1992). The exact interaction the cannabinoids have with these neurotransmitters to cause an effect is not clearly known. It is possible that the effects seen are brought about allosterically via the cannabinoid receptor; a mechanism that would allow some sort of selectivity and action only where there was a link between the two types of receptor. It could be by affecting absorption, distribution or fate of a transmitter or even synthesis, storage and release. Some actions of the cannabinoids could be explained by an effect on drug metabolism like cannabidiol which is a known potent inhibitor of drug metabolism (Narimatsu et al., 1990). There is also a report of cannabis increasing the permeability of the blood brain barrier (Agrawal et al., 1989).

In summary, it is likely that in regard to analgesic effects, the cannabinoids have more than one action on any particular system.

CANNABINOID RECEPTORS

So far, two types of cannabinoid receptor, CB1 and CB2, have been identified. The CNS responses to the cannabinoids are likely to be via the CB1 receptor, as evidence for the presence of the CB2 receptor has only been found in the spleen. The CB1 receptor was the first to be identified and has since been cloned. It has been found in rat brain, with the greatest abundance being in the cortex, cerebellum, hippocampus and striatum, with a lesser concentration in the brain stem and spinal cord (Bidaut-Russell et al., 1990).

Certain of the in-vitro effects seen with the cannabinoids may not be mediated by a receptor mechanism. The lipophilic nature of the cannabinoid compounds results in significant changes in the "fluidity" of phospholipid containing membranes and this may be the property responsible for the altered responses of membrane-associated enzymes and proteins. The mechanism of action is comparable to the steroid anaesthetic, alpha-xolone and the volatile anaesthetic halothane. However THC produces considerably less fluidization than alphaxolone, thus explaining the lack of clinical anaesthesia. The psychotropically inactive, cannabidiol produces an opposite effect: a decrease in the molecular disorder of the lipid bilayer. Apart from the evidence that cannabinoids can alter the physical properties of membranes there is also evidence that THC can alter the composition of the membranes within the brain and affect the biosynthesis of mem-brane lipids (Pertwee, 1988).

MEDICINAL CANNABINOID PROTOTYPES

The structure–activity relationships of cannabinoids have been investigated in consider-able depth (Razdan, 1986). Minor changes in structure have been shown to cause major changes in activity. For example 2-methyl, delta-8-THC is a potent cannabimimetic, but 4-methyl, delta-8-THC is inactive. Such major changes as a result of relatively small chemical modifications are characteristics seen with compounds which act via receptors. Reports that the 11-hydroxy metabolites of D9 THC (Figure 1(a)) and D8 THC (Figure 1(b)) were more potent in the mouse hot plate tests than the parent compounds led to the development of HHC (9-nor-9 beta–hydroxy hexahydrocan-nabinol); see Figure 1(c).

Pfizer Inc. examined the structure–activity relationships for analgesia based on HHC, determining that the c-3 alkyl side chain could be optimised by making it longer and that the phenolic hydroxyl was critical for activity. However, because the pyran ring could be modified without extensive loss of potency, the analgesic and antiemetic drug nantradol was developed by replacing the pyran oxygen with nitrogen and removing the axial methyl substituent. The levo enantiomer of nantradol was found to have twice the potency of the dextro enantiomer. In the battery of animal model tests for analgesia (see below), levonantradol (Figure 1(d)) was found to be up to 100 times more potent than THC (Milne et al., 1980). In a controlled study in humans with acute moderate to severe post-operative pain, levonantradol was significantly superior to placebo in terms of analgesic activity (Jain et al., 1981). Drowsiness was the most reported side effect (40% of responses); fewer than 10% reported other effects such as dry mouth, dizziness, strange dreams, nervousness, headache, hallucinations and dysphoria.

Nabilone (Figure 1(e)) is a successful outgrowth of a cannabinoid research program at Lilly Laboratories. Using the usual approach of pharmaceutical industry, the plan was to discover new therapeutic drugs through synthesis and pharmacological evaluation in animals of hundreds of new chemical entities. Nabilone is a non-THC cannabinoid (i.e., a 9-keto analogue of (+ / −)-hexahyrocannabinol-dimethylheptyl), albeit with a spec-trum of activity closely related to that of (−)-THC.

CP 55940 (Figure 1(f)) is another prototype developed by Pfizer with a spectrum of activity similar to levonantradol but about three times more potent. In comparison to morphine CP55940 has between 8 and 25 times the potency of morphine in a variety of animal analgesic tests (Razdan, 1986). HU-210 (Figure 1(g)) is a dimethylheptyl analogue of 11 hydroxy, delta-8-THC which also has a spectrum of activity similar to levonantradol. Win 55212-2 (Figure 1(h)) is a prototype of a novel series of aminoalkylindole analgesics, synthesised by the research group at Sterling Drug Inc. This compound and its congeners are structurally different from all other known cannabinoids. Nevertheless, studies with tritium labelled Win 55212-2 indicate that this compound binds very strongly to the cannabinoid receptor.

THE ENDOGENOUS CANNABINOID – ANANDAMIDE

Arachidonyl ethanolamine amide, generically named anandamide (Figure 1(i)), is an eicosanoid derivative that was initially isolated from porcine brain. It was independently isolated from calf brain and identified as a regulator of L-type calcium channels. Subsequently, other ethanolamine amides were identified in porcine brain having the same affinity for CB1 as the ethanolamine amide of arachidonic acid. Mechoulam and colleagues proposed that the family of unsaturated fatty acid ethanolamine amides that bind to the cannabinoid receptor be referred to collectively as anandamides (Mechoulam et al., 1995). Anandamide shows analgesic activity in the hot plate test (Fride and Mechoulam, 1993) and has tranquillising effects in animals (Musty et al., 1995). This does support the theory of endogenous cannabinoids having a role in the control of pain and anxiety.

It has been proposed that anandamide is produced and released from neurones in a calcium ion-dependent manner, when they are stimulated with membrane depolarising agents. Devane and Axelrod (1994) propose that anandamide is formed by an enzymatically catalysed condensation reaction between arachidonic acid and ethanolamine. But because endogenous levels of free arachidonic acid and ethanolamine are very low in the brain, Di Marzo et al. (1994) propose that the formation of anandamide occurs through a phosphodiesterase mediated cleavage of a novel phospholipid precursor and that the degradation of anandamide involves hydrolysis to ethanolamine and arachidonic acid.

Anandamide possesses cis double bonds at carbons 5, 8, 11, and 14 and is structurally different from other cannabinoid receptor agonists, such as THC, CP-55940 and HU-210. Anandamide, like other cannabinoids, inhibits forskolin-stimulated cAMP production in cells expressing the cannabinoid receptor and inhibits N-type calcium currents.

Anandamide mimics many of the pharmacological properties of THC, but has a shorter duration of action. Following IV administration of anandamide, the pharmacological effects, with the exception of analgesia, are almost completely dissipated within 30 minutes (Smith et al., 1994). In contrast THC has a long half life and produces effects for hours. Vela et al. (1995) have shown that anandamide, like THC, can decrease naloxone-precipitated withdrawal signs in mice chronically treated with morphine. This further supports the role of anandamide as an endogenous cannabinoid agonist and provides additional support for a link between endogenous opioid and cannabinoid systems.

The discovery of the "anandamide system" is important as it may provide possibilities for new drugs to be developed and even provide targets for existing drugs. These targets may be receptors or the processes of synthesis, storage and release of anandamide itself.

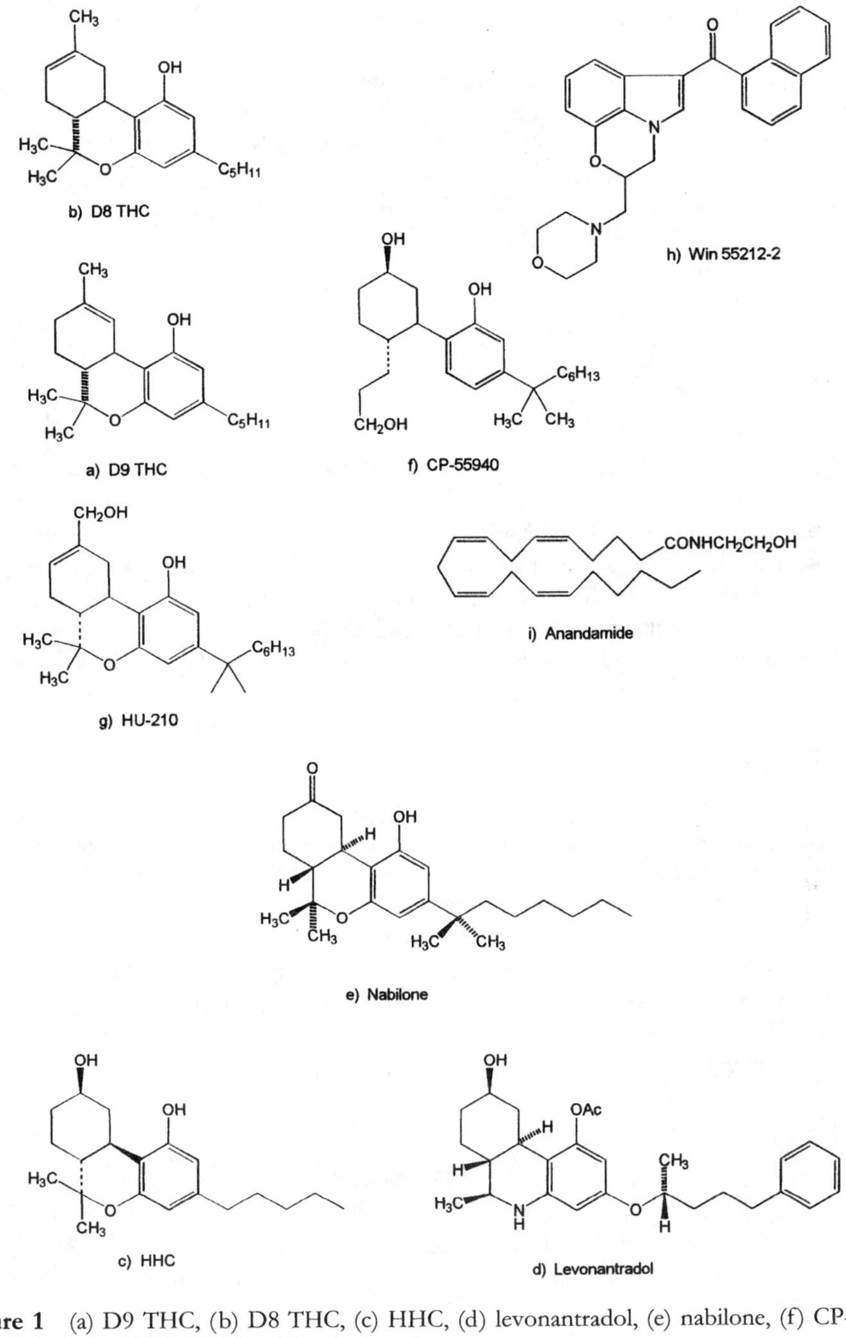

Figure 1 (a) D9 THC, (b) D8 THC, (c) HHC, (d) levonantradol, (e) nabilone, (f) CP-55940, (g) HU-210, (h) WIN 55212-2, (i) anandamide

CANNABINOID RECEPTOR ANTAGONISTS

Sanofi Recherche have made a cannabinoid antagonist, SR141716A, that displays a nanomolar affinity for CB1 but micromolar affinity for CB2 in ligand binding assays. SR141716A antagonises responses of the potent cannabinoid analogues CP-55940, WIN-55212-2 and anandamide in the mouse vas deferens and rat brain adenylate cyclase assays *in vitro*. When administered orally to animals, SR141716A antagonises the analgesic effects produced by WIN-55212-2.

Another antagonist AM630 (iodopravadoline) is a more potent antagonist of THC and CP55940 than of WIN 55212-2 in the mouse vas deferens; unlike SR141716A which is equipotent. This suggests the presence of more than one cannabinoid receptor in the mouse vas deferens (Pertwee *et al.*, 1995).

LABORATORY EVIDENCE FOR CANNABINOID ANALGESIC ACTIVITY

From the large number of methods available for evaluating the effectiveness of analgesics, it is clear that the optimal tool for estimating pain and pain perception is lacking; however, a comprehensive picture can be obtained by using several testing procedures. Experiments with rats and mice have shown that some cannabinoids are effective analgesics in a number of standard tests which are used to evaluate drug analgesic activity, examples of which are:

The Tail-Flick Test

This involves shining a ray of light on the tail of a mouse and measuring the time taken before the mouse moves its tail out of the way. Analgesics would increase the time before the tail would be flicked away (D'Amour *et al.*, 1941). Buxbaum *et al.* (1969) have reported that with intraperitoneal administration in male Sprague–Dawley rats, THC was comparable to morphine in the rat tail-flick test.

Bisher and Mechoulam (1968) reported that 20 mg/kg THC intraperitoneally produced activity in the mouse tail-flick test equivalent to that produced by 10 mg/kg morphine sulphate administered subcutaneously. Dewey *et al.* (1972) found no activity with THC below 100 mg/kg.

The Hot Plate Test

Mice are placed on a plate maintained at 55°C and the time taken before they lick their paws or jump is measured. Analgesics increase this time interval. The mice are not left on the plate for more than 30 seconds (Eddy and Leimbach, 1953). Sofia *et al.* (1975) found oral administration of THC to be equivalent to morphine in the hot plate test.

The Abdominal Stretching Test

Mice are injected intraperitoneally with p-phenylquinone and the number of stretches recorded over a one minute period. Analgesics would tend to decrease the number of stretches (Dewey *et al.*, 1970). THC reduces the number of abdominal stretches but is not as effective as morphine (Dewey *et al.*, 1972).

Carrageenan Induced Oedema Test

In this study, carrageenan is injected into the paws of rats to create oedema. THC was found to be 20 times more potent than aspirin and twice the potency of hydrocortisone in reducing the volume of the oedema. (Sofia *et al.*, 1973).

Acetic Acid Abdominal Constriction Test

0.25 ml of acetic acid 0.5% is injected intraperitoneally in rats and the number of constrictions over 5 minutes counted. THC was found to be 10 times more potent than aspirin in reducing the number of constrictions (Sofia *et al.*, 1973).

Haffner's Tail Pinch Test

An artery clip is placed on the tail of a rat and the time taken for the rat to bite at the clip measured. THC was found to be a very effective analgesic at 11 mg/kg orally where as no analgesia was seen with aspirin at 300 mg/kg (Sofia *et al.*, 1973).

There are many reports of analgesic tests in animals using cannabis or THC and all have varying results (Mechoulam, 1986; Martin, 1985). This may be due to the species used in the experiment or even the housing conditions before and during the experiment. In each case the end point depends upon the expertise of the assessor in evaluating a certain reaction made by the animal in response to the stimulus, whether it be a squeal, head jerk, tail flick, licking of paws or jumping.

SYNERGISTIC ANALGESIC EFFECT OF CANNABINOIDS WITH OPIATES

Despite the differing mechanisms of action between the cannabinoids and opiates there are reports of crude cannabis extract (Ghosh *et al.*, 1979), orally administered delta-6-THC and THC enhancing the analgesic effects of morphine (Mechoulam *et al.*, 1984). Intrathecal administration of numerous cannabinoids with intrathecal morphine as been shown to have synergistic analgesia in mice (Welch *et al.*, 1992). An interesting note from Wirral Hospital, Mersyside, mentions that in the recovery room, patients who had received metoclopramide before an operation needed significantly more opiates in the post-operative period than patients who had received nabilone (Williams and Higgs, 1995).

CLINICAL EVIDENCE FOR CANNABINOID ANALGESIC ACTIVITY

Noyes *et al.* (1975) have shown that THC given orally can reduce pain in patients suffering from advanced cancer. In a double blind study, patients received either placebo (sesame seed oil) or randomly allocated doses of THC in sesame seed oil varying between 5 and 20 mg. The analgesic effect developed gradually and lasted for several hours. The higher doses of THC (15–20 mg) were significantly superior to the placebo but were accompanied by substantial sedation and mental clouding. In comparison to oral codeine, 20 mg THC orally was comparable to both 60 and 120 mg codeine. A dose of 10 mg THC was better tolerated but less effective than the 60 mg dose of codeine.

In contrast to the analgesic effects seen in cancer patients, Hill *et al.* (1974) were unable to detect any analgesic activity after 12 mg doses of THC in healthy volunteers when applying cutaneous electrical stimulation to the fingers. In fact in some instances, THC heightened the amount of pain experienced.

THC given intravenously at a dose of 44 mcg/kg in patients undergoing dental extraction did not produce as much analgesia as 157 mcg/kg of diazepam intravenously (Raft *et al.*, 1977). In fact, placebo appeared to be preferred to THC at a dose of 22 mcg/kg intravenously due to the anxiety and dysphoria produced by the latter drug. It could have been that elevated anxiety responses from the THC may have been misinterpreted as an increased pain experience.

In a double blind trial with levonantradol injected intramuscularly, compared to placebo in post operative pain, levonantradol had superior analgesic action but the side effects were not insignificant (Jain *et al.*, 1981). Significant analgesia was obtained with a 1.5 mg dose of levonantradol; the greatest analgesia and side effects were seen with a 2.5 mg dose. The side effects at the effective analgesic dose appeared to be milder and less frequent than 10 and 20 mg doses of THC. Similarly the observed cardiovascular effects of levonantradol i.e., increase in heart rate and decrease in blood pressure, appeared to be milder than those reported with higher doses of nabilone.

ANECDOTAL EVIDENCE FOR CANNABINOIDS IN PAIN RELIEF

There is a growing body of contemporary anecdotal evidence for the benefits of cannabis in pain relief; there are currently four pain problems where cannabinoids seem to be of most benefit.

Multiple Sclerosis

Here there may be widespread damage to the nervous system and alterations to the neurochemistry. The effects of the disease are very variable. Up to 40% of patients with multiple sclerosis experience pain and this is often unresponsive to conventional analgesics including opiates, anti-depressants and anti-convulsants. Many also get painful bladder spasms which seem particularly responsive to cannabis.

Major Spinal Injury

Traumatic spinal injury up to and including tetraplaegia, may cause significant pain below the level of injury or at the level itself. This pain is due to disruption of pain control mechanisms at the level of spinal cord damage. De-afferentation may be leading to a loss of inhibitory control in pain systems (pathways) in higher parts of the central nervous system.

Other Neurogenic Pain

There are other situations where nerves may be damaged, either traumatically or as a result of various disease processes (e.g. diabetes). The pain produced may be uncontrollable with conventional analgesic practise.

Traumatic Spinal Strains/Sprains

Some musculo-skeletal injuries remain resistant to conventional analgesics up to and including morphine. They are often associated with significant muscle spasm.

The evidence that cannabinoids are effective in these groups of patients is drawn from the experience of the patients themselves with the use of cannabis, usually when smoked. Naturally patients are very reluctant to admit to using an illegal drug. It is therefore impossible to establish a profile for use and effectiveness of cannabis at present.

There is also evidence from the use of the drug nabilone, a synthetic cannabinoid. Currently nabilone is only licensed for use as an anti-emetic during chemotherapy and for short-term use. There is no data on its clinical use in pain control nor is there data on its long term use. Some patients who have gained benefit from the illicit use of cannabis have elected to try nabilone. All patients have persistent (chronic) pain and have obtained little or no benefit from their current therapy. Some are taking a "cocktail" of medications and nearly all wish to discontinue their reliance on them for what is often minimal benefit. Many of these drugs have significant dependency potential, for example opiates and anti-depressants.

Long standing (chronic) pain is a common phenomenon affecting possibly as many as one person in twelve. Conventional analgesic practice is by no means universally effective. Cannabis could significantly improve our ability to help these patients.

NABILONE – CLINICAL EXPERIENCE AT THE JAMES PAGET HOSPITAL

The dose per capsule is 1 mg but we found that this could be excessive for some patients. Therefore, some were started at 0.25 mg by opening the capsule and dividing the resultant powder into four.

The initial time for nabilone use has been at night to reduce the potential discomfort of any side effects. Once the patient's confidence has been developed, the dosage has been increased where appropriate. Those patients who have benefited from nabilone have been through a period of discontinuation to help evaluate the benefits of this drug.

The age range of the 43 patients who have used nabilone is from 25–82 years with 75% between the ages of 30 and 50. More women than men were treated, mainly reflecting a large sex difference in the group with multiple sclerosis.

The diagnoses of the patients were categorised into 6 groups as the most convenient method of presenting the information from such an heterogeneous group. No attempt has been made to do anything more than describe the effects of using nabilone on each individual patient and thereby evaluate whether it might be of value in pain control.

Multiple Sclerosis

Multiple Sclerosis is characterised by widespread and varied damage to the central nervous system. Part of the pain experienced is neuropathic (i.e. pain that results from damage to the nerves) and is often resistant to conventional analgesics. However, patients also experience nociceptive pain (i.e. arising from damaged or dysfunctional tissues where the nervous system is intact). This may be secondary to muscle spasm, mechanical strain or bladder spasm. Patients often present with a cluster of pain symptoms in a variety of sites of their bodies.

Patients will often receive a wide variety of agents to try and treat their pain. They may also have had to come to terms with the progressive and often relentless deterioration of their physical health. The psychological and social effects of this process may add to or complicate the pain experience.

Six out of 13 patients obtained benefit from nabilone and 3 have continued for periods between 2 and 3 years. Two of these long term patients have a varied cluster of sites and types of pain. They have achieved substantial pain relief with nabilone. Their intake of other analgesics and psychotropic agents has become minimal. The third patient with mainly leg pain has maintained an improvement in pain and in sleep.

Three patients have discontinued in spite of benefit. One, with mainly spinal pain, had obtained much better pain relief with smoked cannabis and also improvement in his appetite and well-being. As he had a reliable supply, he has opted to continue with this mode of administration. He is able to provide a precise control of his pain, smoking just as much as he needs. A second patient found that cannabis provided better pain relief than nabilone with less side effects. The final patient was discontinued from nabilone due to difficulties with assessment and drug compliance. However, it was clear that she had gained some benefit with her sleeping.

Seven patients found nabilone either ineffective or suffered side effects without any observed benefit.

Spinal Problems

Back pain is the commonest problem presenting to Pain Relief Clinics. Many of these patients have major problems with their pain, complicated by a multitude of factors. Surgery and other invasive treatment have not only a very limited scope but also a significant failure rate.

The group included a mixture of neuropathic and nociceptor pains in a broad group of patients but with one common factor. There was no other treatment to offer.

Two patients have continued with nabilone. One has both lumbar arachnoiditis and sympathetically mediated pain in her limbs (complex regional pain syndrome). For her, nabilone has acted as an adjuvant to her other analgesics, relieving pain and misery and improving sleep. However, it is dysphoria that prohibits the use of nabilone when she wants to drive. The second patient has scarring of her S1 nerve root, established at 2 operations. She gets pain in the S1 dermatome and has been unable to control it adequately with opioids and a variety of other agents. Nabilone produces a substantial reduction in her pain and improvement in her sleep.

Five other patients with a variety of spinal problems have benefited from using nabilone but have had to stop either because of side effects or because of difficulties in assessment.

Two patients found nabilone completely ineffective to manage the pain. Both of them had had multiple spinal operations and suffered both mechanical spinal pain and neurogenic pain from nerve scarring.

Peripheral Neuropathy

The pain arising from damage to peripheral nerves can be extremely difficult to control satisfactorily. Patients will commonly describe their pain as a constant burning, tearing or pricking sensation. Diabetic neuropathy is a good example and for some patients the use of tricyclic antidepressants or anticonvulsants is satisfactory. However, for others, these may be ineffective. Conventional analgesics are often of little use.

One patient with anaesthesia dolorosa of her face, a condition that can occur after therapeutic damage to the trigeminal nerve, seems to have been cured of her problem. The severe burning, thumping pain that she experienced in her face was uncontrollable with analgesics and a range of psychotropic drugs. Not only did her pain recede completely but she was able to wean herself off all her medication. At 15 months she also discontinued nabilone and a year later has not had any recurrence of the pain in her face. She still has numbness though. Unfortunately, at the point at which nabilone was discontinued, she was discovered to have a metastatic carcinoma of the kidney. There is no evidence that this is anything more than coincidental. She subsequently went on to use nabilone to control the pain from a metastatic lesion in her spine (see below).

Two patients with diabetic neuropathy achieved good pain relief with nabilone but only one continues to use it. The other has found that the dysphoric side effects interfered too much with her daily life. She therefore opted for less satisfactory pain relief, using tramadol. However, from a functional point of view, she was better.

Four other attempts to treat neuropathies with nabilone have not been successful (sensori-motor, diabetic, post-chemotherapy, post herpetic).

Central Neurogenic Pain

This group of 5 patients had pain arising from within the central nervous system. For each of them the underlying cause is very different.

A patient who was tetraplegic following a spinal injury, had some benefit from nabilone. Previously he was smoking cannabis to control pain in his scapula region and in his sacral dermatomes. However, he achieved much better pain control for his distal body pain from regular cannabis use and consequently opted to continue smoking.

Four other patients obtained no benefit (post trigeminal tractotomy for post herpetic neuralgia, thalamic pain post CVA, cervical myelopathy, central cord syndrome post trauma).

Malignancy

Three patients with malignant disease have used nabilone for pain management. The first, a young woman with advanced cervical cancer, supplemented her opiate analgesia with nabilone before dying a few weeks later. A second patient using morphine to control the pain of a spinal metastasis reached the limit with opiates due to side effects. Her pain control substantially improved with a small amount of nabilone. She had used nabilone previously for anaesthesia dolorosa (see above). A third patient with bronchial carcinoma and a tracheo-oesophageal fistula and severe chest pain has used nabilone as an adjuvant to opiate analgesia. He improved sleep and well-being. Previously he was a frequent cannabis user.

Heartsink Patients

Most Pain Relief Clinics have their collection of patients who are best described as "Heartsinks" from the feelings engendered in the clinician trying to support them. Hence they are often ready and willing candidates for any new treatment.

Of six patients with a wide variety of complex physical, psychological and social problems, 1 definitely obtained benefit from nabilone, whilst 3 others probably were helped. The former preferred smoking cannabis for his neck pain and insomnia opted for this approach. The other 3 were discontinued due to difficulties with assessment.

Benefits and Side Effects of Nabilone and Cannabis

As time has passed we have observed many patients using nabilone and/or cannabis. The effects and side effects have been the same although the effects of cannabis have always been preferred.

From our observations the benefits can probably be divided into 3 categories with possibly a variety of mechanisms:

1. For pain directly arising from nerve damage (neuropathic) pain a direct reduction in pain was observed. Therefore effects on the receptors within the pain pathways can be proposed.
2. For muscle spasms and bladder cramps there has been a reduction, presumably by a direct effect on the neural control of muscle tension.
3. The adjuvant effects of improved sleep, "pain compressing", "pain distancing", anxiolysis, relaxation, euphoria, and relief of depression all have benefits on the patient's perception and management of pain. These will be mediated through cortical and limbic system mechanisms.
4. There are probable peripheral effects on inflammation which would influence pain although all of the patients here have visited NSAID therapy in the past.

Side effects seem to have occurred with a similar frequency to many other psychoactive drugs. Drowsiness and dysphoria have been the major problems for patients. A dry mouth has been reported. No other side-effects of nabilone have been noted.

Summary of Nabilone Experience in 43 Patients

Benefitted	Continued or cured	9
Benefitted	Stopped, cannabis instead	6
Benefitted	Stopped, side effects	3

Benefitted ?	Stopped	7
No Benefit	Stopped	6
No Benefit	Side effects, stopped	12
		43

Cannabis Use in Chronic Pain

Many patients self-report their use of cannabis for the relief of their chronic pain and a record of these has been kept recently. Two specific observations have been made. Firstly, a significant number of these patients have spinal pain as a result of trauma. The injuries have never been severe enough to cause neurological damage. However, all have experienced soft tissue damage and including one with a vertebral crush fracture. A common feature has been a combination of muscle spasm, excessive superficial spinal tenderness (hyperalgesia, allodynia) and lack of sleep.

Secondly, a patient with an established prolapsed intervertebral disc could relieve all his pain with cannabis. His prolapsed disc was later surgically treated with substantial relief of symptoms.

Occasionally patients in the clinic will ask for advice on their use of cannabis and its interaction with other drugs. In parts of London where cannabis is widely used socially, discussion of this issue is the norm for certain health workers.

Costs

Nabilone is a very expensive drug because of its current limited use. It is producing budgeting problems within our Hospital. Cannabis grown commercially is extremely cheap. Nabilone possibly costs 10 times as much as the equivalent amount of cannabis bought on the street. Cannabis can be grown to yield a cocktail of cannabinoids of known and repeatable concentrations. The illogicalities are evident.

Future Uses

There are many possible uses for cannabinoids in pain relief. Not only is the whole field of chronic pain open to such a new potential analgesic, but also its place in palliative care needs exploring. Acute pain, particularly acute back pain may be another valuable use of it. Sedation in Intensive Care is another possibility although this area is littered with agents that have come and gone. The anti-inflammatory effects may preclude its use here. Perhaps the traditional use of benzodiazepines for premedication prior to surgery could be challenged!

Future Studies

Leaving aside the flights of fancy described above, and focusing just on chronic pain, the issue of drug trials must be addressed. All trials of treatment for chronic pain are fraught with difficulty and commonly yield conflicting or ambiguous results. There are many reasons but two will suffice. Firstly, pain cannot be directly measured. Secondly these patients are a heterogenous group often with a multitude of physical, psychological and social problems. No two patients with multiple sclerosis are the same. Therefore treating these patients as a group can be a statistical nightmare when one comes to analyse the results. Conducting trials on cannabinoids will be worse due to the variety of effects they produce. It may be that the trial of the drug(s) in single individuals is the only way forward at the current time (an "N of 1" trial).

Trials of cannabinoids will require not only the current oral forms that are essentially of slow onset and long acting, but also a fast release variant. The nasal or inhalational route may be optimal providing a route for breakthrough pain and also a method of "fine tuning" the drug. In the past asthma has been treated with inhaled cannabinoids.

The experiences described above have been just that. No attempt has been made to try and prove benefit or compare effects. The purpose has been to define some areas for future studies. The challenge is now for clinicians to start clinical research in this area.

However, we must not lose sight of the fact that there are a large number of patients with chronic pain who might benefit from this group of drugs. Currently their options for analgesia are limited or non-existent. This is particularly poignant when one considers the history and safety of cannabis.

Of the patients continuing on nabilone evidence relating to their tolerance of the side effects of dysphoria and drowsiness was conflicting, this was also true of their tolerance of the analgesic capability of the drug, and it is on this basis why further investigative work is needed.

THE EXPERIENCE OF AN MS PATIENT WHO USES CANNABIS

CH (a multiple sclerosis patient) writes of her experiences:

"It was hard enough to deal with all the problems that multiple sclerosis has brought me during the past 10 years, but it became much more difficult when I realised I had to cope with the unpleasant side-effects of my medication as well. I was being prescribed a whole range of medicines. There were pills to stop me feeling sick. These made me clumsy and drowsy. There were pills to relieve my bladder spasms but they made me feel sick and gave me blurred vision. There were pills to help me sleep but they made me anxious and were habit forming.

Then a friend showed me an American article about the growing number of people with MS who had found safe and effective relief from their symptoms by taking cannabis. I could not see what I had to lose, so I decided to see if it would work for me too.

I am a middle-aged suburban housewife living in the north of England and have two young children. I am not remotely involved in the "drug scene". It took me some time, difficulty and expense to lay my hands on this illegal substance. When I had got it, I had no idea how to find out how much I should take and when. The doctors who see me were all interested and supportive but they didn't know much about it either, so I had to work everything out for myself. For about a year now, I have been regularly taking a small amount of cannabis resin – less than the size of a pea – late at night. I used to smoke it with dried herbs (not tobacco), but I worried that my children might see me smoking, so now I eat it. After a short time my body completely relaxes, which relieves my tension and spasms. During the day I have to use a catheter when ever I want to empty my bladder and most notably, cannabis relieves the discomfort and difficulty I have controlling it. It has also stopped the nausea that kept me awake at night. It is hard to regulate the dose. The quality varies, and, I suppose, like alcohol, the same amount on different occasions can have different effects.

Over the months I found I was able to reduce the doses of standard medication and now take none at all. I don't often take enough to "get high". When I do, I'm sure the feeling of calm and euphoria does my spirits a lot of good. Like many people who are ill, it is only too easy to become introspective and self-pitying. It is neither easy nor pleasant for me to obtain supplies, but I am happy to carry on as it does me so much good. I don't like breaking the law, I would far rather be able to obtain it from the chemist's".

SIDE EFFECTS OF CANNABIS AND NABILONE

Cannabis produces a very wide range of effects in human behavior. It produces euphoria, anxiety, lethargy, drowsiness, impaired performance, memory defects, changes in perception of time and depersonalization. These effects do vary between subjects and can also depend on factors such as the subject's pre-existing mood state, the social setting in which the cannabis is taken and whether the patient has taken cannabis before. The most common physical side effects of cannabis are a reddening of the eyes and a slight increase in the heart rate. Neither of these is reported to be uncomfortable or dangerous. Propranolol can prevent the tachycardia but does not affect the subjective and behavioral effects.

Balanced against these side effects it must be remembered that after so many years of use through history there is still no credible evidence that cannabis has ever caused a single death.

Clinical trials with nabilone showed that nearly all patients experienced at least one side effect. The most common being drowsiness followed by dizziness, euphoria, dry mouth, ataxia, visual disturbances, concentration difficulties, sleep disturbance, dysphoria, hypotension, headache and nausea (in decreasing order of incidence). Other reported side effects include confusion, disorientation, hallucinations, psychosis, depression, decreased co-ordination, tremors, tachycardia, decreased appetite and abdominal pain. Tolerance to drowsiness and euphoria develops rapidly without any noticeable drop in analgesic capability.

Patients who have taken nabilone 2 mg each day for several months and then stopped, have not reported any withdrawal symptoms; when restarted, tolerance to the drowsiness has to be re-acquired. Nabilone may impair mental and/or physical abilities when operating machinery or driving a car, and so patients must be made aware of this. Some patients at the James Paget Hospital have found nabilone to be excellent at reducing their pain but have been unable to continue treatment because they need to drive. The effect of nabilone may persist for a variable and unpredictable period of time following oral administration and side effects have been known to persist up to 72 hours after stopping the drug. Slight changes in mood and personality can occur in a few patients. This is usually noticed by the patient's partner.

Laboratory studies have so far shown no evidence of teratogenicity, but no controlled studies have been carried out in pregnant women. Nabilone can elevate supine and standing heart rates and cause postural hypotension. It should therefore be used cautiously in the elderly and in patients with hypertension and heart disease.

PHARMACOKINETICS OF THC

Absorption

About 50% of the THC available in cannabis is present in the smoke inhaled from a whole cannabis cigarette. This produces an almost immediate effect, with a peak effect within about 20–30 minutes and lasting for three to four hours (Huestis et al., 1992; Seth, 1991).

The analgesic effects are also experienced quickly but peak plasma concentrations of THC do not quite coincide with the "high" or the "high" with peak changes in heart rate indicating that the pharmacokinetics and dynamics of THC are still not completely understood. The time taken to peak concentration following oral administration of THC is between 2 and 4 hours and the effects may persist for 3–5 hours. About 90–95% of THC is absorbed from the gut but only between 10% and 20% reaches the systemic circulation due to significant metabolism with the "first pass" through the liver, and high solubility in fat. (Data sheet for Dronabinol, Roxane Laboratories, Columbus, Ohio, USA.)

Following a single injection of radio labelled THC, the concentration in fat was found to be 10 times greater than in any other tissue examined (Lemberger et al., 1970). It is most probable that the action of a single dose is terminated not by metabolism but by redistribution into sites like fat where it cannot act.

Half-Life

The elimination of THC is in two stages. The half-life of the first stage is about 4 hours and the half-life of the second about 25–36 hours. The effects last in most cases about 4 hours but sometimes for up to 24 hours. This is important to consider especially if patients believe that their performance is no longer impaired after the "high" has worn off. Traces of THC and its metabolites persist in the plasma of man for several days and can be detected in the fat and brain of animals for days after a single dose. Metabolites can be found in the urine for weeks after a dose. Studies do not show evidence of accumulation after repeated doses over several weeks but it may occur with very high doses over a longer period. People who have used cannabis for a long time metabolise THC more rapidly than novice users.

Elimination

THC is rapidly converted into an active metabolite, 11-hydroxy-THC, which produces effects identical to THC, and an inactive metabolite; 11-nor-9-carboxy delta 9 tetrahydrocannabinol (Huestis et al., 1992). The active 11-hydroxy-THC then gets converted to a more polar inactive metabolite 8,11-dihydroxy-THC which is then excreted in the faeces and urine. After 72 hours, 50% of orally administered THC is excreted in the faeces as unchanged drug and metabolites, and in the urine, 10–15% as metabolites only.

PHARMACOKINETICS OF NABILONE

Absorption

Nabilone is readily absorbed after oral administration and is rapidly distributed into the tissues. It is metabolized by reduction of the 9-keto group to form a mixture of isomeric alcohols together with several unidentified metabolites. (Mechoulam *et al.*, 1986)

Half-Life

The plasma half-life of nabilone is about 2 hours but the half-lives of the metabolites are very much longer.

Elimination

The major excretory pathway is the biliary system; about 65% of the nabilone dose is excreted in the faeces and 20% in the urine as polar, acidic metabolites.

THE FUTURE FOR CANNABIS AS A DRUG FOR PAIN RELIEF

Pharmacologically, cannabis is referred to as a "dirty drug". It contains many active compounds that have many effects, the mechanisms of which are barely known. It is tempting to consider all these compounds a hindrance to development; but they must be thought of as exciting opportunities. A major advantage of cannabis is its safety and low potential for physical dependency. The disadvantages are the side effects, its reputation and its legal status, which inhibits research.

Separating the analgesic activity of the cannabinoids from their significant side effects has been long sought. In the 1970s, Wilson and May suggested that separation was possible which spurred plenty of interest (Wilson and May, 1975). Unfortunately the only separation achieved so far is to produce derivatives that still maintain the psychopharmacological effects but have no analgesic properties. (Reggio *et al.*, 1991)

It is known that THC has an analgesic effect, but there are many other cannabinoids in cannabis that can contribute to its pain relieving effects. Cannabidiol, which is devoid of cannabimimetic effects is a potent cyclo-oxygenase inhibitor and analgesic but has been overlooked as studies have concentrated on THC and its derivatives. Metabolites of the cannabinoids may be contributing to the analgesia by an indirect route, e.g. by increasing the permeability of the blood brain barrier to other mediators.

There are problems associated with the design of clinical trials of these agents. Pain relief is a subjective sensation and is more difficult to evaluate than for example, blood pressure. Obtaining consistent samples of cannabis is a problem.

The ideal derivative would be crystalline, stable at room temperature, soluble in water and have the beneficial effects of the cannabinoids but none of the side effects. A suitable pharmaceutical form would preferably have the rapid effects of smoking cannabis. Possibilities could be a cannabinoid in an aerosol preparation which could incorporate Patient Controlled Analgesia technology, thereby regulating the rate of delivery.

CONCLUSION

There is only one route forward for the advancement of the medicinal use of cannabis for pain relief and it is the slow route followed by all other prospective drugs; to identify all the active components of the cannabis plant (smoked and unsmoked), isolate them and carry out properly planned clinical trials with a suitable route of delivery. Cannabis and nabilone are unlikely to have a dramatic part in the physician's repertoire of analgesics as a whole, but it will provide pain relief in some cases when all else has failed.

REFERENCES

Agrawal, A.K., Kumar, P., Galati, A. and Seth, P.K. (1989) Cannabis induced neurotoxicity in mice: effect on cholinergic receptors and blood brain barrier permeability. *Res. Commun. Subst. Abuse*, **10**, 155.

Bhattacharya, S.K. (1986) The antinociceptive effect of intracerebroventricularly administered prostaglandin D2 in the rat. *Psychopharmacology*, **89**, 121.

Bidaut-Russell, M., Devane, W.A. and Howlett, A.C. (1990) Cannabinoid receptors and the modulation of cyclic AMP accumulation in the rat brain. *J. Neurochem.*, **55**, 21.

Bisher, H.I. and Mechoulam, R. (1968) Pharmacological effects of two active constituents of marihuana. *Arch. Int. Pharmacodyn.*, **172**, 24.

Bloom, A.S. and Dewey, W.L. (1978) A comparison of some pharmacological actions of morphine and THC in the mouse. *Psychopharmacology*, **57**, 243.

Burstein, S., Levin, E. and Varanell, C. (1973) Prostaglandins and cannabis – II. Inhibition of biosynthesis by the naturally occurring cannabinoids. *Biochem. Pharmacol.*, **22**, 2905.

Buxbaum, D., Sanders-Bush, E. and Efron, D. (1969) Analgesic activity of tetrahydrocannabinol (THC) in the rat and mouse. *Fed. Proc.*, **28**, 735.

Chesher, G.B., Dahl, C.J., Everingham, M., Jackson, D.M., Marchant-Williams, H. and Stamer, G.A. (1973) *Br. J. Pharmac.*, **49**, 588–594.

Chesher, G.B. (1980) Cannabis and the opiate receptor – any link? *Proc. Austr. Physiol. Pharmacol. Soc.*, **11**, 24.

Christison, A. (1851) On the natural history, action and uses of Indian hemp. *Mon. J. Med. Sci.*, **13**, 26 and 177.

D'Amour, F.E. and Smith, D.L. (1941) A method for determining loss of pain sensation. *J. Pharmacol. Exp. Ther.*, **72**, 74.

Devane, W.A., Hanus, L., Breuer, A., Pertwee, R.G., Stevenson, L.A., Griffin, G., Gibson, D., Mandelbaum, A., Etinger, A. and Mechoulam, R. (1992) Isolation and structure of a brain constituent that binds to the cannabinoid receptor. *Science*, **258**, 1946–1949.

Devane, W.A., Dysarz, F.A., Johnson, M.R., Melvin, L.S. and Howlett, A.C. (1988) Determination and characterisation of a cannabinoid receptor in rat brain. *Mol. Pharmacol.*, **34**, 605–613.

Devane, W.A. and Axelrod, J. (1994) Enzymatic synthesis of anandamide, an endogenous ligand for the cannabinoid receptor, by brain membranes. *Proc. Nat. Acad. Sci.*, **91**, 6698–6701.

Dewey, W.L., Harris, L.S., Howes, J.F. and Nuite, J.A. (1970) The effect of various neurohumoral modulators on the activity of morphine and the narcotic antagonists in the tail-flick and phenylquinone tests. *J. Pharmacol. Exp. Ther.*, **175**, 435.

Dewey, W.L., Harris, L.S. and Kennedy, J.S. (1972) Some pharmacological and toxicological effects of 1-trans-8 and 1-trans-9 tetrahydrocannabinol in laboratory rodents. *Arch. Int. Pharmacodyn.*, **196**, 133–145.

Di Marzo, V., Fontana, A., Cadas, H., Schinelli, S., Cimino, G., Schwartz, J.C. and Piomelli, D. (1994) Formation and inactivation of endogenous cannabinoid anandamide in central neurons. *Nature*, **372**, 686–691.

Donovan, M. (1845) On the physical and medicinal qualities of Indian hemp (Cannabis indica), *Dublin J. Med. Sci.*, **26**, 368 and 459.

Eddy, N.B. and Leimbach, D. (1953) Synthetic analgesics. II. Dithienylbuteryl and dithienyl-butylamines. *J. Pharmacol. Exp. Ther.*, **107**, 385.

Fride, E. and Mechoulam, R. (1993) Pharmacological activity of the cannabinoid receptor agonist, anandamide, a brain constituent. *Eur. J. Pharmacol.*, **231**, 313–314.

Gaoni, Y. and Mechoulam, R. (1964) Isolation, structure and partial synthesis of an active constituent of hashish. *J. Am. Chem. Soc.*, **86**, 1646–1647.

Ghosh, P. and Bhattacharya, S.K. (1979): Cannabis-induced potentiation of morphine analgesia in rat – Role of brain monoamines. *Ind. J. Med. Res.*, **70**, 275.

Hill, S.Y., Scherwin, R., Goodwin, D.W. and Powell, B.J. (1974) Marihuana and pain. *J. Pharmacol. Exp. Ther.*, **88**, 415–418.

Huestis, M. A., Henningfield, J. E. and Cone, E. J. (1992) Blood cannabinoids. 1. Absorption of THC and formation of 11-OH-THC and THCCOOH during and after smoking Marijuana. *J. Anal. Toxicol.*, **16**, 276–82.

International Cannabis Research Society Meeting Summary, Keystone, CO (June 19–20, 1992) In: *Drug and Alcohol Dependence (1993)*, **31**, 219–227.

Jain, A.K., Ryan, J.R., McMahon, F.G. and Smith, G. (1981) Evaluation of intramuscular levonantradol and placebo in acute post-operative pain. *J. Clin. Pharmacol.*, **21**, S320–326.

Lemberger, L., Silberstein, S.D., Axelrod, J. and Kopin, I.J. (1970) Marihuana: Studies of the disposition and metabolism of delta-9-tetrahydrocannabinol in man. *Science*, **17**, 1320–1321.

Lichtman, A.H. and Martin, B.R. (1991) Cannabinoid-induced antinociception is mediated by a spinal alpha-2 noradrenergic mechanism. *Brain Research*, **559**, 309–314.

Martin, B.R. (1985) Characterisation of the antinociceptive activity of intravenously administered THC in mice. In: D.J. Harvey (ed.), *Marihuana 84*. IRL Press, Oxford, 685.

Mechoulam, R. (1984) Recent advances in the use of cannabinoids as therapeutic agents. In: W. Agurell, W.L. Dewey and D. Willette (eds.), *The Cannabinoids: Chemical, Pharmacologic and Therapeutic Aspects*, Academic Press Inc., New York, pp. 777–793.

Mechoulam, R. (1986) *Cannabinoids as Therapeutic Agents*. CRC Press. Boca Raton Florida. pp. 106–120.

Mechoulam, R. (1995) Pharmacology of cannabinoid receptors. *Ann. Rev. Pharmacol. Toxicol.*, **35**, 607–34.

Milne, G.M., Koe, B.K. and Johnson, M.R. (1980) Stereospecific and potent analgetic activity for nantradol – a structurally novel, cannabinoid – related analgetic. *NIDA Res. Monograph Ser.*, **27**, 84–92.

Milne, G.M. and Johnson, M.R. (1981) Levonantrodol: a role for central prostanoid mechanisms. *J. Clin. Pharmacol.*, **21**, 367.

Musty, R.E., Reggio, P. and Consroe, P. (1995) A review of recent advances in cannabinoid research and the 1994 international symposium on cannabis and the cannabinoids. *Life Science*, **56**(23/24), 1933–1940.

Narimatsu, S., Watanabe, K., Matsunaga, T., Yamamoto, I., Imaoka, S., Funae, Y. and Yoshimura, H. (1990) Inhibition of hepatic microsomal cytochrome P450 by cannabidiol in adult male rats. *Chem. Pharm. Bull.*, **38**, 1365.

Noyes, R., Brunk, S., Baram, D. and Canter, A. (1975). Analgesic properties of delta-9-THC and codeine. *Clin. Pharmacol. Ther.*, **18**, 84–89.

O'Shaugnessy, W.B., (1841) *Cannabis in the Bengal Dispensatory and Pharmacopoeia*, Bishop's College Press, Calcutta, p. 579.

O'Shaugnessy, W.B., (1843) On the Cannabis indica or Indian hemp, *Pharmacol. J. Trans.*, **2**, 594.

Pertwee, R.G. (1988) The central neuropharmacology of psychotropic cannabinoids. *Pharmac. Ther.*, **36**, 189–261.

Pertwee, R., Griffin, G., Fernando, S., Li, X., Hill, A. and Markriyannis, A. (1995) AM630, a competitive cannabinoid receptor antagonist. *Life Sciences*, **56**(23/24), 1949–1955.

Raft, D., Gregg, J., Ghia, J. and Harris, L. (1977) Effect of intravenous tetrahydrocannabinoids on experimental and surgical pain. *Clin. Pharmacol. Ther.*, **21**, 26–33.

Razdan, R.K. (1986) Structure activity relationships in cannabinoids. *Pharmacol. Rev.*, **38**, 75–149.

Reggio, P.H., McGaughey, G.B., Odear D.F., Seltman, H.H., Compton, D.R. and Martin, B.R. (1991) A Rational Search for the Separation of Psychoactivity and Analgesia in Cannabinoids. *Pharmacol. Biochem. Behaviour*, **40**, 479–486.

Seth, R. (1991) Chemistry and pharmacology of cannabis. *Progr. Drug Res.*, **36**, 71–115.

Smith, P.B., Compton, D.R., Welch, S.P., Razdan, R.K., Mechoulam, R. and Martin, B.R. (1994) The pharmacological activity of anandamide, a putative endogenous cannabinoid, in mice. *J. Pharmacol. Exp. Ther.*, **270**, 219–227.

Sofia, R.D., Nalepa, S.D., Harakal, J.J. and Vassar, H.B. (1973) Anti-edema and analgesic properties of Δ^9 THC. *J. Pharmacol. Exp. Ther.*, **186**, 646–655.

Sofia, R.D., Vasser, H.B. and Knoblock, L.C., (1975) Comparative analgesic activity of various naturally occuring cannabinoids in mice and rats. *Psychopharmacologia (Berlin)*, **40**, 285.

Vela, G., Ruiz-Gayo, M. and Fuentes J.A. (1995) Anandamide decreases naloxone-precipitated withdrawal signs in mice chronically treated with morphine. *Neuropharmacology*, **34**(6), 665–668.

Welch, S.P. and Stevens, D.L. (1992) Anti-nociceptive activity of intrathecally administered cannabinoids alone, and in combination with morphine, in mice. *J. Pharmacol. Exp. Ther.*, **262**, 10–18.

Williams, P.I. and Higgs, A. (1995) Effect of nabilone on nausea and vomiting. *Br. J. Anaesth.*, **74**(1), 111–112.

Wilson, R.S. and May, E.L. (1975) Analgesic properties of the tetrahydrocannabinoids, their metabolites and analogues. *J. Med. Chem.*, **18**, 700–703.

9.   CANNABIS ADDICTION AND WITHDRAWAL: ATTITUDES AND IMPLICATIONS

DAVID E. SMITH and RICHARD B. SEYMOUR

Haight Ashbury Free Clinics Inc., San Francisco, USA

INTRODUCTION

For a variety of reasons, cannabis, or marijuana is the most controversial illicit drug in use today. Its defenders maintain that marijuana is harmless, or at least less harmful than other drugs, including alcohol and tobacco, while its more virulent attackers claim that it causes numerous sequelae, including birth defects more horrendous than those that resulted from thalidomide use. Much rhetoric has been expounded by both sides of the controversy. Unfortunately, even though there is ample evidence that cannabis is a dangerous substance, the extreme claims from both ends of the spectrum have tended to cloud the clinical reality of marijuana use, pharmacology and neurochemistry.

The confusion of claims about marijuana are also inhibiting prevention efforts in the United States, resulting in increased use among teenagers. The 1994 National Household Survey on Drug Abuse, compiled by the National Institute on Drug Abuse (NIDA) reported that about 7.3% of American teenagers, about 1.3 million between the ages of 12 and 17, smoked marijuana last year. This figure is up 4% from the 1992 Survey. It was noted that until 1992, marijuana use had declined every year since 1979 (NIDA, 1994). President Clinton's national drug policy co-ordinator reported that marijuana accounts for 81% of the United States' illicit drug use, and its rise among teenagers reflects a growing sense that marijuana is benign. Only 42% of teenagers consider marijuana a dangerous drug. This change in attitude, combined with the reversal in teen marijuana use, suggests that marijuana use will remain a significant public health and policy issue for the foreseeable future.

The danger of marijuana use has been further exacerbated in recent years by its increase in potency since the 1960s and 1970s. In the past, the most common marijuana cigarette smoked in the United States contained approximately 10 mg of delta-9-tetrahydrocannabinol (THC). More recently, with the advent of selective breeding, there has been an increase in the THC content and potency of marijuana cigarettes. In addition, the use of "hash oil," a concentrated tincture of hashish, with the marijuana leaf significantly increases the amount of THC in the combination cigarette (Gold, 1994, 1989). This change in marijuana potency has been a primary factor in transforming the low-dose, self-experimentation of marijuana use typical in the 1960s to high-potency, high-reward/reinforcement marijuana use and dependence prevalent in the 1990s. Properties that increase reinforcement potential, such as rapid absorption, high intrinsic pharmacological activity of a drug, and rapid entry into specific regions in the brain, are present in the high potency THC of the 1990s. Similarly, the factors that favour physical dependence, such as long half-life, low clearance, cumulative drug load, and high intrinsic pharmacological activity, are also present.

THE NEUROPHARMACOLOGY OF MARIJUANA

While marijuana can be eaten, the most common mode of marijuana self-administration is by smoking and inhalation. Marijuana smoke contains more than 150 compounds in addition to the major psychoactive component, THC. Many of the cannabinoids and other complex organic compounds appear to have psychoactive properties and others have not been tested for long- or short-term safety in animals or human beings.

Following volatalisation of THC by burning of a cigarette and deep inhalation, the pharmacokinetics of marijuana are complex. Cannabinoids are rapidly absorbed from the lungs and THC and major metabolites can be traced throughout the body and brain. In the past, there had been some debate over whether marijuana's main constituent, THC, acts directly on the central nervous system (CNS) or whether it must first be metabolized to 11-hydroxy THC. It appears that THC is directly psychoactive. (Lemberger et al., 1976)

When marijuana is smoked, it appears to produce its psychoactive effects through specific binding with endogenous "THC" receptors. Radioligand binding studies with a water-soluble cannabinoid have revealed high-affinity sites in the brain that are specific for cannabinoids and that can be inhibited by myelin-basic protein in the rat. (Nye et al., 1988). Anandamide (the name given to the structure of arachidonylethanolamide, an arachidonic acid derivative in the porcine brain) has recently been shown to inhibit the specific binding of a radiolabeled cannabinoid probe to synaptosomal membranes in a manner typical of competitive ligands. This effect produces a concentration-dependent inhibition of the electrically evoked twitch response of the mouse vas deferens, a characteristic effect of psychotropic cannabinoids. These properties suggest that anandamide may function as a natural ligand for the cannabinoid receptor (DeVane et al., 1992). In the first in vivo examination of anandamide, Fride and Mechoulam (1993) reported that it produced hypothermia and analgesia, effects that parallel those caused by psychotropic cannabinoids. As with previous endogenous endorphin research, the definition of the endogenous neurochemical process via the identification of a THC receptor and its ligand helps to explain marijuana's analgesic, antinausea, concentration and amnesic effects, by showing that the drug has an affinity to areas in the brain involving pain and nausea control, and cerebral activities such as memory and concentration. Thomas et al. (1992) reported that the cannabinoid binding of two ligands was densest in the basal ganglia and cerebellum (molecular layer), with intermediate binding in Layers I and VI of the cortex, and the dentate gyrus and CA-pyramidal cell regions of the hippocampus. The identification of a THC receptor and its ligand also suggests the possible development of new pharmacological treatments for marijuana abuse. Recent research with a THC receptor antagonist helps establish that cannabis dependence exists and may lead to a therapeutic agent, similar to the development of naltrexone, an opioid receptor antagonist agent which is both an opioid blocker and an anticraving agent for alcohol.

Further verification of the THC receptor has recently been announced by the American National Institute on Drug Abuse (NIDA) (Swan, 1995; Aceto *et al.*, 1995). Experiments involving the application of a THC antagonist, SR 141716A produced a dramatic withdrawal syndrome in rats. According to senior investigator Billy Martin, M.D., "The fact that people do seek treatment for marijuana dependence is evidence of marijuana withdrawal in humans." Noting that withdrawal in humans is usually long and drawn out, he added, "But with rats, using SR 141716A as an effective antagonist, we compress and accentuate that withdrawal process."

THC is highly lipid-soluble and a complex relationship exists between THC, which can be measured in the blood after self-administration, and that which is transferred rapidly into lipid and other areas of the central and peripheral nervous system. Direct correlations between self-reports of euphoria and blood levels have been hindered by this relationship and the metabolism of THC in the liver to 11-hydroxy-THC and 11-nor-carboxy THC (Wall *et al.*, 1983) and tens of other metabolites with psychoactive properties. THC and THC metabolites are primarily excreted in the feces. The slow release of THC and active metabolites from lipid stores and other areas may explain the so-called carry-over effects on driving and other reports of behavioural changes over time. THC is stored in body fat and its slow excretion may make the urine test positive for more than 30 days, particularly if the individual is a chronic abuser. If the person is subject to drug testing in industry, the urine test can be positive, thereby putting the person's job in jeopardy, since the cut-off levels in industry are relatively low, i.e. 50 µg to 100 ng per ml. Even relatively low levels of use can be detected. This issue is further complicated by the fact that a prescription drug, Marinol (synthetic THC), used for glaucoma and the nausea associated with cancer chemotherapy, can make the urine screen positive for THC. In drug testing wherein an employee has been referred to the company's Employee Assistance Program for evaluation, including urinalysis, a medical review officer (MRO), usually a physician retained by the company to review all cases in which a potentially positive urine is reported, is required to make the distinction between medical use and illicit use. (Seymour and Smith, 1990, 1994; Clark, 1990)

EFFECTS OF MARIJUANA

Intoxication is similar to other drugs. Marijuana is taken for the euphoria or high. Marijuana is self-administered by laboratory animals and appears to have effects on the putative reward neuroanatomy similar to that of other drugs of abuse (Gardner and Lowinson, 1991). One recent study found that THC treatment (like dopamine (DA) agonists) caused a decline in plasma prolactin levels accompanied by a decreased DOPAC/DA ratio in the medial basal hypothalamus, indicating that acute exposure to THC can augment brain DA neurotransmission (Rodriguez de Fonseca *et al.*, 1992). In addition, THC binds with the mu-receptor, or an opioid receptor subtype stimulated by morphine. Chronic mu-decreased activity could cause locus ceruleus hyperactivity during withdrawal (Gold and Miller, 1992). Furthermore, the opiate antagonist naloxone has been shown in animals to attenuate the enhanced dopamine (DA) levels associated with THC administration (Chen *et al.*, 1989). Again, opioid receptor interactions appear to be important for marijuana to exert its effects. While future studies of THC and its receptors in the brain will change our understanding of marijuana reinforcement, reward and withdrawal, naloxone's alteration of THC effects suggest that marijuana engages endogenous brain opioid circuitry, causing an association between these endogenous opioids and DA neurones that appears fundamental to marijuana's euphoric effects.

MARIJUANA WITHDRAWAL SYNDROME

The identification of a THC receptor and ligand, and the recognition of the effect retarded release of THC and its active metabolites may have on carry-over effects, may be of help in identifying and understanding a marijuana withdrawal syndrome. As recently as the publication of the American Society of Addiction Medicine's Principles of Addiction Medicine in 1994, authors in that work were reporting that, aside from mild increases in heart rate, blood pressure, and body temperature, no clinically significant physiological withdrawal syndrome associated with discontinuation of marijuana use had been identified (Wilkins and Gorelick, 1994).

These same authors point out that "psychological manifestations" of marijuana withdrawal may include anxiety, depression, irritability, insomnia, tremors, and chills. They add that these symptoms usually only last a few days, but subtle symptoms can persist for weeks. It seems curious to us that such symptoms as tremors and chills should be seen as psychological manifestations with no physiological basis, and one would hope that an increasing understanding of receptor site science, as it applies to marijuana, will help clarify the nature of marijuana withdrawal.

Ciraulo and Shader (1991) state that marijuana withdrawal syndromes are likely to resemble those commonly associated with ethanol withdrawal. They add that individuals chronically using cannabis are at high risk for such serious physical illnesses as pulmonary disease and also at high risk for developing concomitant polydrug abuse problems.

Our own experience at the Haight Ashbury Free Clinics and clinical discussions with colleagues who are treating marijuana users suggests the presence of a prolonged withdrawal syndrome, primarily characterised by anxiety and insomnia. We have also seen the onset of depression during withdrawal, particularly in adolescents suffering from motivation impairment manifested in learning difficulty and family relation problems during their marijuana use (Smith and Seymour, 1982). Ongoing experience at the Haight Ashbury Free Clinics has verified these findings.

Chronic marijuana users often fit the profile for addictive disease, characterised by compulsion, loss of control, and continued use in spite of adverse consequences. In recovery, these individuals may respond well to such supported recovery fellowships as Alcoholics Anonymous (AA) and Marijuana Anonymous (MA). Subtle withdrawal symptoms may persist for extended periods of time, however, and it is not uncommon to hear chronic marijuana smokers in long-term recovery comment that it was several years into abstinence and sobriety before they were truly aware of the adverse effects marijuana had on their thinking and behavior.

CONCLUSION: ON THE QUESTION OF MARIJUANA LEGALISATION

Having reviewed the neuropharmacology and other evidence pointing to the existence of a THC withdrawal syndrome, we will conclude by turning to the issue of marijuana legalisation and present our views on that volatile subject as well as our reasons for arriving at the conclusion that marijuana should not be legalised.

The proponents of legal reform pertaining to marijuana maintain that its use among adults should not be curtailed any more than that of alcohol or tobacco. Our own opinion that no currently illicit drug, including marijuana, should be placed on the open market as a legal substance for non-medical purposes, is based on public health considerations, and is in no small way influenced by the fact that the currently legal drug "tobacco" is responsible for over 400,000 deaths a year in the United States. In addition, we have great concern that, if marijuana was to be legalised, it would be distributed by the tobacco industry and marketed aggressively to youth in a fashion similar to that in which cigarettes are currently marketed. According to the National Institute on Drug Abuse (NIDA), "Scientists at the University of California, Los Angeles, found that daily use of one to three marijuana joints appears to produce approximately the same lung damage and potential cancer risk as smoking five times as many cigarettes" (NIDA Capsules, 1993). Is it wise to give ourselves permission to freely use yet another drug that is a known pulmonary toxin and potential carcinogen?

While the effects of chronic marijuana use on the lungs and pulmonary system are the most obvious physical threat from cannabis, these may just be the tip of the iceberg. Numerous medical studies suggest that marijuana may act as a general immunosuppressant, while clinical observation confirms a potential for young people who start using tobacco and alcohol at an early age are at greater risk for moving on to marijuana use and addiction to such drugs as heroin and cocaine (DuPont, 1984).

While we feel that the escalation of criminal penalties is not the answer for discouraging the use of marijuana among young people, we strongly support the concepts of education regarding the dangers of marijuana and other psychoactive drugs, early intervention, diversion to treatment from the criminal justice system, and treatment on demand for marijuana dependence. Marijuana withdrawal is less intense but more prolonged than alcohol or opiate withdrawal. It is characterised by anxiety, sleep dysfunction and drug craving. The Haight Ashbury Free Medical Clinics utilises its symptomatic medication protocol using non-addictive medications such as trazadone for sleep. Symptomatic medication is used in conjunction with individual and group counselling, as well as group recovery support. Understanding the dangers, education, prevention, and treatment represent for us the true course of a realistic programme of "harm prevention" based on clinical and public health realities.

REFERENCES

Aceto, M.D., Scates, S.M., Lowe, J.A. and Martin, B.R. (1995) Cannabinoid-precipitated withdrawal by a selective antagonist: SR 141716A. *Eur. J. Pharmacol.* **282**(1–3), Rl–R2.

Chen, J., Paredes, W., Li, J., Smith, D. and Gardner, E.L. (1989) *In vivo* brain microdialysis studies of delta 9-tetrahydrocannabinol on presynaptic dopamine efflux in nucleus accumbens of the Lewis rat. *Soc. Neurosci. Abs.*, **15**, 1096.

Ciraulo, D.A. and Shader, R.I. (1991) *Clinical Manual of Chemical Dependence*, American Psychiatric Press, Inc., Washington/London, p. 186.

Clark, H.W. 1990 The medical review officer and workplace drug testing. *J. Psychoact. Drugs*, **22**(4), 435–445.

DeVane, W.A., Hanu, L. and Bruer, A. (1992) Isolation of a brain constituent that binds to the cannabinoid receptor. *Science*, **258**, 1946–1949.

DuPont, R.L.(1984) *Gateway Drugs*. American Psychiatric Press, Inc., Washington/London.

Fride, E. and Mechoulam, R. (1993) Pharmacological activity of the cannabinoid receptor agonist anandamide, a brain constituent. *Eur. J. Pharmacol.*, **23**(2), 313–314.

Gardner, E.L. and Lowinson, J.H. (1991) Marijuana's interaction with brain reward systems: Update 1991. *Pharmacol. Biochem. Behav.*, **40**, 571–580.

Gold, M.S. (1994) Marijuana. In: Miller, N.S. (ed.) *Principles of Addiction Medicine*, American Society of Addiction Medicine, Washington.

Gold, M.S. (1989) *Marijuana. Drugs of Abuse: A Comprehensive Series for Clinicians*, Vol. I, Plenum Publishing, New York.

Gold, M.S. and Miller, N.S. (1992) Seeking drugs/alcohol and avoiding withdrawal: The neuro-anatomy of drive states and withdrawal. *Psychiat. Ann.*, **22**(8), 430–435.

Lemberger, L., McMahon, R. and Archer, R. (1976) The role of metabolic conversion on the mechanisms of action of cannabinoids. In Braude, M.S. and Szara, S. (eds.) *Pharmacology of Marihuana*, Vol. 1, Raven Press, New York, pp. 125–133.

National Institute on Drug Abuse (1994) *The 1994 National Household Survey*, The Institute, Rockville, MD.

National Institute on Drug Abuse (1993) *Marijuana update*. NIDA Capsules, Revised September 1993, The Institute, Rockville, MD.

Nye, J.S., Voglmaier, S., Martenson, R.E. and Snyder, S.H. (1988) Myelin basic protein is an endogenous inhibitor of the high-affinity cannabinoid binding site in brain. *J. Neurochem.*, **50**(4), 1170–1178.

Rodriguez De Fonseca, F., Fernandez-Ruiz, J.J., Murphy, L.L., Cebeira, M., Steger, R.W., Bartke, A. and Ramos, J.A. (1992) Acute effects of delta-9-tetrahydrocannabinol on dopa-minergic activity in several rat brain areas. *Pharmacol. Biochem. Behaviour*, **42**(2), 269–275.

Seymour, R.B. and Smith, D.E. (1994) Controlling substance abuse in the workplace. In: LaDou, J. (ed.) *Occupational Health & Safety*. 2nd Edition, National Safety Council, Itasca, IL, pp. 287–306.

Seymour, R.B. and Smith, D.E. (1990) Identifying and responding to drug abuse in the work-place. *J. Psychoact. Drugs*, **22**(4), 383–405.

Smith, D.E. and Seymour, R.B. (1982) Clinical Perspectives on the Toxicity of marijuana: 1967-1981. In: *Marijuana and Youth: Clinical Observations on Motivation and Learning*, National Institute on Drug Abuse, Rockville, MD, pp. 61–72.

Swan, N. (1995) Marijuana antagonist reveals evidence of THC dependence in rats. *NIDA Notes*, **10**(6), 1 and 6.

Thomas, B.F., Wei, X. and Martin, B.R. (1992) Characterization and autoradiographic localization of the cannabinoid binding site in rat brain using [3H]-OH-delta 9-THC-DMH. *J. Pharmacol. Exp. Res.* **263**(3), 1383–1390.

Wall, M.E., Sadler, B.M. and Brine, D. (1983) Metabolism, disposition and kinetics of delta-9-tetrahydrocannabinol in men and women. *Clin. Pharm. Ther.*, **34**, 352–363.

Wilkins, J.N. and Gorelick, D.A. (1994) Management of phencyclidine, hallucinogen and marijuana intoxication and withdrawal. In: Miller, N.S. (ed.) *Principles of Addiction Medicine*, American Society of Addiction Medicine, Washington.

10. SIDE EFFECTS OF CANNABIS USE AND ABUSE

SIMON WILLS

Head of Drug Information Service, St Mary's Hospital, Portsmouth, UK

Information on the adverse effects of cannabis has accrued over the past few decades in a piecemeal fashion. Unfortunately, systematic, well-conducted research on the toxicology of the drug in humans is comparatively rare. In most cases, animal studies are of dubious relevance to human toxicology, and this review concentrates almost exclusively on investigations involving humans. Most information has come from small scale investigations, case reports and anecdotal evidence. Many studies assume that the adverse effects of smoking cannabis and the adverse effects of oral or intravenous delta-9-tetrahydrocannabinol or THC (the main psychoactive constituent) are equivalent. This is not necessarily correct since, as discussed in other chapters, there are many other cannabinoids present in the plant. In addition, certain adverse effects of cannabis may depend upon the route of administration. Smoking gives rise to very rapid, and high, plasma and CNS levels of THC. This, for example, produces a different quality of psychoactive effect to that produced by oral administration which has a slow onset, but provides more sustained plasma levels of THC. Smoked cannabis produces adverse effects upon the lungs which do not occur when THC is given orally.

Cannabis varies considerably in potency and purity. These two factors can have a marked affect upon the pharmacological properties of a particular sample. Even users are unlikely to know the dose that they have taken, and dosage can significantly affect the likelihood of adverse effects. In addition, many of those who use cannabis take other substances at the same time which can be hard to identify. Even when these are known, assessing the role of other street drugs in the aetiology of a particular patient's symptoms can be difficult. All of this makes determination of the side effect profile of cannabis less than easy. The picture is further complicated by the fact that the subject of the adverse effects of cannabis can form part of a political or moral agenda, with inevitable bias in interpretation of data.

ACUTE INTOXICATING EFFECTS

Cannabis is used socially for its intoxicating effects, which can develop quickly. THC is particularly lipophilic and when smoke laden with it enters the lungs, THC dissolves rapidly in pulmonary surfactant. This, in turn, facilitates fast absorption across blood vessels into plasma. The high lipophilicity also allows unbound THC to penetrate the blood brain barrier readily. Consequently THC reaches the brain within minutes of drawing on a cannabis cigarette and, as a result, the psychoactive effects begin very shortly after smoking has commenced.

In the case of oral administration, the onset is delayed by one to three hours after ingestion. The onset is more rapid if the stomach is empty. Extensive first pass metabolism and rapid distribution into adipose tissue, markedly reduce the availability of oral THC to the brain. Whatever the route of administration, the initial amounts of THC which reach the CNS are surprisingly small as a proportion of the total dose administered. This is because of the extensive protein binding of THC (about 97%) which impedes crossing of the blood–brain barrier.

The subjective effects of cannabis which are sought by users are those which affect the brain. The intensity, duration and precise nature of these effects varies widely between individuals, and even within the same individual, depending upon several factors. The principal determinants involved are the dose, the method of administration, the user's emotional state and the concomitant use of other mind-altering substances.

The psychotropic actions of cannabis are dose-dependent and the dose itself is, in turn, affected by the form of cannabis used and its purity. The proportion of THC in samples of street cannabis varies. Dried cannabis herb typically contains approximately 0.5–2.5% THC, while cannabis resin taken directly from the plant with no further treatment can contain up to 8–10%. Concentrated liquid resin may be over 60% THC.

Other factors affecting the subjective experience of cannabis intoxication include the route of administration. As discussed above, the pharmacokinetics of smoked and oral cannabis affect the time course and intensity of the drug's effects. Cannabis, like many psychoactive substances can amplify emotions which are prevalent at the time of intoxication. This is in itself determined by such factors as the environment of abuse, the presence of others, the expectations of the user, the user's previous experience of cannabis (if any), and the degree of apprehension or anxiety. The symptoms of pre-existing psychiatric illness can be exacerbated (see below). It is often stated that the paranoia which is characteristic of many dysphoric experiences, is related to the individual's apprehension about using an illicit substance (i.e. breaking the law).

The concomitant use of other psychoactive drugs with cannabis can affect the nature of the intoxication produced. The interactions between cannabis and other specific drugs are discussed in more detail below. However, it is impossible to predict the subjective experience of intoxication with combinations of psychotropic substances. The outcome of mixing psychotropic drugs is very variable; knowledge of the effects of each drug individually may lead one to anticipate or identify the effects within an individual which have been produced by a single agent, but often this is not possible.

Features of cannabis intoxication vary widely, but some common experiences are described briefly below:

(1) Dizziness, nausea, tachycardia, facial flushing, dry mouth and tremor can occur initially. These features are similar to sympathetic nervous system arousal.
(2) Merriment, happiness, and even exhilaration at high doses.
(3) This is succeeded by disinhibition, relaxation, increased sociability and talkativeness.
(4) Enhanced sensory perception giving rise, for example, to increase appreciation of music, art and touch.
(5) Heightened imagination leading to subjectively increased creativity.
(6) Time distortions may occur such that time appears to have been slowed down.

(7)   Illusions, delusions and hallucinations are rare except at high dosage.

(8)   Impaired judgement, reduced co-ordination and ataxia, which could impede driving ability and render accidents and risk-taking more likely.

(9)   Emotional lability, incongruity of affect, dysphoria, disorganised thinking, inability to converse logically, agitation, paranoia, confusion, restlessness, anxiety, drowsiness and panic attacks can occur.

(10) Increased appetite and short-term memory impairment are common.

Intoxication persists for a variable length of time, largely determined by dose and method of administration. Whilst the psychoactive effects of oral cannabis can last for as long as eight hours, smoking cannabis typically produces a shorter intoxication of up to four hours.

Initially THC distributes rapidly into adipose tissue, with a half-life for plasma clearance of approximately 4 hours. This is followed by a much longer phase of elimination during which THC slowly partitions out of fatty tissues with a beta half-life of 25–36 hours. The complete clearance of one dose can require many weeks.

CARDIOVASCULAR EFFECTS

One of the most common adverse effects of cannabis is sinus tachycardia which can last for several hours. Sometimes users complain of palpitations. Heart rate is typically increased by 20–50% depending upon individual circumstances. The exact mechanism is not clear, but the effect is antagonised by the beta-adrenergic blocker, propranolol. Cannabis causes release of adrenaline, has antimuscarinic properties and can increase peripheral blood flow. Tachycardia could be mediated by these mechanisms acting in concert – all of them involve the adrenergic system. Tolerance to tachycardia does occur to some extent after chronic use, but the response is never completely lost. There are remarkably few reports of cannabis-induced arrhythmias, although premature ventricular beats are occasionally observed.

Tachycardia could be harmful to those suffering from cardiac disorders such as angina or heart failure. Two studies have assessed the effects of cannabis smoking on the exercise tolerance of patients with angina (Aronow and Cassidy, 1974; Aronow and Cassidy, 1975). Both concluded that cannabis cigarettes decreased the amount of exercise needed to trigger anginal attacks by an average of 50%. By comparison, placebo cigarettes reduced exercise tolerance by only 9%, and high nicotine content cigarettes by 23%.

Acute cardiac toxicity has been described very rarely. There are a very small number of cases of myocardial infarction attributed to cannabis in the medical literature (Collins et al., 1985; Macinnes and Miller, 1984). However, infarction and cannabis smoking are both common, and their occurrence in the same person at the same time is probably no more than coincidence. Usually cannabis falls suspect because the patient experiencing infarction is younger than average, and investigators then begin to look for a cause.

Orthostatic hypotension is a well-known adverse effect of cannabis. It has been suggested that this occurs when cannabinoids bind to the peripheral cannabinoid receptor (CB1) and inhibit sympathetic tone (Varga *et al.*, 1995). Sometimes the hypotensive effect is dramatic enough to cause syncope (Merritt *et al.*, 1982). High doses produce a reddening of the eyes due to vasodilation. Facial flushing is also a known effect.

However, the dizziness reported by many cannabis users when standing upright after smoking cannabis may not be entirely due to hypotension. A study of ten young men revealed that reduced middle cerebral artery blood velocity could be identified in all subjects who reported moderate to severe dizziness after smoking cannabis and standing upright. Only those with the most severe dizziness also showed reduced blood pressure. The authors suggested that impaired cerebral autoregulation of bloodflow was important in causing dizziness (Mathew *et al.*, 1992).

ENDOCRINE EFFECTS

The effects of cannabis upon sexual performance are not clearly understood, and the published research available often yields conflicting results. Traditionally, high doses have been claimed to reduce libido and cause impotence, whilst more moderate intake has been associated with heightened pleasure from sexual contact (Buffman, 1982; Abel, 1981). If the drug has the ability to enhance the subjective experience of sex, then this might be due to cannabis-induced relaxation or increased tactile sensitivity.

Several animal studies have shown that THC reduces the secretion of testosterone (Abel, 1981). Kolodny *et al.* (1974) recruited 20 young men aged between 18 and 28 years who had used cannabis on at least four days per week for a minimum of six months. The number of cannabis cigarettes smoked per week ranged from 5 to 13; the average was 9.4. The average plasma testosterone level in the men under study was 416 ng/100 ml, compared to 742 ng/100 ml in 20 matched controls who had never used cannabis. This effect of cannabis appeared to be dose-dependent in that those who consumed ten or more cigarettes per week had a lower average testosterone (309 ng/100 ml) than those consuming nine or less (503 ng/100 ml). Human chorionic gonadotrophin was given to four of the men who continued to use cannabis. This produced an increase in plasma testosterone of between 121% and 269%. Six of the seventeen cannabis users who were tested exhibited markedly reduced sperm counts, and two were impotent. Three users discontinued cannabis for two weeks, and this resulted in a rapid rise in plasma testosterone.

Other researchers have failed to verify the findings of Kolodny's group. Seven separate studies involving over one hundred chronic cannabis smokers have individually demonstrated an absence of significant effect upon human testosterone levels (Block *et al.*, 1991; Cushman, 1975; Erdolu *et al.*, 1985; Hembree *et al.*, 1976; Mendelson *et al.*, 1974, 1978; Schaefer *et al.*, 1975). In 1976, Kolodny and colleagues studied the effects of cannabis upon human testosterone levels for a second time. On this occasion, the investigation was more rigorous. Plasma testosterone levels were significantly reduced at 30, 120 and 180 minutes after smoking cannabis, compared to levels measured in the same individuals during a non-smoking period. However, a study by Mendelson and co-workers in 1978, using a similarly rigorous design, again failed to reproduce these results.

It is unclear why Kolodny and associates have consistently reported that cannabis can reduce testosterone levels whilst other groups have not confirmed this finding. Testosterone plasma levels do show a very wide diurnal variation in man. Perhaps particularly high doses and/or very long duration of administration of cannabis is required. Even if cannabis can reduce testosterone levels by as much as Kolodny estimated, then the functional significance of this is unclear, since the levels reported are within the normally observed human male range for this hormone.

THC is able to trigger the growth of breast tissue in rats (Harmon and Aliapoulios, 1974). In 1972 the same authors reported three cases of human gynaecomastia which they attributed to heavy cannabis use. In 1980, Olusi reported three further cases in which the plasma prolactin levels were almost doubled in each case prompting the author to suggest a mechanism for the reaction. However, no other cases of cannabis-induced human gynaecomastia have been described in the medical literature, suggesting that the reaction is probably rare and that cannabis use might even be a coincidental finding.

Kolodny et al. (1974) reported that 6 of 17 men who habitually smoked cannabis had a reduced sperm count. This was also described in five men who were chronic users (Hembree et al., 1976). These individuals were required to discontinue cannabis for 14–21 days, before resuming for 4 weeks. Sperm count did not fall during the smoking phase, but did drop by an average of 58% during a two week post-exposure recovery phase. This was attributed to a direct action on spermatogenesis since reproductive hormone levels were unaffected; the delayed response was assumed to be due to the long time required for sperm to form in man. Another study has also reported decreased spermatogenesis in cannabis smokers (Erdolu et al., 1985). In none of these small studies was sperm count reduced to the levels of male infertility. By contrast, the research of Close and co-workers failed to find a link between cannabis use and decreased sperm count or motility (Close et al., 1990).

There is even less information available concerning the effect of cannabis upon human female reproductive function. In animals, cannabis can impair ovulation. Kolodny et al. (1979) studied the adverse effects of cannabis upon the human female reproductive cycle. Cannabis users had a greater proportion of cycles in which either ovulation did not take place, or in which the luteal phase was very short (38.3%) compared to non-users (12.5%). The duration of the human menstrual cycle was also measured. Cannabis smokers had an average cycle length of 26.8 days, compared to 28.8 days in non-users. However, cannabis smokers consumed significantly greater amounts of alcohol compared to controls, which is known to have detrimental effects upon the human menstrual cycle. Cannabis users were also more sexually active. Unexpectedly, another study has suggested that regular use of cannabis can significantly shorten the time to conception amongst fertile women (Joesoef et al., 1993).

Studies suggest that cannabis does not affect plasma levels of follicle-stimulating hormone or luteinizing hormone in men or women (Block et al., 1991; Erdolu et al., 1985; Kolodny et al., 1974). Most studies have found that cannabis does not usually affect human prolactin levels, although cases of human male hyperprolactinaemia attributed to cannabis are reported (Erdolu et al., 1985; Olusi, 1980). Cannabis does not seem to affect plasma levels of thyroid hormones, or glucocorticosteroids. The effects of cannabis upon glycaemic control are discussed below under interactions with anti-diabetic drugs.

NEUROLOGICAL AND BEHAVIOURAL EFFECTS

One of the most famous papers concerned with the adverse effects of cannabis was written by Campbell and colleagues, and published in 1971. This group investigated the structure of the CNS in chronic cannabis smokers, who had taken the drug regularly for at least three years. Ten males, with an average age of 22 years, were subjected to air encephalography to measure the size of brain ventricles. The results showed significant enlargement of the lateral and third ventricles in cannabis users compared to thirteen controls of a similar age. The changes were suggestive of cerebral atrophy.

The changes reported by Campbell and colleagues can be caused by many factors including birth trauma, alcoholism, head injury, psychomotor epilepsy, and CNS disease. In many cases, findings of this kind are apparently without an identifiable cause.

Contrary to the findings of Campbell and associates, studies of CNS structure using the more sophisticated and accurate method of computerised tomography, did not reveal physical evidence of brain damage in cannabis smokers (Co et al., 1977; Kuehnle et al., 1977). Cannabis is hence no longer suspected to be a cause of cerebral atrophy. Notwithstanding these findings, chronic use may be associated with a range of psychiatric or psychological problems (see below), which Campbell and colleagues had pointed to as supportive evidence for the widespread existence of cerebral atrophy amongst cannabis smokers.

In animals, THC has varying effects upon seizure frequency depending upon the dose given, the species involved, and whether the animal is prone to epilepsy or not. There is no consistent pattern of effect between species – THC being pro-convulsant in some animals and anticonvulsant in others. Cannabidiol seems only to have anticonvulsant actions.

Despite its very widespread usage for four decades, cannabis has been reported as the cause of human seizures in only one published case report (Keeler and Reifler, 1967). The authors described a young male epileptic who had not experienced a fit for 6 months, but who suffered three during a 21 day period in which he smoked seven cannabis cigarettes. The convulsions were not temporally linked to the administration of the drug. In 1976, Feeney described thirteen young epileptics who regularly smoked cannabis. Eleven of them had not noticed a change in seizure frequency since smoking cannabis, but one patient reported that the drug seemed to reduce the number of convulsions that he experienced; a second stated it increased seizure frequency. No further significant details were provided.

In the absence of supporting evidence, cannabis is unlikely to cause convulsions in humans at normal levels of consumption. The two cases above, where cannabis intake was cited as the cause of convulsions are probably coincidental. Consroe and colleagues reported a case in 1975 in which cannabis smoking was apparently necessary to prevent convulsions occurring. The patient was taking pentobarbitone and phenytoin to control epilepsy, with only partial success. He discovered that fit frequency was reduced significantly by smoking cannabis regularly. The authors hailed this as a direct effect of cannabis but, as discussed below, this could have been caused by THC reducing the metabolism of the barbiturate, thus increasing its therapeutic effect.

Some evidence suggests that if cannabis has any effect at all on human epilepsy, it may protect against convulsions (Ng *et al.*, 1990). Cannabinol has even been used for this purpose therapeutically (Carlini and Cunha, 1981).

Cannabis is known to have detrimental effects upon human short-term memory function. When intoxicated with cannabis, short-term memory is impaired in a similar manner to that observed following ingestion of alcohol. However, detailed investigation of the phenomenon is bedevilled by a host of methodological difficulties (Deahl, 1991). The first rigorous study of the effects of cannabis upon human memory was conducted by Schwartz *et al.* (1989). Ten chronic abusers of cannabis were matched to seventeen controls in terms of IQ and age. All were 14–16 years of age, middle class, well-educated and none had a history of learning disability. Eight of the controls were drug abusers who had not used cannabis chronically, and the remaining nine did not abuse any drugs. All of the participants were subjected to six neuropsychological tests. The chronic cannabis users had significantly impaired memory for two of the tests which assessed auditory and visual memory. This memory deficit had not significantly improved when repeated after six weeks abstinence from cannabis. The permanence of this effect, and the duration of cannabis usage required to attain it, is not known.

The practical significance of these "laboratory" observations to a cannabis user's quality of life is not clear. Although differences between cannabis users and non-users are apparent from testing, the tests are, of course, not "real life", and understanding what a small under-performance in a neuropsychological test really means is the key question. In most studies, memory is only impaired in certain tests, not all of them, and the retrieval of memories stored outside of the testing environment is not generally affected.

Cannabis usage may impair the ability to maintain concentration on certain tasks and avoid distractions, and the effect seems to persist in ex-users (Solowij, 1995). It has been suggested that cannabis could cause persistent effects upon attention and learning via one of three basic mechanisms:

(1) by a continuing pharmacological action caused by accumulation of drug in the brain accompanied by very slow excretion;
(2) as a symptom of a chronic withdrawal reaction;
(3) by direct neurotoxicity (Pope and Yurgelun-Todd, 1996).

There have been suggestions that cannabis can cause personality changes in chronic users. However, it has not been proven that cannabis can increase violence, reduce intelligence or cause amotivation. The majority of studies have failed to make an association between cannabis usage and any of these parameters. It is suggested, particularly often, that cannabis can cause amotivation – a lack of interest in achievement, inability to organise, and disinterest in life events. As with psychiatric illness (see below), it may be that those with a pre-existing predisposition to this are more likely to abuse cannabis. Obsession with regular drug-taking has been reported with many other drugs of abuse. A recent study has suggested that the apparent amotivation in chronic cannabis users is due to depression (Musty and Kaback, 1995).

A knowledge of the effects of cannabis in humans, leads one to suspect that the drug should impair driving performance. Common sense dictates that this must be so, and one should assume that this is the case unless proven otherwise. However, although in acute intoxication cannabis can impair concentration, and affect judgement and co-ordination, experiments have not provided proof that it impairs real driving. A succinct, but detailed review by Adams and Martin (1996) summarised the evidence available at that time. Although laboratory studies suggest impairment should occur, driving simulations and supervised negotiation of driving courses have not demonstrated a deleterious effect. Doses as large as 250 µg per kg body weight have been reported as having surprisingly little effect. When under the influence of cannabis, users can retain remarkable awareness of their state of intoxication. Consequently it has been suggested that, when driving, users are able to recognise the potential impairment of performance and that they overcompensate by concentrating more diligently and driving with extra care. Limited studies entailing the analysis of cannabis levels in drivers involved in road accidents have not supported an association. Many studies of cannabis and driving ability have investigated the acute effect of cannabis in intermittent users. In chronic regular users any potentially detrimental effects upon performance should be less, since a degree of tolerance develops.

One study has demonstrated that smoking cannabis decreased cerebral blood flow in those who had little or no previous exposure to the drug (Matthew and Wilson, 1992). The authors attributed this to the anxiety of novices about drug administration. In experienced users, cannabis increased blood flow in both hemispheres but especially to the frontal and left temporal areas. The authors suggested that this was a direct effect of cannabis, since the degree of increased blood flow corresponded to the extent of intoxication.

Like alcohol, cannabis can cause ataxia during intoxication if large enough doses are taken.

PSYCHIATRIC EFFECTS

First-time users, in particular, may be especially prone to adverse mental effects, perhaps due to a lack of experience in dealing with the psychotropic effects of the drug. Some of these acute adverse effects, confined to the period of intoxication, are described above. Panic attacks are the most frequently observed serious adverse psychiatric effects of cannabis, and are often characterised by a high level of anxiety, fear, and concern about losing control. Usually, removal of the patient to a quiet location, accompanied by calming words, will terminate a panic attack relatively quickly. It is rarely necessary to administer anxiolytics.

Acute psychotic episodes can occur after using cannabis, but are not common. In 1995, McBride and Thomas briefly described the results of two studies in which the prevalence of psychotic symptoms amongst cannabis users was estimated. Neither study has ever been published in full. In the first of these, 18 individuals from a total of 100 attending alcohol and drug abuse clinics in South Wales reported psychotic symptoms, in clear consciousness. In the second investigation, 15% of 199 cannabis users identified via a survey in New Zealand, reported psychotic symptoms after smoking the drug. No details of severity, history of psychiatric illness or concomitant use of other psychotropic substances were divulged.

Typically users presenting with psychotic symptoms have taken large doses. The reaction can present in a variety of ways, but it is often characterised by paranoid delusions, agitation, disordered thinking and hypomania. Hallucinations can also occur. Sometimes these episodes are described as schizophreniform reactions, and they can occur in those already suffering from psychosis (or latent psychosis), as well as those with no apparent tendency to mental illness. Usually the sufferer exhibits marked or complete improvement within a week of abstinence from cannabis, often with minimal treatment. There has been a reluctance to label this reaction officially as "acute cannabis psychosis", perhaps because the presentation is very variable, but this is in the nature of the action of psychoactive substances. The effects of acute intoxication vary widely between individuals, and for the same individual, depending on the factors already discussed. Several studies have shown that there is little, if any, difference between symptoms of cannabis-induced psychosis and other forms of acute psychosis. Many psychiatrists, whilst acknowledging that the drug can cause acute psychosis-like symptoms, have concluded that acute cannabis psychosis should not be labelled as such because, in terms of presentation, it is difficult or impossible to differentiate from other forms (Thomas, 1993; Thornicroft et al., 1992; McGuire et al., 1992). However, clinical presentation is not the only method by which diseases can be classified, and it would seem entirely reasonable to differentiate cannabis related acute psychosis by virtue of aetiology at least (i.e. drug-induced). The situation is complicated by the fact that many individuals who suffer from chronic psychotic disorders abuse cannabis. In these subjects, differentiating a natural relapse from a drug-induced one may be impossible clinically.

Cannabis certainly can exacerbate pre-existing schizophrenia, by precipitating a general relapse, or more specifically causing or worsening hallucinations and delusions. Three mechanisms have been proposed, by which cannabis might exacerbate schizophrenia (Martinez-Arevalo et al., 1994):

(1) A direct adverse effect of cannabis upon normal cognitive mechanisms, which results in precipitation of psychotic symptoms in those that are predisposed.
(2) A cannabis-induced psychosis superimposed upon schizophrenia.
(3) An antagonism of anti-psychotic medication. As mentioned below, smoking cannabis accelerates the elimination of the archetypal antipsychotic, chlorpromazine, from the body. But cannabis might also possess pharmacological properties which antagonise those of antipsychotic drugs. This area requires further study.

It has not been proven that cannabis can cause chronic psychotic illness. Indeed, in practice it would seem ethically impossible to construct an experiment to prove or disprove this hypothesis. Several studies have shown that those suffering from schizophrenia are more likely to have abused cannabis. For example, one group conducted a 15 year follow-up of 45,570 Swedish conscripts and revealed that those who had used cannabis more than fifty times were six times more likely to have been diagnosed as suffering from schizophrenia than those who had not used the drug (Andreasson *et al.*, 1987). The greater the frequency of cannabis use, the greater was the likelihood of developing schizophrenia. However, studies of this nature do not establish a link between cannabis and chronic psychotic illnesses. An equally valid interpretation of the results would be that those with a predisposition to mental illness are more likely to consume drugs of abuse. In this respect, it is interesting to note that 430 of the 730 conscripts who had consumed cannabis on more than fifty occasions had been suffering from a non-psychotic mental illness prior to conscription. It may be the case that such victims of mental illness seek out street drugs as a form of self-medication for their condition. A similar interpretation could be placed upon studies which have suggested that cannabis is an important predictor of relapse in those with existing psychotic illness (Martinez-Arevalo *et al.*, 1994).

Flashbacks have also been attributed to cannabis; these episodes involve re-experiencing an aspect of a cannabis intoxication at a later date. These phenomena can recur for months after exposure to cannabis. They are probably more common in chronic cannabis users, and usually diminish in intensity and frequency with continued abstinence.

Although the drug can cause anxiety during acute intoxication in the inexperienced user, cannabis also seems to have anxiolytic effects. A synthetic cannabinoid derivative, nabilone, was found to have some anxiolytic effects at a dose of 2 mg orally. Nabilone was not as potent as diazepam 5 mg (Nakano *et al.*, 1978).

PULMONARY EFFECTS

Cannabis is associated with a range of adverse effects upon the human lung. This is wholly attributable to the usual method of administration of the drug – viz. smoking. Regular smoking of either cannabis or tobacco is associated with an increased incidence of upper and lower respiratory tract infections, chronic bronchitis, cough, wheeze, expectoration, pharyngitis, and decreased exercise tolerance. Smoke particles can also trigger acute asthma attacks in those who are predisposed.

The pulmonary effects of cannabis specifically are difficult to isolate because most users mix cannabis with tobacco before smoking, and/or are regular tobacco smokers at the same time. This can make separation of the effects caused by the two agents difficult. Notwithstanding this limitation, the adverse respiratory effects of cannabis itself are gradually emerging.

Cannabis produces a mild bronchodilation after acute exposure, which has led to speculation that purified forms of the drug might have therapeutic potential (Vachon *et al.*, 1976). However, chronic exposure causes increased airflow resistance, largely due to an action on large airways (Tashkin *et al.*, 1987). This is in contrast to tobacco smokers, where small airways disease is characteristic. Minor, but significant, increase in airway obstruction has been observed after only 6 weeks of heavy usage (Tashkin *et al.*, 1976).

Both cannabis and tobacco cause structural changes to pulmonary macrophages, although whether this results in decreased functional ability is not clear. Macrophages are an important defence mechanism against microbial infection. Cannabis may be the direct cause of pulmonary aspergillosis in the immunocompromised, since the drug may harbour spores of *Aspergillus* spp. Cases of aspergillosis attributed to smoking cannabis have been described in AIDS patients, those undergoing chemotherapy and transplant recipients (Sutton *et al.*, 1986; Chusid *et al.*, 1975; Marks *et al.*, 1996). It has been recommended that immunocompromised individuals should heat the drug to 150°C for 15 minutes before using it, in order to kill spores (Levitz and Diamond, 1991).

When smoked, cannabis gives rise to a greater blood level of carbon monoxide, and a greater lung deposition of tar than tobacco. This is probably largely due to the fact that cannabis smokers tend to take longer puffs from their cigarettes than tobacco smokers, as well as deeper inhalations. They also retain smoke in the lungs for greater periods of time by holding the breath for substantially longer than tobacco smokers (Wu *et al.*, 1988). One group has shown that the increased breath-holding time is the most important determinant of carbon monoxide levels and tar retention, and that it does increase THC absorption from the lungs (Tashkin *et al.*, 1991a). Significantly, unlike tobacco smokers, most users smoke their cannabis cigarettes to the smallest possible butt length before discarding them. This is an important observation because the proximal end of the cigarette apart from delivering more THC, also delivers more tar and more carbon monoxide (Tashkin *et al.*, 1991b). Smoking cannabis to a longer butt length would probably be less detrimental to the lung therefore.

Cannabis cigarettes are always home-made and so more loosely wrapped than commercially produced tobacco cigarettes. Consequently the amount of particles and tar that is filtered out by the actual shaft of the cigarette is reduced. Cannabis cigarettes also do not contain filters. These features make the practice potentially more damaging to the airways. The consumption of 3 or 4 cannabis cigarettes daily can damage the pulmonary epithelium to the same extent as twenty or more tobacco cigarettes (Gong *et al.*, 1987).

Cannabis may be associated with the development of lung cancer, although this has not been proven. A single cannabis cigarette deposits more tar in the lungs than one of the tobacco variety, but tobacco smokers consume more cigarettes per day than those who use cannabis. Annual lung exposure to tar is likely to be greater in the average cigarette smoker than in the average cannabis smoker. The consensus is that since the risk of tobacco causing lung cancer is dose-related – and the tar produced from cannabis is at least as toxic as that from tobacco – there must be at least some risk of lung cancer from smoking cannabis. If tobacco is used as a carrier for smoking cannabis then there is a definite risk of developing lung cancer.

Histological studies of the pulmonary tissues of chronic cannabis smokers reveal evidence of changes which could be pre-cancerous (Tennant, 1980; Tashkin *et al.*, 1990). In the report from Tashkin's group all of those with histopathological changes were young and asymptomatic, whereas Tennant's subjects were studied because they had already exhibited changes in lung function. Tashkin and associates investigated eleven separate pre-determined adverse microscopical changes affecting epithelium, basement membrane and submucosa. These were all more common in tobacco smokers than non-smokers, and nine of these changes were even more common in those who smoked cannabis alone. Those who smoked both drugs had the greatest extent of change of all indicators. Sherman and co-workers in 1995 showed that the smoking of cannabis was associated with an increased incidence of DNA damage in human alveolar macrophages – another change that could be pre-cancerous.

A series of small studies suggest that cannabis smoking may be associated with an increased risk of cancer of the upper respiratory tract and mouth (Nahas and Latour, 1992).

MISCELLANEOUS EFFECTS

Walter *et al.* (1996) reported a case of suspected cannabis-induced hyperthermia. A male patient presented with a body temperature of 41$C, delirium, flushing, and hot and dry skin. He had been jogging on a warm day after smoking cannabis. The only link with cannabis was a temporal one. This reaction has never been reported in humans before, despite widespread use over three decades, suggesting the link with cannabis in this case was probably coincidental.

Lambrecht *et al.* (1995) reported a single case of renal infarction which was identified in a 29 year old man who had smoked cannabis regularly for ten years. His last consumption of cannabis was the day prior to admission. The authors suggested that the vasodilatation caused by cannabis, together with the drug's adrenergic effects, and anaemia, might have triggered renal artery thrombosis. They also speculated that cannabinoid-induced damage to the arterial endothelium might have resulted in thrombus formation. However, the patient had smoked cannabis for ten years uneventfully, and there is no proven link between cannabis consumption and any form of thrombosis. Renal infarction has not been associated with cannabis before. There was thus no temporal, pharmacological or independent association between cannabis and the development of symptoms. The only association was that the patient happened to smoke cannabis. It seems more likely that the link was coincidental. No other adverse effects of cannabis upon the kidney have been reported.

Two studies of the effects of cannabis upon liver function tests have yielded opposing results. In 1974, Kolodny and colleagues showed that chronic smoking of cannabis had no ill effect upon alkaline phosphatase (ALP) or serum glutamic oxalacetic transaminase (SGOT). However, in 1976, Frank and associates reported that during a 28 day period of smoking one cannabis cigarette per day, both ALP and SGOT levels increased in the 25 men under study. In those smoking 2% THC cigarettes, SGOT rose continually until the end of the study (day 29), then rapidly returned to normal. The value never exceeded the upper limit of the normal range. SGOT was unaltered throughout the study for subjects who smoked 1% THC cigarettes. Individuals using both strengths of cannabis developed increased plasma ALP levels. Levels rose continually until day 29 in both groups, but ALP almost reached the upper limit of the normal range in those consuming 2% THC cigarettes. Levels returned to pre-exposure levels in both groups after day 29, but more slowly in the 2% THC cigarettes group. Cannabis has not been shown to cause acute liver damage.

SUMMARY OF ADVERSE EFFECTS

1. Symptoms of intoxication are very variable but include initial adrenergic-like effects, followed by merriment, disinhibition, increased sensory perception and time distortions.

2. Negative psychotropic effects during acute intoxication include emotional lability, dysphoria, anxiety, confusion and panic attacks. Hallucinations are rare, as are flashbacks.

3. Cannabis can cause short-lived, psychotic reactions and also exacerbation of pre-existing schizophrenia. It has not been proven that cannabis can cause chronic psychosis.

4. During acute intoxication, cannabis can cause reduced co-ordination, impaired judgement and ataxia.

5. Common sense leads one to believe that cannabis will impair driving ability, but this has not been proven.

6. Cannabis can impair short-term memory and ability to concentrate, which may be a persistent effect lasting several weeks or more. But the relevance of laboratory measured parameters to everyday living is not clear. A link with amotivation has not been demonstrated.

7. Cannabis does not cause cerebral atrophy.

8. Cannabis is not likely to cause convulsions in humans, but there is some evidence that it might have an anticonvulsant action.

9. Cannabis can cause sinus tachycardia, which could be detrimental to those suffering from angina. Reported association with myocardial infarction should be regarded as coincidental based on present evidence.

10. The drug can cause vasodilatation, giving rise to orthostatic hypotension, reddening of the eyes, dizziness and facial flushing.

11. Cannabis does not cause reduced plasma levels of testosterone in humans, in most studies. Where decreased levels have been reported, these are within the normal human range.

12. Reports of gynaecomastia and decreased human sperm count are small in number and so a link to cannabis cannot be regarded as proven. Decreased sperm count has not been associated with infertility.

13. The effects of cannabis upon human female fertility are not clear.

14. Cannabis does not have significant effects upon human plasma levels of LH, FSH, thyroxine, glucose, or glucocorticosteroids.

15. Cannabis is not associated with frank adverse effects upon human renal or hepatic function, although one study suggested that raised values for liver function tests could occur.

16. The regular smoking of cannabis causes chest infections, bronchitis, pharyngitis, wheeze, expectoration, cough and increased airways resistance.

17. Cannabis smoking deposits larger amounts of tar in the lungs than tobacco smoking. It is associated with histopathological and DNA damage that could be pre-cancerous.

18. Spores of *Aspergillus* spp. in cannabis can cause infection in the immunocompromised.

DRUG INTERACTIONS INVOLVING CANNABIS

Information on the interactions between any illicit drug of abuse and other pharmacologically active substances is notoriously difficult to find. Most of the information available is derived from small-scale studies and case reports, and so should be interpreted with caution. The text below provides a critical appraisal of the reported interactions between cannabis and therapeutic drugs, as well as other drugs of abuse.

Amphetamines

It is unusual for cannabis and amphetamines to be taken together socially, since they have very different, and to some extent, opposing effects. Studies of co-administration in man, using oral dexamphetamine and smoked cannabis, tend to show that each drug exerts its characteristic effects independent of the presence of the other drug. For example, in one investigation 15 mg dexamphetamine largely failed to counteract the negative effects of cannabis on tests of cognition, except for a small effect on tests of mathematical addition (Zalcman et al., 1973). The cannabis used contained approximately 15 mg THC. In another study dexamphetamine failed to reverse the deleterious effects of smoked cannabis on motor function (Evans et al., 1974). Evans' group utilised comparatively smaller doses of both dexamphetamine (10 mg per 70 kg body weight) and cannabis (equivalent to 50 mcg THC per kg). Perhaps surprisingly, amphetamine has no effect upon cannabis induced tachycardia.

In mice, the stimulant effects of methamphetamine result in increased locomotor activity of various kinds. If intravenous cannabis extract or pure THC is given prior to administration of methamphetamine, the locomotor activity is reduced significantly. The lethality of a fixed high dose of methamphetamine is also increased by pretreatment with THC, but not by pretreatment with cannabis extract. None of these interactions with methamphetamine occur in mice that are tolerant to cannabinoids (Yamamoto et al., 1988).

Antidiabetic Drugs

Cannabis increases appetite and so when taken regularly it may appear to reduce the effectiveness of antidiabetic medication by increasing carbohydrate intake.

An investigation of healthy non-diabetic adults showed that smoked cannabis had no ill effects upon carbohydrate metabolism (Permutt et al., 1976; Weil et al., 1968). Monitoring of blood glucose levels revealed no episodes of hypoglycaemia amongst seven patients who had fasted for 24–72 hours beforehand. The smoking of cannabis was also found to have no adverse effects upon the response to the glucose tolerance test in ten healthy volunteers. Most other researchers have found that cannabis does not affect blood glucose, or that it produces a small, short-lived increase (Jones and Benowitz, 1976; Hollister and Reaven, 1974; Podolsky et al., 1971).

By contrast, a case of diabetic ketoacidosis supposedly caused by oral administration of cannabis has been described (Hughes et al., 1970). The patient exhibited the classic signs and symptoms of diabetic coma within 24 hours of eating an unspecified quantity of cannabis. The link with cannabis is exceedingly tenuous for the following reasons: he had a family history of diabetes; he had smoked cannabis before on numerous occasions without problems; and he was still frankly diabetic, requiring insulin, one month after this event. The temporal link between florid symptoms of diabetes and the ingestion of cannabis was undoubtedly coincidental.

Antimuscarinic Drugs

When atropine and cannabis are co-administered there is an additive effect upon heart rate. One study of two volunteers showed that smoking cannabis alone increased heart rate by about 20 beats per minute. The same increase in heart rate was demonstrated after 600 mcg atropine injection. But the smoking of cannabis 30 minutes after receiving atropine increased the pulse by about 50 beats per minute (Beaconsfield *et al.*, 1972). Antimuscarinic drugs cause sinus tachycardia by competitively antagonising the acetyl-choline released from parasympathetic neurones of the vagus nerve on the myocardium. The tachycardia induced by cannabis is mediated via adrenergic receptors since it is blocked by propranolol.

Barbiturates

Administration of both THC and cannabidiol results in inhibition of the metabolism of barbiturates in man and in experimental animals. In an investigation of the subjective effects of giving the combination, intravenous THC (27–134 µg per kg body weight) was given to seven volunteers who had already received a single dose of intravenous pentobarbitone (100 mg per 70 kg body weight). Five of the volunteers experienced intense psychotropic side effects such as hallucinations and anxiety; as a result four of them could not participate in the study further (Johnstone *et al.*, 1975). Similar additive effects were observed when secobarbitone was co-administered with THC (Lemberger *et al.*, 1976). The combination of the barbiturate and THC was found not to affect ventilation; tidal volume and plasma carbon dioxide levels remained unaltered (Johnstone *et al.*, 1975).

These additive psychotropic effects of barbiturates and cannabis can be exacerbated by cannabinoid inhibition of barbiturate metabolism, which has been demonstrated in several human studies (Benowitz and Jones, 1977; Benowitz *et al.*, 1980; Paton and Pertwee, 1972). Both THC and cannabidiol have this effect.

Chlorpromazine

The rate of elimination of chlorpromazine was investigated in a study of 31 patients taking the drug. The average clearance in eleven smokers of tobacco was increased by 38% compared to controls, and by 50% in five smokers of cannabis. In those who smoked both, the clearance was increased by 107% (Chetty *et al.*, 1994).

Cocaine

Cocaine and cannabis are rarely taken together at street level. The combination does not seem to have been studied in humans, but in rats THC did not alter the pharmacokinetics of intravenous cocaine (Vadlamani *et al.*, 1984).

Cytotoxic Agents

In vitro studies using human cancer cell lines reveal that tetrahydrocannabinol does not potentiate or antagonise the cytotoxic actions of actinomycin D, adriamycin, methotrexate, cisplatin, nitrogen mustard or velban (Harbell and DiBella, 1982). The cancer cell lines used were for human breast, uterus, ovary and melanoma.

By contrast, one study has suggested that tetrahydrocannabinol has additive immuno-suppressant effects when administered at the same time as cyclophosphamide in rats (Ader and Grota, 1981). The researchers measured the rat antibody response to the injection of sheep red blood cells. The administration of THC in a dose of 100 mg/kg reduced the antibody titre to about 80% of the value obtained in controls. Cyclophosphamide 30 mg/kg reduced the titre to a little over 50% of the control value. The combination produced an antibody titre of barely 20% of the control value. However, the importance of this in humans has not been investigated, and remains speculative. The amount of THC used on a mg/kg basis was very large compared to doses which humans use socially; it was also given as an intra-peritoneal injection – an unrepresentative mode of administration.

It has long been contended that cannabis has immunosuppressive effects in humans. If this were so, then an additive or synergistic effect with cytotoxics would be expected. However, the evidence for this immunosuppressive effect is not very convincing. *In vitro* studies using high doses of cannabis or cannabinoids, have demonstrated various adverse effects upon immunity in several mammalian species, but clinical studies in man do not support these findings (Hollister, 1992). For example, two studies of drug abuse in HIV positive individuals have shown that cannabis use is not associated with increased likelihood of progression to AIDS (DiFranco *et al.*, 1996; Ronald *et al.*, 1994).

Disulfiram

A single case report has described an interaction between cannabis and disulfiram (Lacoursiere and Swatek, 1983). A 28 year old man taking disulfiram 250 mg daily exhibited signs of hypomania following the smoking of cannabis. He likened this to the effects of amphetamine. He was hyperactive, euphoric and suffered from pressure of speech, irritability and insomnia. He had used cannabis prior to commencing disulfiram and had not experienced adverse reactions. Furthermore, when disulfiram was discontinued he resumed cannabis use with no ill effects. Several months later, however, the patient was prescribed disulfiram again and, although free from adverse effects initially, he again developed hypomanic symptoms upon smoking cannabis during disulfiram treatment.

This interaction has not been reported to affect other patients receiving disulfiram and cannabis concurrently. Rosenberg *et al.* (1978) investigated the ability of the combination to induce alcoholics to enter or remain in treatment; no cases of hypomania were described. The patient described above had a long history of substance abuse and he may have abused other substances and not disclosed them to the authors of the case report. Alternatively, the cannabis that he used may have been adulterated with other street drugs.

Ethanol

Human performance in tests of mental, motor and perceptual ability is significantly reduced when THC or cannabis is taken with ethanol before testing (Manno *et al.*, 1971; Bird *et al.*, 1980). This detrimental effect is greater with the combination than when either agent is taken alone.

In 1992, Lukas *et al.* studied the interaction between alcohol and cannabis in 15 human volunteers. After consuming 0.7 g per kg body weight of ethanol, peak plasma levels of 78 mg/dl occurred 50 minutes later. However, on a separate occasion the subjects smoked a cannabis cigarette containing approximately 2.5% THC 30 minutes after ingesting the same amount of ethanol as previously. The peak plasma levels of ethanol were lower (55 mg/dl), and the peak occurred much later (105 minutes after drinking). These changes to plasma ethanol pharmacokinetics were mirrored by a decrease in the subjective duration of both ethanol and cannabis effects.

This work appears to conflict with the earlier work of Benowitz and Jones (1977), who demonstrated that administration of oral THC at a dose of 60–180 mg per day for 10–17 days caused only a slight decrease in the rate of metabolism of ethanol. However, it may be that the metabolic effects of the whole cannabis plant, as studied by Lukas' group, differ from THC alone as studied by Benowitz and Jones. But a research group headed by Bird in 1980 reported that none of the major individual cannabinoids THC, cannabinol and cannabidiol affected blood ethanol levels when given prior to drinking. The amount of THC given by Benowitz and Jones was significantly greater than that used by Lukas; the cannabis was also given orally as opposed to being smoked, and theirs was not an acute, single dose study. All of these factors may have influenced the different results obtained by the two groups.

Fluoxetine

THC causes aggressive behaviour in rats that have been selectively deprived of REM sleep. The administration of fluoxetine or tryptophan was found to potentiate this aggression. These drugs boost CNS levels of serotonin, but drugs with anti-serotonin effects prevented THC causing aggression (Carlini and Lindsey, 1982).

In 1991, Stoll and colleagues reported the case of a 21 year old woman who developed mania subsequent to taking both fluoxetine and cannabis. She had smoked cannabis on several occasions in the past without experiencing any unusual side effects. After taking fluoxetine 20 mg daily for 4 weeks, she smoked two cannabis cigarettes within 36 hours. Within 24 hours she became euphoric, and exhibited signs of increased energy, hypersexuality, pressure of speech, delusions of grandeur, agitation and manic excitement. These abated over 3 to 4 days after admission to hospital, cessation of fluoxetine, and treatment with lorazepam and perphenazine. Twenty-nine days after the original episode, fluoxetine was re-introduced but the patient discontinued the drug two weeks later because she could not sleep and felt "hyper".

An obvious criticism of this purported interaction is that fluoxetine alone can cause mania, and suggestive symptoms were even reported by the patient after the second exposure to fluoxetine alone. The main reason for suggesting that an interaction took place was the temporal link between administration of cannabis and the onset of mania. As the authors point out, both THC and fluoxetine inhibit serotonin reuptake and an additive effect cannot be ruled out as the cause of mania. However, the patient did not exhibit other signs of the serotonin syndrome and, despite the widespread use of both fluoxetine and cannabis, no other cases have been reported.

Indomethacin

Pretreatment of human subjects with indomethacin before smoking cannabis was found to reduce the elevation of plasma prostaglandin levels which are associated with administration of THC. Indomethacin significantly, but modestly, blunted the subjective intensity of euphoria induced by cannabis, and also slightly reduced THC-induced cardiac acceleration. The most interesting finding was that indomethacin prevented cannabis causing its characteristic, subjective, distortions of time perception. Indomethacin did not reduce the detrimental effects of cannabis on recall (Perez-Reyes *et al.*, 1991).

Lithium

Ratey *et al.* (1981) reported a single case of cannabis apparently elevating serum lithium concentrations. The authors suggested that cannabis inhibited peristalsis, allowing a greater proportion of the lithium dose to be absorbed. This seems an unlikely mechanism because drugs that slow gut motility tend to also slow the rate of absorption such that peak plasma levels of drug are reduced. The patient in question had a complex psychiatric history & lithium levels had fluctuated widely during the preceding year. This makes interpretation of the case report, and the role of cannabis, very difficult.

Opioids

Predictably opioids and THC seem to have some additive CNS effects. In human volunteers, intravenous THC (27–134 µg per kg) potentiated the sedative actions of intravenous oxymorphone (1 mg/70 kg). More surprising was the discovery that THC could potentiate the respiratory depressant effects of oxymorphone. In eight volunteers, ventilation fell from an average of 24.9 litres/minute before drug administration, to 14.1 litres/minute after injection of oxymorphone. The subsequent administration of varying doses of intravenous THC progressively reduced this parameter – ultimately to only 6.6 litres/minute after 134 µg/kg THC. Cardiovascular effects of THC were unaltered by the presence of an opioid (Johnstone *et al.*, 1975).

Phencyclidine

Cannabis and phencyclidine are purported to have additive effects when consumed together at street level. In dogs, the THC component of cannabis inhibits the metabolism of phencyclidine – the clearance is reduced, but the half-life and volume of distribution are unaltered (Godley *et al.*, 1991).

Physostigmine

This drug is an inhibitor of cholinesterase, the enzyme which destroys acetylcholine and as such, has opposite effects to antimuscarinic drugs such as atropine (see above). This was investigated by Freemon and co-workers in 1975. Five volunteers were given 20–40 mg of oral THC, and then 0.75–1.25 mg intravenous physostigmine two hours later. Predictably, physostigmine reduced the tachycardia associated with THC, presumably by increasing the persistence of acetylcholine at synapses of the vagus nerve on the myocardium. Physostigmine also reduced THC-induced conjunctival injection, and increased the lethargy and sleepiness induced by THC during the late phase of intoxication. Physostigmine did not affect the peak psychotropic effects of cannabis.

Propranolol

This beta-blocker inhibits the cardiac acceleration caused by cannabis (Beaconsfield *et al.*, 1972; Hillard and Vieweg, 1983; Sulkowski *et al.*, 1977). Propranolol may also attenuate the reddening of the eyes that is so common after smoking cannabis. It would not be anticipated that propranolol would affect the ability of cannabis to impair cognition, and this was confirmed in one study (Drew *et al.*, 1972). However, another study of six experienced cannabis smokers revealed that pre-treatment with propranolol prevented cannabis from impairing performance at certain learning tests (Sulkowski *et al.*, 1977). Propranolol also had a slight blunting effect upon the subjective experience of cannabis intoxication.

Theophylline

Jusko studied the effects of cannabis on the clearance of theophylline in two studies. In the first of these, fourteen cannabis smokers were recruited (Jusko *et al.*, 1978). Cannabis users had smoked the drug at least twice weekly for several months. Seven of them smoked cannabis only, and seven smoked tobacco regularly as well. The half-life and clearance of aminophylline varied in each of the groups studied:

	Half-life (hrs)	Clearance (ml/kg/hr)
Non-smokers ($n=19$)	8.1	52
Cannabis alone ($n=7$)	5.9	73
Tobacco alone ($n=24$)	5.7	75
Cannabis plus tobacco ($n=7$)	4.3	93

In the second study, the clearance of theophylline was compared in three different groups (Jusko *et al.*, 1979). One hundred and seventy-seven patients that did not smoke cannabis had a clearance of 56 ml/hr/kg. Nine participants who smoked cannabis less than once a week had similar clearance values to non-users (54 ml/hr/kg). However, those who smoked cannabis at least twice per week had markedly elevated clearance of theophylline (83 ml/hr/kg).

Smoking cannabis accelerates the elimination of theophylline, unless usage is very low. Jusko suggests that the method of administration of cannabis is important if this interaction is to occur. Smoking cannabis (and tobacco) causes the production of polycyclic aromatic hydrocarbons, which can induce liver metabolism, thus accelerating clearance. Presumably there would be no interaction between oral cannabis and theophylline, but this has not been studied.

Tricyclic Antidepressants

The combined use of cannabis and certain tricyclic antidepressants (TCAs) has been reported to cause sinus tachycardia and adverse psychotropic effects. An account of four male adolescents who received this combination has been published (Wilens *et al.*, 1997). The four young men, aged from 15 to 18 years, suffered from attention deficit hyperactivity disorder. Three were taking desipramine (150–200 mg daily) and one nortriptyline (75 mg daily). All four subjects experienced tachycardia after smoking cannabis, and this was accompanied by confusion. In addition, various other psychotropic effects were documented in individual cases, including lightheadedness, delirium, labile mood, and hallucinations. One adolescent reported that these effects had not occurred on occasions when he had smoked cannabis prior to starting a TCA.

A case report in 1983 described a 21 year old woman who, before receiving any treatment for depression, had smoked cannabis on several occasions without apparent ill effects (Hillard and Vieweg, 1983). She was treated for depression with nortriptyline 30 mg daily for nine months which she tolerated well. She did not smoke cannabis during this time. However, whilst continuing TCA treatment she smoked a cannabis cigarette and developed severe sinus tachycardia of 160 beats per minute. This required emergency hospital treatment with propranolol because the patient was so alarmed. She also felt a tightness in her chest and a lump in her throat. Subsequent to the successful termination of the arrhythmia she continued nortriptyline for 6 months with no further episodes of tachycardia. When nortriptyline administration was terminated she resumed occasional cannabis smoking with no adverse cardiac effects.

Similarly, in 1980, Kizer described a 25 year old man with a sinus tachycardia of 120 beats per minute who smoked cannabis whilst taking imipramine 25 mg twice daily. The patient described feeling restless and dizzy.

Cannabis and certain TCAs can both cause mild tachycardia. In the case of TCAs this is mainly due to their antimuscarinic actions. In some subjects this effect of the two drugs seems to be at least additive. Those receiving TCAs which are known to accelerate the heart rate should ideally not smoke cannabis. For those with depression that want to continue smoking cannabis, it would be preferable to prescribe an antidepressant which does not have cardiac effects (e.g. a serotonin reuptake inhibitor) or choose a TCA which is unlikely to cause tachycardia (e.g. lofepramine, trazodone).

REFERENCES

Abel, E.L. (1981) Marihuana and sex: a critical survey. *Drug Alcohol Dependence* **8**: 1–22.
Adams, I.B. and Martin, B.R. (1996) Cannabis: pharmacology and toxicology in animals and humans. *Addiction* **91**: 1585–1614.
Ader, R. and Grota, L.J. (1981) Immunosuppressive effect of tetrahydrocannabinol plus cyclophosphamide (letter). *New Eng. J. Med.* **305**: 463.
Andreasson S., Ergstrom, A., Allebeck, P. and Rydberg, U. (1987) Cannabis and schizophrenia. A longitudinal study in Swedish conscripts. *Lancet* **2**: 1483–86.
Aronow, S. and Cassidy, J. (1974) Effect of marihuana and placebo-marihuana smoking on angina pectoris. *New Eng. J. Med.* **291**: 65–67.
Aronow, S. and Cassidy, J. (1975) Effect of smoking marihuana and of a high nicotine cigarette on angina pectoris. *Clin. Pharmacol. Ther.* **17**: 549–54.
Beaconsfield, P., Ginsburg, J. and Rainsbury, R. (1972) Marijuana smoking: cardiovascular effects in man and possible mechanisms. *New Eng. J. Med.* **287**: 209–12.

Benowitz, N.L. and Jones, R.T. (1977) Effects of delta-9-tetrahydrocannabinol on drug distribution and metabolism: antipyrine, pentobarbital and ethanol. *Clin. Pharmacol. Ther.* **22**: 259.

Benowitz, N.L, Trong-Lang, N., Jones, R.T., Herning, R.I. and Bachman, J. (1980) Metabolic and psychophysiologic studies of cannabidiol-hexobarbital interaction. *Clin. Pharmacol. Ther.* **28**: 115–19.

Bird, K.D., Boleyn, T., Chesher, G.B., Jackson, D.M., Starmer, G.A. and Teo, R.K. (1980) Intercannabinoid and cannabinoid-ethanol interactions and their effects on human performance. *Psychopharmacol.* **71**: 181–88.

Block, R.I., Farinpour, R. and Schlechte, J.A. (1991) Effects of chronic marijuana use on testosterone, luteinizing hormone, follicle stimulating hormone, prolactin and cortisol in men and women. *Drug Alcohol Dependence* **28**: 121–28.

Buffman, J. (1982) Pharmacosexology: the effects of drugs on sexual function a review. *J. Psychoactive Drugs* **14**: 5–44.

Campbell, A.M.G., Evans, M., Thomson, J.L.G. and Williams, M.J. (1971) Cerebral atrophy in young cannabis smokers. *Lancet* **ii**: 1219–25.

Carlini, E.A. and Cunha, J.M. (1981) Hypnotic and antiepileptic effects of cannabinol. *J. Clin. Pharmacol.* **21**(Suppl): 417S–27S.

Carlini, E.A. and Lindsey, C.J. (1982) Effect of serotonergic drugs on the aggressiveness induced by delta-9-tetrahydrocannabinol in REM-sleep-deprived rats. *Braz. J. Med. Biol. Res.* **15**: 281–3.

Chetty, M., Miller, R. and Moodley, S.V. (1994) Smoking and body weight influence the clearance of chlorpromazine. *Eur. J. Clin. Pharmacol.* **46**: 523–6.

Chusid, M.J., Gelfand, J.A., Nutter, C. and Fauci, A.S. (1975) Pulmonary aspergillosis, inhalation of contaminated marijuana smoke, and chronic granulomatous disease. *Ann. Int. Med.* **82**: 682–83.

Close, C.E., Roberts, P.L. and Berger, R.E. (1990) Cigarettes alcohol and marijuana are related to pyospermia in infertile men. *J. d'Urologie* **144**: 900–03.

Co, B., Goodwin, D.W., Gado, M., Mikhael, M. and Hill, S.Y. (1977) Absence of cerebral atrophy in chronic cannabis users by computerized transaxial tomography. *J. Am. Med. Assoc.* **237**: 1229–30.

Collins, J.S.A., Higginson, J.D.S., Boyle, D.M.C. and Webb, S.W. (1985) Myocardial infarction during marijuana smoking in a young female. *Eur. Heart J.* **6**: 637–38.

Consroe, P.F., Wood, G.C. and Buchsbaum, H. (1975) Anticonvulsant nature of marihuana smoking. *J. Am. Med. Assoc.* **234**: 306–7.

Cushman Jr, P. (1975) Plasma testosterone levels in healthy male marihuana smokers. *Am. J. Drug Alcohol Abuse* **2**: 269–75.

Deahl, M. (1991) Cannabis and memory loss. *Br. J. Addiction* **86**: 249–52.

DiFranco, M.J., Sheppard, H.W., Hunter, D.J., Tosteson, T.D. and Ascher, M.S. (1996) The lack of association of marijuana and other recreational drugs with progression to AIDS in the San Francisco men's health study. *Annals Epidemiol.* **6**: 283–89.

Drew, W.G., Kiplinger, G.F., Miller, L.L. and Marx, M. (1972) Effects of propranolol on marijuana-induced cognitive dysfunctioning. *Clin. Pharmacol. Ther.* **13**: 526–33.

Erdolu, C., Saglam, R. and Harmankaya, C. (1985) The effects of marihuana and tranquilizers on male sexual functions. *Bull. Gulhane Mil. Med. Acad.* **27**: 77–82.

Evans, M.A., Martz, R., Lemberger, L., Rodda, B.E. and Forney, R.B. (1974) Clinical effects of marihuana dextroamphetamine combination. *Pharmacologist* **16**: 281.

Feeney, D.M. (1976) Marihuana use among epileptics (letter). *J. Am. Med. Assoc.* **235**: 1105.

Frank, I.M., Lessin, P.J., Tyrrell, E.D., Hahn, P.M. and Szara, S. (1976) Acute and cumulative effects of marihuana smoking in hospitalized subjects: a 36-day study, in Braude, M.C. and Szara, S. (Eds.) *The Pharmacology of Marihuana*, Raven Press, New York, pp. 673–79.

Freemon, F.R., Rosenblatt, J.E. and El-Yousef, M.K. (1975) Interaction of physostigmine and delta-9-tetrahydrocannabinol in man. *Clin. Pharmacol. Ther.* **17**: 121–26.

Godley, P.J., Moore, E.S., Woodworth, J.R. and Fineg, J. (1991) Effects of ethanol and delta-9-tetrahydrocannabinol on phencyclidine disposition in dogs. *Biopharmaceutics Drug Disposition* **12**: 189–99.

Gong, H., Fligiel, S., Tashkin, D.P. and Barbers, R.G. (1987) Tracheobronchial changes in habitual, heavy smokers of marijuana with and without tobacco. *Am. Rev. Resp. Dis.* **136**: 142–9.

Harbell, J.W. and DiBella, N.J. (1982) Studies on the interaction of tetrahydrocannabinol (THC) with chemotherapeutic agents against human tumors *in vitro. Proc. Am. Assoc. Cancer Res.* (*AACR Abstracts No. 891*) **23**: 226.

Harmon, J. and Aliapoulios, M.A. (1972) Gynecomastia in marihuana users (letter). *New Eng. J. Med.* **287**: 936.

Harmon, J. and Aliapoulios, M.A. (1974) Marihuana-induced gynecomastia: clinical and laboratory experience. *Surg. Forum* **25**: 423–25.

Hembree III, W.C., Zeidenberg, P. and Nahas, G.G. (1976) Marihuana's effects on human gonadal function, in Nahas, G.G. (Ed.) *Marihuana: Chemistry, Biochemistry and Cellular Effects*, Springer-Verlag, New York, pp. 521–32.

Hillard, J.R. and Vieweg, W.V.R. (1983) Marked sinus tachycardia resulting from the synergistic effects of marijuana and nortriptyline. *Am. J. Psychiatry* **140**: 626–7.

Hollister, L.E. and Reaven, G.M. (1974) Delta-9-tetrahydrocannabinol and glucose tolerance. *Clin. Pharmacol. Ther.* **16**: 297–302.

Hollister, L.E. (1992) Marijuana and immunity. *J. Psychoactive Drugs* **24**: 159–64.

Hughes, J.E., Steahly, L.P. and Bier, M.M. (1970) Marihuana and the diabetic coma. *J. Am. Med. Assoc.* **214**: 1113–14.

Joesoef, M.R., Beral, V., Aral, S.O., Rolfs, R.T. and Cramer, D.W. (1993) Fertility and use of cigarettes, alcohol, marijuana and cocaine. *Ann. Epidemiol.* **3**: 592–94.

Johnstone, R.E., Lief, P.L., Kulp, R.A. and Smith, T.C. (1975) Combination of delta-9-tetrahydrocannabinol with oxymorphone or pentobarbital: effects on ventilatory control and cardiovascular dynamics. *Anesthesiology* **42**: 674–84.

Jones, R.T. and Benowitz, M. (1976) The 30-day trip – clinical studies of cannabis tolerance and dependence, in Braude, M.C. and Szara, S. (Eds.) *The Pharmacology of Marihuana*, Raven Press, New York, pp. 627–42.

Jusko, W.J., Schentag, J.J., Clark, J.H., Gardner, M. and Yurchak, A.M. (1978) Enhanced biotransformation of theophylline in marijuana and tobacco smokers. *Clin. Pharmacol. Ther.* **24**: 406–10.

Jusko, W.J., Gardner, M.J., Mangione, A., Schentag, J.J., Koup, J.R. and Vance, J.W. (1979) Factors affecting theophylline clearances: age, tobacco, marijuana, cirrhosis, congestive heart failure, obesity, oral contraceptives, benzodiazepines, barbiturates, and ethanol. *J. Pharmaceutical Sci.* **68**: 1358–66.

Keeler, M.H. and Reifler, C.F. (1967) Grand mal convulsions subsequent to marijuana use. *Dis. Nerv. Syst.* **18**: 474–5.

Kizer, K.W. (1980) Possible interaction of TCA and marijuana (letter). *Ann. Emerg. Med.* **19**: 444.

Kolodny, R.C., Masters, W.H., Kolodner, R.M. and Toro, G. (1974) Depression of plasma testosterone levels after chronic intensive marihuana use. *New Eng. J. Med.* **290**: 872–74.

Kolodny, R.C., Lessin, P., Toro, G., Masters, W.H. and Cohen, J. (1976) Depression of plasma testosterone with acute administration, in Braude, M.C. and Szara, S. (Eds.) *The Pharmacology of Marihuana*, Raven Press, New York, pp. 217–25.

Kolodny, R.C., Webster, S.K., Tullman, G.D. and Dornbush, R.I. (1979) Chronic marihuana use by women: menstrual cycle and endocrine findings. Presented at the New York Postgraduate Medical School Second Annual Conference on Marihuana: "Marihuana – Biomedical Effects and Social Implications", June 28–29, 1979, *per* Abel, E.L. (1981) Marihuana and sex: a critical survey. *Drug Alcohol Dependence* **8**: 1–22.

Kuehnle, J., Mendelson, J.H., Davis, K.R. and New, P.F.J. (1977) Computed tomographic examination of heavy marihuana smokers. *J. Am. Med. Assoc.* **237**: 1231–32.

Lacoursiere, R.B. and Swatek, R. (1983) Adverse interaction between disulfiram and marijuana: a case report. *Am. J. Psychiatry* **140**: 243.

Lambrecht, G.L.Y., Malbrain, M.L.N.G., Coremans, P., Verbist, L. and Verhaegen, H. (1995) Acute renal infarction and heavy marijuana smoking. *Nephron* **70**: 494–96.

Lemberger, L., Dalton, B., Martz, R., Rodda, B. and Forney, R. (1976) Clinical studies of the interaction of psychopharmacologic agents with marihuana. *Ann. New York Acad. Sci.* **281**: 219–28.

Levitz, S.M. and Diamond, R.D. (1991) Aspergillosis and marijuana (letter). *Ann. Int. Med.* **115**: 578–79.

Lukas, S.E., Benedikt, R., Mendelson, J.H., Kouri, E., Sholar, M. and Amass, L. (1992) Marihuana attenuates the rise in plasma ethanol levels in human subjects. *Neuropsychopharmacology* **7**: 77–81.

McBride, A.J. and Thomas, H. (1995) Psychosis is also common in users of "normal" cannabis (letter). *Br. Med. J.* **311**: 875.

McGuire, P.K., Jones, P., Harvey, I., Bebbington, P., Toone, B., Lewis, S. and Murray, R.M. (1992) Cannabis and acute psychosis. *Schizophrenia Res.* **13**: 161–67.

Macinnes, D.C. and Miller, K.M. (1984) Fatal coronary artery thrombosis associated with cannabis smoking. *J. Roy. Coll. Gen. Prac.* **34**: 575–76.

Manno, J.E., Kiplinger, G.F., Scholz, N., Forney, R.B. and Haine, S.E. (1971) The influence of alcohol and marihuana on motor and mental performance. *Clin. Pharmacol. Ther.* **12**: 201–11.

Marks, W.H., Florence, L., Lieberman, J., Chapmna, P., Howard, D., Roberts, P. and Perkinson, D. (1996) Successfully treated invasive pulmonary aspergillosis associated with smoking marijuana in a renal transplant recipient. *Transplantation* **61**: 1771–74.

Martinez-Arevalo, M.J., Calcedo-Ordonezz, A. and Varo-Prieto, J.R. (1994) Cannabis consumption as a prognostic factor in schizophrenia. *Br. J. Psychiatry* **164**: 679–81.

Mathew, R.J., Wilson, W.H., Humphreys, D., Lowe, J.V. and Wiethe, K.E. (1992) Middle cerebral artery velocity during upright posture after marijuana smoking. *Acta Psychiatr. Scand.* **86**: 173–78.

Mendelson, J.H., Ellingboe, J., Kuehnle, J.C. and Mello, K. (1978) Effects of chronic marihuana use on integrated plasma testosterone and luteinizing hormone levels. *J. Pharmacol. Exp. Ther.* **207**: 611–17.

Mendelson, J.H., Kuehnle, J.C., Ellingboe, J. and Babor, T.F. (1974) Plasma testosterone levels before, during and after chronic marihuana smoking. *New Eng. J. Med.* **291**: 1051–55.

Merritt, J.C., Cook, C.E. and Davis, K.H. (1982) Orthostatic hypotension after delta-9-tetrahydrocannabinol marihuana inhalation. *Ophthalm. Res.* **14**: 124–28.

Musty, R.E. and Kaback, L. (1995) Relationship between motivation and depression in chronic marijuana users. *Life Sci.* **56**: 2151–58.

Nahas, G. and Latour, C. (1992) The human toxicity of marijuana. *Med. J. Aust.* **156**: 495–97.

Nakano, S., Gillespie, H.K. and Hollister, L.E. (1978) A model for evaluation of antianxiety drugs with the use of experimentally induced stress: comparison of nabilone and diazepam. *Clin. Pharmacol. Ther.* **23**: 54–62.

Ng, S.K.C., Brust, J.C.M., Hauser, W.A. and Susser, M. (1990) Illicit drug use and the risk of new-onset seizures. *Am. J. Epidemiol.* **132**: 47–57.

Olusi, S.O. (1980) Hyperprolactinaemia in patients with suspected cannabis-induced gynaecomastia. *Lancet* **1**: 255.

Paton, W.D.M. and Pertwee, R.G. (1972) Effect of cannabis and certain of its constituents on pentobarbitone sleeping time and phenazone metabolism. *Br. J. Pharmacol.* **44**: 250–61.

Perez-Reyes, M., Burstein, S.H., White, W.R., McDonald, S.A. and Hicks, R.E. (1991) Antagonism of marihuana effects by indomethacin in humans. *Life Sci.* **48**: 507–15.

Podolsky, S., Pattavina, C.G. and Amaral, M.A. (1971) Effect of marijuana on the glucose-tolerance test. *Ann. N. Y. Acad. Sci.* **191**: 54–60.

Pope, H.G. and Yurgelun-Todd, D. (1996) The residual cognitive effects of heavy marijuana use in college students. *J. Am. Med. Assoc.* **275**: 521–27.

Ratey, J.J., Ciraulo, D.A. and Shader, R.I. (1981) Lithium and marijuana. *J. Clin. Psychopharmacol.* **1**: 32–33.

Permutt, M.A., Goodwin, D.W., Schwin, R. and Hill, S.Y. (1976) The effect of marijuana on carbohydrate metabolism. *Am. J. Psychiatry* **33**: 220–24.

Ronald, P.J.M., Robertson, J.R. and Elton, R.A. (1994) Continued drug use and other cofactors for progression to AIDS among injecting drug users. *AIDS* **8**: 339–43.

Rosenberg, C.M., Gerrein, J.R. and Schnell, C. (1978) Cannabis in the treatment of alcoholism. *J. Stud. Alcohol* **39**: 1955.

Schaefer, C.F., Gunn, C.G. and Dubowski, K.M. (1975) Normal plasma testosterone concentrations after marihuana smoking. *New Eng. J. Med.* **292**: 867–68.

Schwartz, R.H., Gruenewald, P.J., Klitzner, M. and Fedio, P. (1989) Short-term memory impairment in cannabis-dependent adolescents. *Am. J. Dis. Child.* **143**: 1214–19.

Sherman, M.P., Aeberhard, E.E., Wong, V.Z., Simmons, M.S., Roth, M.D. and Tashkin, D.P. (1995) Effects of smoking marijuana, tobacco or cocaine alone or in combination on DNA damage in human alveolar macrophages. *Life Sci.* **56**: 2201–07.

Solowij, N. (1995) Do cognitive impairments recover following cessation of cannabis use? *Life Sci.* **56**: 2119–26.

Stoll, A.L., Cole, J.O. and Lukas, S.E. (1991) A case of mania as a result of fluoxetine-marijuana interaction (letter). *J. Clin. Psychiatry* **52**: 280–1.

Sulkowski, A., Vachon, L. and Rich, E.S. (1977) Propranolol effects on acute marijuana intoxication in man. *Psychopharmacol.* **52**: 47–53.

Sutton, S., Lum, B.L. and Torti, F.M. (1986) Possible risk of invasive aspergillosis with marijuana use during chemotherapy for small cell lung cancer. *Drug Intelligence Clin. Pharm.* **20**: 289–91.

Tashkin, D.P., Shapiro, B.J., Lee, Y.E. and Harper, C.E. (1976) Subacute effects of heavy marihuana smoking on pulmonary function in healthy men. *New Eng. J. Med.* **294**: 125–29.

Tashkin, D.P., Coulson, A.H., Clark, V.A., Simmons, M., Bourque, L.B., Duann, S., Spivey, G.H. and Gong, H. (1987) Respiratory symptoms and lung function in habitual heavy smokers of marijuana alone, smokers of marijuana and tobacco, smokers of tobacco alone, and non-smokers. *Am. Rev. Resp. Dis.* **135**: 209–16.

Tashkin, D.P., Fligiel, S., Wu, T-Z, Gong, H., Barbers, R.G. Coulson, A.H. *et al.* (1990) Effects of habitual use of marijuana and/or cocaine on the lung, In Chiang, C.N. and Hawks, R.L. (Eds.) *Research Findings on Smoking of Abused Substances*, NIDA Research Monograph No.99, US Dept Health and Human Services, New York, pp. 63–87.

Tashkin, D.P., Gliederer, F., Rose, J., Chang, P., Hui, K.K., Yu, J.L. and Wu, T-Z (1991a) Effects of varying marijuana smoking profile on deposition of tar and absorption of CO and delta-9-THC. *Pharmacol. Biochem. Behav.* **40**: 651–56.

Tashkin, D.P., Gliederer, F., Rose, J., Chang, P., Hui, K.K., Yu, J.L. and Wu, T-Z (1991b) Tar, CO and delta-9-THC delivery from the 1st and 2nd halves of a marijuana cigarette. *Pharmacol. Biochem. Behav.* **40**: 657–61.

Tennant Jr, F.S. (1980) Histopathologic and clinical abnormalities of the respiratory system in chronic hashish smokers. *Substance Alcohol Actions Misuse* **1**: 93–100.

Thomas, H. (1993) Psychiatric symptoms in cannabis users. *Br. J. Psychiatry* **163**: 141–49.

Thornicroft, G., Meadows, G. and Politi, P. (1992) Is 'cannabis psychosis' a distinct category? *Eur. Psychiatry* **7**: 277–82.

Vachon, L., Mikus, P., Morrissey, W., FitzGerald, M. and Gaensler, E. (1976) Bronchial effect of marihuana smoke in asthma, In Braude, M.C. and Szara, S. (Eds.) *The Pharmacology of Marihuana*, Raven Press, New York, pp. 777–84.

Vadlamani, N.L., Pontani, R.B. and Misra, A.L. (1984) Effect of diamorphine, delta-9-tetrahydrocannabinol and ethanol on intravenous cocaine disposition. *J. Pharm. Pharmacol.* **36**: 552–54.

Varga, K., Lake, K., Martin, B. and Kunos, G. (1995) Novel antagonist implicates the CB1 cannabinoid receptor in the hypotensive action of anandamide. *Eur. J. Pharmacol.* **278**: 279–83.

Walter, F.G., Bey, T.A., Ruschke, D.S. and Bemowitz, N.L. (1996) Marijuana and hyperthermia. *J. Toxicol. Clin. Toxicol.* **34**: 217–21.

Weil, A.T., Zinberg, N.E. and Nelsen, J.M. (1968) Clinical and psychological effects of marihuana in man. *Science* **162**: 1234–42.

Wilens, T., Biederman, J. and Spencer, T.J. (1997) Case study: effects of smoking marijuana while receiving tricyclic antidepressants. *J. Am. Acad. Child Adolesc. Psychiatry* **36**: 45–48.

Wu, T-Z, Tashkin, D.P., Djahed, B. and Rose, J.E. (1988) Pulmonary hazards of smoking marijuana as compared with tobacco. *New Eng. J. Med.* **318**: 347–51.

Yamamoto, I., Umebayashi, H., Watanabe, K. and Yoshimura, H. (1988) Interactions of cannabis extract, delta-9-tetrahydrocannabinol and 11-hydroxy-delta-8-tetrahydrocannabinol with methamphetamine in mice. *Res. Commun. Substances Abuse* **9**: 107–16.

Zalcman, S., Liskow, B., Cadoret, R. and Goodwin, D. (1973) Marijuana and amphetamine: the question of interaction. *Am. J. Psychiatry* **130**: 707–8.

INDEX